Water and Nutrient Management in Natural and Constructed Wetlands

Jan Vymazal
Editor

Water and Nutrient Management in Natural and Constructed Wetlands

Springer

Editor
Jan Vymazal
ENKI, o.p.s.
Dukelská 145
379 01 Trebon
Czech Republic

and

Czech University of Life Sciences Prague
Faculty of Environmental Sciences
Department of Landscape Ecology
Námesti Smirickych 1
281 63 Kostelec nad Cernymi lesy
Czech Republic
vymazal@yahoo.com

ISBN 978-90-481-9584-8 e-ISBN 978-90-481-9585-5
DOI 10.1007/978-90-481-9585-5
Springer Dordrecht Heidelberg London New York

Library of Congress Control Number: 2010935493

Printed on acid-free paper

Springer is part of Springer Science+Business Media (www.springer.com)

Preface

Wetlands provide a wide variety of ecosystem services within the landscape and their importance is commonly accepted. Among the most important are regulating services, i.e. benefits obtained from the regulation of ecosystem processes. For example, wetlands contribute to climate regulation. Land cover can affect local temperature and precipitation, wetland ecosystems may affect greenhouse gas sequestration and emissions, or affect the timing and magnitude of runoff and flooding, for example. Wetlands also improve water quality through mechanical, physical, physico-chemical, biological and biochemical processes. These abilities are also used in constructed wetlands for wastewater treatment but within a more controlled environment. In addition, wetlands provide the supporting services necessary for the production of all other ecosystem services such as soil formation and retention, nutrient cycling, primary production or water cycling. In short, wetlands are clearly among the most valuable ecosystems on Earth.

In 1993, we started to think about a meeting that would bring together scientists and researchers studying natural and constructed wetlands. At that time, both groups usually met during separate meetings. Researchers in constructed wetlands for water treatment were intensively developing technology in the early 1990s, while natural wetlands had been studied for several decades by that time. We believed that "constructed wetlands people" would benefit from the knowledge coming from "natural wetlands people" and vice versa, and that these two groups should share their knowledge. Therefore, in 1995 we organized a workshop "Nutrient Cycling in Natural and Constructed Wetlands" at Třeboň in South Bohemia. The meeting was a success and we were encouraged to continue organizing future meetings. The first workshop was primarily aimed at constructed wetlands but during the following meetings more papers dealing with natural wetlands, such as peatlands, floodplains or wet meadows were submitted. Papers dealing with processes occurring in wetlands, such as evapotranspiration or energy dissipation, and ecosystem services such as biomass production, nutrient and sediment storage or flood retention have become more frequent.

The seventh meeting took place again at Třeboň, for the first time in spring (April 22–25, 2009). The workshop was attended by participants from 16 countries in Europe, North America and Australia. This volume contains a selection from the papers presented during the workshop and which were subsequently peer-reviewed. In the first part of the book, chapters dealing with the use of constructed wetlands are found, while the second part of the book deals with natural wetlands.

The organization of the workshop was partially supported by grant No. 2B06023 "Development of Mass and Energy Flows Evaluation in Selected Ecosystems" from the Ministry of Education, Youth and Sport of the Czech Republic.

Praha, Czech Republic
December 2009

Jan Vymazal

Contents

Contributors

Edit Ágoston-Szabó Hungarian Danube Research Station of the Hungarian Academy of Sciences, 2163 Vácrátót, Hungary, edit@botanika.hu

Herbert John Bavor Centre for Water and Environmental Technology – Water Research Laboratory, University of Western Sydney – Hawkesbury, Locked Bag 1797, Penrith South DC, NSW 1797, Australia, j.bavor@uws.edu.au

Geert Bogaert Department of Applied Ecology and Environmental Biology, Ghent University, Coupure Links 653, B-9000 Gent, Belgium

Jakub Brom Faculty of Agriculture, Department of Landscape Management, University of South Bohemia in České Budějovice, Studentská 13, 370 05 České Budějovice, Czech Republic, jbrom@zf.jcu.cz

David Cooper ARM Ltd, Rydal House, Colton Road, Rugeley, Staffordshire, WS15 3HF, UK

Philip A. Davies Sustainable Environment Research Group, Aston University, Birmingham B4 7ET, UK

Mária Dinka Hungarian Danube Research Station of the Hungarian Academy of Sciences, 2163 Vácrátót, Hungary

Nataša Dolinar Biotechnical faculty, Department of Biology, University of Ljubljana, Večna pot 111, 1000 Ljubljana, Slovenia

Gijs Du Laing Laboratory of Analytical Chemistry and Applied Ecochemistry, Ghent University, Coupure Links 653, B-9000 Gent, Belgium, Gijs.DuLaing@UGent.be

Jiří Dušek Institute of Systems Biology and Ecology, v.v.i., Academy of Sciences of the Czech Republic, Na Sádkách 7, České Budějovice, Czech Republic, jiri.dusek@usbe.cas.cz

Ivet Ferrer Environmental Engineering Division, Department of Hydraulic, Maritime and Environmental Engineering, Technical University of Catalonia (UPC), c/ Jordi Girona 1-3, Building D1, E-08034 Barcelona, Spain

Nat Fonder Epuvaleau, npa, Hydrology and Rural Hydraulics Department, University of Liege, Gembloux Agro-Bio Tech, Passage des déportés, 2-5030, Gembloux, Belgium, fonder@epuvaleau.eu

Alenka Gaberščik Biotechnical faculty, Department of Biology, University of Ljubljana, Večna pot 111, 1000 Ljubljana, Slovenia, alenka.gaberscik@bf.uni-lj.si

Magdalena Gajewska Faculty of Civil & Environmental Engineering, ul, Gdansk University of Technology, Narutowicza 11/12, 80-233 Poznaň, Poland, mgaj@pg.gda.pl

Joan García Environmental Engineering Division, Department of Hydraulic, Maritime and Environmental Engineering, Technical University of Catalonia (UPC), c/ Jordi Girona 1-3, Building D1, E-08034 Barcelona, Spain, joan.garcia@upc.edu

David Güell Depuradores d'Osona, S.L. c/Historiador Ramon d'Abadal i de Vinyals 5, 3r, E-08500 Vic, Spain

Tom Hanssen Laboratory of Analytical Chemistry and Applied Ecochemistry, Ghent University, Coupure Links 653, B-9000 Gent, Belgium

Tom Headley Helmholtz Centre for Environmental Research (UFZ), Permoserstrasse 15, D-04318, Leipzig, Germany

Petra Hesslerová ENKI, o.p.s., Dukelská 145, 379 01 Třeboň, Czech Republic and Department of Applied Geoinformatics and Spatial Planning, Faculty of Environmental Sciences, Czech University of Life Sciences Prague, Kamýcká 1166, Praha, CZ-165 00, Czech Republic, M42barta@seznam.cz

Hanna Huryna Institute of Physical Biology, University of South Bohemia, Zámek 136, 37 333 Nové Hrady, Czech Republic

Paul R Knowles Sustainable Environment Research Group, Aston University, Birmingham B4 7ET, UK, knowlesp@aston.ac.uk

Lenka Kröpfelová ENKI, o.p.s., Dukelská 145, 379 01 Třeboň, Czech Republic; Faculty of Environmental Sciences, Department of landscape Ecology, Czech Republic and Czech University of Life Sciences, náměstí Smiřických 1, 281 63 Kostelec nad Černými lesy, Czech Republic

Günter Langergraber Institute of Sanitary Engineering and Water Pollution Control, University of Natural Resources and Applied Life Sciences, Vienna (BOKU), Muthgasse 18, A-1190 Vienna, Austria, guenter.langergraber@boku.ac.at

Piet N.L. Lens Department of Environmental Resources, UNESCO-IHE Institute for Water Education, PO Box 3015, 2601 DA Delft, The Netherlands

Stephanie Liey Ecole Supérieure d'Agriculture, 55, rue Rabelais, F 49000 Angers, France

Esther Llorens Environmental Engineering Division, Department of Hydraulic, Maritime and Environmental Engineering, Technical University of Catalonia (UPC), c/ Jordi Girona 1-3, Building D1, E-08034 Barcelona, Spain

Clodagh Murphy ARM Ltd, Rydal House, Colton Road, Rugeley Staffordshire, WS15 3HF, UK, clodagh.murphy@armreedbeds.co.uk

Jaime Nivala Environmental and Biotechnology Center (UBZ), Helmholtz Center for Environmental Research (UFZ), Permoserstrasse 15, 04318 Leipzig, Germany, jaime.nivala@gmail.com

Hanna Obarska-Pempkowiak Faculty of Civil & Environmental Engineering, ul, Gdansk University of Technology, Narutowicza 11/12, 80-233 Poznaň, Poland

Arkadiusz Ostojski Department of Sanitary Engineering, Faculty of Civil and Environmental Engineering, Gdansk University of Technology, 11/12 Narutowicza Str, 80-952 Gdansk, Poland, arkusost@pg.gda.pl

Tomáš Picek Department of Ecosystem Biology, University of South Bohemia, Faculty of Science, Branišovská 31, 370 05 České Budějovice, Czech Republic

Jan Pokorný ENKI o.p.s., Dukelská 145, 379 01 Třeboň, Czech Republic, pokorny@enki.cz

Paula Pongrac Biotechnical faculty, Department of Biology, University of Ljubljana, Večna pot 111, 1000 Ljubljana, Slovenia

Jan Procházka Faculty of Agriculture, Department of Landscape Management, University of South Bohemia in České Budějovice, Studentská 13, 370 05, České Budějovice, Czech Republic

Georges Reeb Atelier Reeb, 13 Quai des Bateliers, 67000, Strasbourg, France, georges.reeb@atelier-reeb.fr

Marjana Regvar Biotechnical faculty, Department of Biology, University of Ljubljana, Večna pot 111, 1000 Ljubljana, Slovenia

Alžběta Rejšková ENKI o.p.s., Dukelská 145, 379 01 Třeboň, Czech Republic

Diederik P.L. Rousseau Department of Environmental Resources, UNESCO-IHE Institute for Water Education, PO Box 3015, 2601 DA Delft, The Netherlands, d.rousseau@unesco-ihe.org

Miloš Rozkošný Výzkumný ústav vodohospodářský T.G.M., v.v.i., Mojmírovo nám.16, 612 00, Brno, Czech Republic, milos_rozkosny@vuv.cz

Saleh A.A.E. Saleh Department of Environmental Resources, UNESCO-IHE Institute for Water Education, PO Box 3015, 2601 DA Delft, The Netherlands

Christian B. Sekomo Department of Chemistry, National University of Rwanda, PO Box 117, Butare, Rwanda; Department of Environmental Resources, UNESCO-IHE Institute for Water Education, PO Box 3015, 2601 DA Delft, The Netherlands

Nina Šraj Biotechnical faculty, Department of Biology, University of Ljubljana, Večna pot 111, 1000 Ljubljana, Slovenia

Filip M.G. Tack Laboratory of Analytical Chemistry and Applied Ecochemistry, Ghent University, Coupure Links 653, B-9000 Gent, Belgium

Enrica Uggetti Environmental Engineering Division, Department of Hydraulic, Maritime and Environmental Engineering, Technical University of Catalonia (UPC), c/ Jordi Girona 1-3, Building D1, E-08034 Barcelona, Spain

Jan Vymazal Faculty of Environmental Sciences, Department of Landscape Ecology, Czech University of Life Sciences in Prague, Prague, Czech Republic; ENKI, o.p.s., Dukelská 145, 379 01 Třeboň, Czech Republic, vymazal@yahoo.com

Scott Wallace Naturally Wallace Consulting, LLC. 4276 Greenhaven Court, Vadnais Heights, MN 55127, USA

Ewa Wojciechowska Faculty of Civil & Environmental Engineering, ul, Gdansk University of Technology, Narutowicza 11/12, 80-233 Poznaň, Poland

Chapter 1
Application of Constructed Wetlands in Recycling, Agriculture and Agroforestry: Water Management for Changing Flow Regimes

Herbert John Bavor

Abstract Increasingly, constructed wetland systems are being utilized for treatment and buffering of effluent and runoff water, functioning in nutrient removal, disinfection and also as transitional environments in recycling applications. Agriculture and agroforestry opportunities are abundant for water treated by wetland systems. Progressively, wetland use as a component in potable recycling may be more acceptable in areas as changing flow regimes impact on water supply. This paper examines an innovative wetland application, the McGraths Hill Wetland Complex in the Hawkesbury District of NSW, Australia. The system has operated for more than a decade, founded on a concept to develop approximately 100 ha of land as an integrated reuse and wetland system with the objectives of (a) minimising effluent derived pollutants discharged to receiving waters and increasing water and nutrient reuse while meeting EPA requirements; (b) improving the visual amenity of the entrance to the local township through enhancement of the natural environment; (c) providing an economic return while reducing management costs; (d) providing and improving floodplain wetlands, receiving water riparian zones and other potential habitats; (e) providing a passive recreation resource and educational and research opportunities. The system is considered a success and, in addition to significantly reducing direct nutrient discharge to receiving waters, has produced of the order of 10,000 bales of lucerne hay (approximately 4200 T) per year and provided a woodlot resource. The integrated treatment complex is suggested as a key model for "treatment train" approaches to wastewater management. Changing flow regimes brought about by climate change and increased water abstraction are making multi-function, green technologies critical for community sustainability.

Keywords Agroforestry · Climate change · Recycled water · Surface flow · Sustainability · Water quality · Wetlands

---

H.J. Bavor (✉)
Water Research Laboratory, University of Western Sydney – Hawkesbury, Locked Bag 1797, Penrith South, NSW 1797, Australia
e-mail: j.bavor@uws.edu.au

J. Vymazal (ed.), *Water and Nutrient Management in Natural and Constructed Wetlands*, DOI 10.1007/978-90-481-9585-5_1,

1.1 Introduction

Over much of the world, precipitation fed surface and groundwater utilisation has met or exceeded sustainable levels. Wastewater, via various ameliorating treatments, represents a valuable "new" water resource. The McGraths Hill Sewage Treatment Plant (STP), operated by the local Hawkesbury Council, was originally constructed in 1938 as a traditional oxidation pond system and then upgraded with a trickling filter in 1969 to serve an equivalent population of 9500. The plant was designed to carry out tertiary treatment of the effluent including disinfection and oxygenation in a series of four polishing ponds. For many years, effluent from the polishing ponds was discharged to South Creek, a tributary flowing into the Hawkesbury River, via a natural depression and ephemeral wetland. With the addition of effluent flow, the wetland became a perennial water body covering approximately 5 ha of open water and 3 ha of marshland. In 1988 Council gained approval from the EPA to increase disinfection detention time to 30 days and reuse effluent on approximately 20 ha of cropping land and 6 ha of woodlot adjacent to the STP.

By 1995, the local population and input flows to the STP had more than doubled and water quantity and quality constraints were of concern in the receiving waters. More than 800 ML yr^{-1} (800×10^3 m^3 yr^{-1}) of secondary treated effluent, contributing loads of approximately 6400 kg N yr^{-1} and 1400 kg P yr^{-1}, was discharged to the receiving creek. Council took steps to acquire an additional 50 ha of predominantly floodplain land which surrounded the STP. A steering committee was established comprising representatives from Council, Hawkesbury Nepean Catchment Management Trust, EPA, Greening Australia, University of Western Sydney and the New South Wales Department of Land and Water Conservation, to plan and develop the STP complex as an integrated agriculture/agroforestry and wetland system. The key objectives and performance outcomes of the implementation are presented below.

Concept objectives of:

- Minimising effluent derived pollutants discharged to receiving waters and increasing water and nutrient reuse while meeting EPA requirements;
- improving the visual amenity of the entrance to the local township through enhancement of the natural environment;
- providing an economic return while reducing management costs;
- providing and improving floodplain wetlands, receiving water riparian zones and other potential habitats;
- providing a passive recreation resource and educational and research opportunities.

1.2 Results and Discussion

Implementation of the concept involved soils and geotechnical survey of the site, lime treatment and compaction of sub-soil to ensure low infiltration rates, thus minimising the potential for minimise percolation (Sakadevan, Maheshwar, & Bavor,

2000). A modular free-water-surface wetland design was used to allow for maintenance of individual wetland cells, allow for isolation of wetland cells in the event of high nutrient or toxic spill events which could be captured and managed, allow for increased wet weather storage and to minimise the potential for short circuiting of flows (Bavor & Mitchell, 1994). A schematic of the system is set out in Fig. 1.1.

The existing secondary treatment STP, a traditional trickling filter system, and disinfection ponds were retained in the scheme on the 76 ha site. The ponds provided a minimum of 30 days effluent retention for input flows of about 3250 m^3 d^{-1}. After detention ponding, approximately 25% of the flow was spray-irrigated over approximately 31 ha of *Medicago sativa* (Lucerne or Alfalfa) pasture (Fig. 1.2).

It should be noted that in the secondary-treated effluent used in the irrigation had N concentrations of 20–35 mg L^{-1} and P concentrations of 7–10 mg L^{-1}. Remaining flows were dosed with alum (aluminum sulphate), mixed in a flocculation basin and then settled in phosphorus sedimentation ponds. The flocculation treatment reduced total phosphorus concentration from an average of 5.2–1.0 mg L^{-1}, prior to discharge into a modular constructed wetland system. The constructed wetland system was developed as 15 module cells, each cell approximately 20 m wide, 100 m long and 0.3–0.5 m in depth. The wetland cells further reduced total phosphorus concentration to less than 0.25 mg L^{-1} and total nitrogen to less than 5.0 mg L^{-1}, as shown in Table 1.1.

Some of the wetland outflow was irrigated over woodlots and amenity landscaped areas, with the remainder discharged into a creek adjacent the treatment site. Flows, irrigation and discharge volumes are presented in Fig. 1.3.

The woodlots were planted with *Eucalyptus robusta* (Swamp Mahogany) and *Eucalyptus camaldulensis* (River Red Gum). Additional native species which may be considered in future plantings are *Eucalyptus grandis* (Flooded Gum) and *Eucalyptus nitens* (Shining Gum), which have been reported to perform well in regional effluent irrigation trials (Sharma & Ashwath, 2006; Moroni, Worledge, & Beadle, 2003).

It is considered important to consider a range of pasture forages and forestry options to enhance high productivity and longevity in the agroforestry cropping stands. Winter dormancy and resistance to Anthracnose (*Colletotrichum trifolii* fungal rot) are potential problems for lucerne cropping under flooded or irrigation conditions (Lattimore, 2009). Further, it has been noted that elevated concentrations of soil nitrogen have been strongly associated with eucalypt decline that was in turn correlated to increased foliar N and P (Davidson et al., 2007; Dugald, Close, Davidson, & Watson, 2008). In recycled water agroforestry complexes, managed as production systems, it may be desirable to operate with some periodic reduction in the effluent loading rate. Such management practices would reduce the tendency for irrigated trees to have a larger total biomass with a smaller proportion of below ground biomass and also promote the development of a robust root system throughout the soil profile and reduce susceptibility to toppling (Moroni, Worledge, & Beadle, 2003).

The development of the constructed wetland complex, coupled with phosphorus flocculation and utilization of treated effluent for irrigation, has resulted in very significant reduction of nutrient loads to the receiving water. The STP prior to

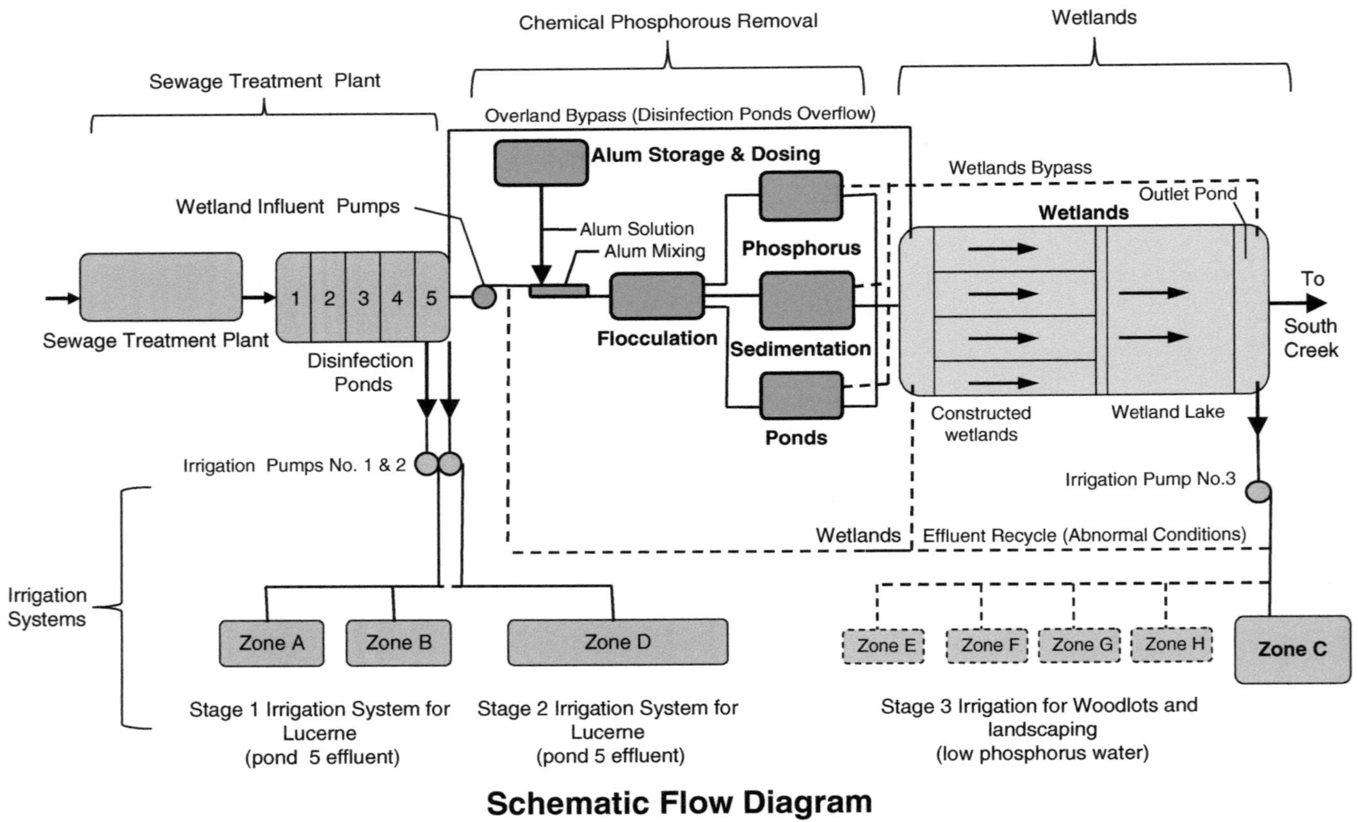

Fig. 1.1 Schematic of STP complex showing integrated system flow through disinfection ponds, phosphorus precipitation dosing, constructed wetland cells, irrigation zones and system discharge

Fig. 1.2 Spray irrigation pasture planted with *Medicago sativa.* Woodlot areas planted with *Eucalyptus robusta* and *Eucalyptus camaldulensis* are shown in the background. (Photo author)

Table 1.1 Nitrogen, phosphorus and BOD_5 reduction following sequential disinfection pond (Pond 5), alum-phosphorus sediment ponding (PSP) and constructed wetland (CW) treatment. Values are presented as mg l^{-1} for mean yearly performance, 2008. Sampling points are shown in Fig. 1.1

Sampling point	Pond 5 outlet (mg L^{-1})	PSP outlet (mg L^{-1})	CW outlet (mg L^{-1})
Total nitrogen	21.89	5.09	3.06
Total phosphorus	7.75	1.07	0.25
BOD_5	7.88	10.69	6.17

implementation of integrated agroforestry reuse and the constructed wetland system discharged 6400 kg N and 1400 kg P yr^{-1} into a downstream creek. Following the upgrade, over all direct treated effluent discharge to the receiving creek was reduced by only about 12%. Significantly, however, the improved nutrient removal achieved by the integrated system resulted in lower nitrogen and phosphorus loads of 3550 kg N yr^{-1} and 1420 kg P yr^{-1}, respectively. These loads represented 44.8% nitrogen and 72.1% phosphorus reductions, on a yearly basis, of direct nutrient input from the STP to the receiving creek. Thus, volumetric environmental flow was only marginally reduced while nutrient inputs were very greatly reduced.

An additional beneficial outcome for the scheme has been realized through the sale of lucerne hay harvested from the irrigated pasture cropping areas (Fig. 1.4).

Yield from the Lucerne crop has been as high as 4.200 T yr^{-1} which, at a sale price of \$125 T yr^{-1}, has brought in over \$500000 per year over good growing years. At that level of return the integrated system development cost of \$2.2 M could be recouped in just over 4 years. Such economic performance has made consideration of such schemes very attractive to regional water management authorities and agroindustries (Davies, Sakadevan, & Bavor, 2001; Davison, Pont, Bolton, & Headley, 2006).

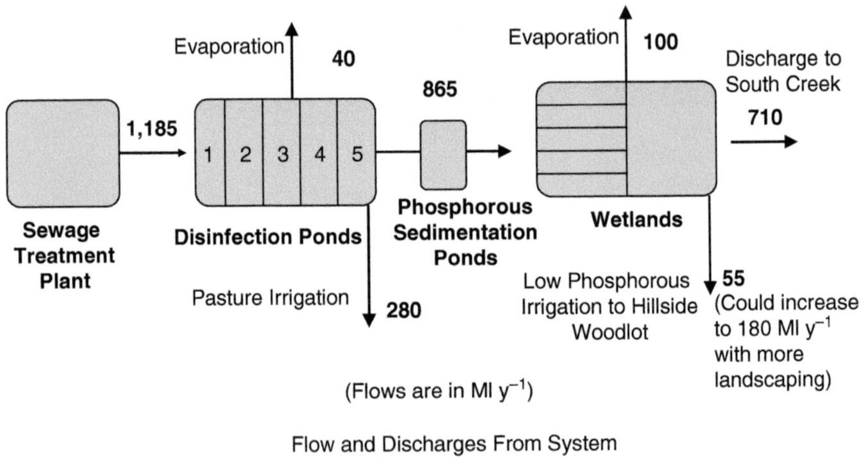

Fig. 1.3 Flow and treated effluent discharges from STP and constructed wetland system. Volumes are presented in ML yr^{-1} (1 ML = 10^3 m^3)

Fig. 1.4 Irrigated pasture and wetland bays. Bales of Lucerne hay are shown, ready for transport to local livestock farmers. (Photo author)

1.3 Conclusions

The planning and implementation of an integrated wastewater treat plant and agroforestry system was considered an environmental and economic success. The system met design objectives and achieved 44.8% nitrogen and 72.1% phosphorus load reductions in direct discharge to receiving waters. In addition to significantly reducing eutrophication potential, the system has produced of the order of 10,000 bales of Lucerne hay (approximately 4200 T) per year and provided a woodlot resource. The integrated treatment complex is suggested as an important model for "treatment train" approaches to wastewater management. Changing flow regimes brought about by climate change and increased water abstraction are making multi-function, green technologies critical for community sustainability.

References

Bavor, H. J., & Mitchell, D. S. (Eds.). (1994). Wetland systems in water pollution control. *Water Science and Techoogy, 29*(4), 1–336.

Davidson, N. J., Close, D. C., Battaglia, M., Churchill, K., Ottenschlarger, M., Watson, T., & Bruce, J. (2007) Eucalypt health and agricultural land management within bushland remnants in the Midlands of Tasmania, Australia. *Biological Conservation, 139*, 439–446.

Davies, C. M., Sakadevan, K., & Bavor, H. J. (2001). Removal of stormwater-associated nutrients and bacteria in constructed wetland and water pollution control pond systems. In J. Vymazal (Ed.), *Transformations of nutrients in natural and constructed wetlands* (pp. 483–495). Leiden: Backhuys Publishers.

Davison, L., Pont, D., Bolton, K., & Headley, T. (2006). Dealing with nitrogen in subtropical Australia: Seven case studies in the diffusion of ecotechnological innovation. *Ecological Engineering, 28*, 213–223.

Dugald C. Close, D. C., Davidson, N. J., & Watson, T. (2008). Health of remnant woodlands in fragments under distinct grazing regimes. *Biological Conservation, 141*, 2395–2402.

Lattimore, M. (2009) Current lucerne varities (2009). New South Wales Department of Primary Industries. *Primefact, 705*, 1–3.

Moroni, M. T., Worledge, D., & Beadle, C. L. (2003). Root distribution of Eucalyptus nitens and E. globulus in irrigated and droughted soil. *Forest Ecology and Management, 177*, 399–407

Sakadevan, K., Maheshwar, B., & Bavor, H. J. (2000). Availability of nitrogen and phosphorus under recycled water irrigation. *Australian Journal of Soil Research, 38*, 653–664.

Sharma, A., & Ashwath, N. (2006). Land disposal of municipal effluents: importance of choosing agroforestry systems. *Desalination, 187*, 361–374.

Chapter 2
Properties of Biosolids from Sludge Treatment Wetlands for Land Application

Enrica Uggetti, Ivet Ferrer, Esther Llorens, David Güell, and Joan García

Abstract Sludge treatment wetlands consist of constructed wetlands which have been upgraded for sludge treatment over the last decades. Sludge dewatering and stabilisation are the main features of this technology, leading to a final product which may be recycled as an organic fertiliser or soil conditioner. In this study, biosolids from full-scale treatment wetlands were characterised in order to evaluate the quality of the final product for land application, even without further post-treatment such as composting. Samples of influent and treated sludge were analysed for pH, Electrical Conductivity, Total Solids (TS), Volatile Solids (VS), Chemical Oxygen Demand (COD), Dynamic Respiration Index (DRI), nutrients (Total Kjeldahl Nitrogen (TKN), Total Phosphorus (TP) and Potasium (K)), heavy metals and faecal bacteria indicators (*E. coli* and *Salmonella* spp.). According to the results, sludge water content and therefore sludge volume are reduced by 25%. Organic matter biodegradation leads to VS around 43–44%TS and COD around 500 g $kgTS^{-1}$. The values of DRI_{24h} (1000–1500 mgO_2 $kgTS^{-1}$ h^{-1}) indicate that treated sludge is almost stabilised final product. Besides, the concentration of nutrients is quite low (TKN~4%TS, TP~0.3%TS and K~0.2–0.6%TS). Both heavy metals and faecal bacteria indicators meet current legal limits for land application of the sludge. Our results suggest that biosolids from the studied treatment wetlands could be valorised in agriculture, especially as soil conditioners.

Keywords Compost · Reed beds · Organic waste · Wastewater

J. García (✉)
Environmental Engineering Division, Department of Hydraulic, Maritime and Environmental Engineering, Technical University of Catalonia (UPC), Barcelona, Spain
e-mail: joan.garcia@upc.edu

J. Vymazal (ed.), *Water and Nutrient Management in Natural and Constructed Wetlands*, DOI 10.1007/978-90-481-9585-5_2,

2.1 Introduction

Sewage sludge is the organic waste generated by wastewater treatment processes, after solid and liquid separation units. The amount of sludge produced and its composition depend on the influent's characteristics and wastewater treatment used. Sludge production in conventional activated sludge processes ranges from 60 to 80 g of total solids per person per day. In Europe, the Urban Wastewater Treatment Directive 91/271/EEC (Council of the European Union, 1991) promoted the implementation of wastewater treatment plants (WWTP) with secondary wastewater treatment in municipalities above 2000 Persons Equivalent (PE); and the Water Framework Directive (Council of the European Union, 2000) encouraged wastewater treatment even in municipalities below 500 PE. As a result sludge production has increased in the European Union by 50% since 2005 (Fytili & Zabaniotou, 2008). For instance, in Catalonia (Spain) around 50% of the WWTP (170) were constructed between 2000 and 2006 (Agencia Catalana del Agua, 2007).

In Spain, in order to manage the increasing amount of sludge produced the following hierarchy was proposed (Consejo de Ministros, 2001): (1) valorisation in agriculture, (2) energetic valorisation, and (3) landfilling. Agricultural valorisation is nowadays preferred to landfilling, since sludge recycling ensures the return of organic constituents, nutrients and microelements to crop fields which eases the substitution of chemical fertilizers (Oleszkiewicz & Mavinic, 2002). Sludge disposal onto agricultural land is regulated by the European Sludge Directive, which sets up land application of sewage sludge based on maximum heavy metals concentrations (Council of the European Union, 1986). Recent regulation proposals are more restrictive in terms of heavy metals, and also consider emerging pollutants and microbial faecal indicators (Environment DG, EU, 2000).

In practice, sludge treatment systems have to provide a final product suitable for land application (fulfilling legislation requirements), with reasonable investment as well as operational and maintenance costs. In this sense, sludge treatment wetlands might be regarded as a recent technology for sludge management, which is particularly appropriate for small communities from both an economical and environmental point of view.

Treatment wetlands (TW) reproduce self-cleaning processes occurring in natural wetlands and are being used for wastewater treatment in many regions of the world (Caselles-Osorio et al., 2007). Since the late 1980s, TW have been adapted for sludge treatment developing a technology that is nowadays used in most European countries and in North America (Uggetti, Ferrer, Llorens, & García, 2010).

Sludge treatment wetlands consist of shallow tanks filled with a gravel layer and planted with emergent rooted wetland plants such as *Phragmites australis* (common reed) (Cole, 1998). Thickened secondary sludge is pumped and spread on the wetland's surface. Here, part of the sludge water content is rapidly drained by gravity through the gravel layer; while another part is evapotranspirated by the plants. In this way, a concentrated sludge residue remains on the surface of the bed where, after the resting time, thickened sludge is a new spread, starting the following feeding cycle.

The roots of the plants ease oxygen transfer to the gravel and sludge layers, creating aerobic microsites that promote sludge mineralization and stabilization (Reed, Crites, & Middlebrooks, 1988). Furthermore, the complex root system maintains pores and small channels within sludge layer that preserve the drainage efficiency through the bed (Nielsen, 2003b). When the sludge is dry, the movement of plant stems by the wind prompts the cracking of the surface, improving the aeration of the sludge layer.

During feeding periods, the sludge layer height increases at a certain rate (around 10 cm yr^{-1}). When the layer approaches the top of the tank, feeding is stopped during a final resting period (from 1 to 2 months to 1 year), aimed at improving sludge dryness and mineralisation. The final product is subsequently withdrawn, starting the following operating cycle.

The quality of this product is the result of both dewatering processes (draining and evapotranspiration) and organic matter biodegradation (Nielsen, 2003b). According to Nielsen and Willoughby (2005) it is suitable for land application; although further post-treatments might be required to improve sludge hygienisation (Zwara & Obarska-Pempkowiak, 2000). Nevertheless, detailed studies on the properties of biosolids from treatment wetlands are still lacking in the literature.

In this study, full-scale treatment wetlands were evaluated with the aim of studying the efficiency of the process in terms of sludge dewatering, mineralization and hygienisation; while assessing the quality of the final product for land application. To this end, physico-chemical and microbiological parameters, together with stability indexes, were considered as proposed in the European Sludge Directive (Council of the European Union, 1986), the 3rd Draft EU Working Document on Sludge (Environment DG, EU, 2000) and the 2nd Draft EU Working Document on Biological Treatment of Biowaste (Environment DG, EU, 2001).

2.2 Materials and Methods

2.2.1 System Description

The studied full-scale treatment wetlands are located at the WWTP of Seva (1500 PE), in the province of Barcelona (Catalonia, Spain). The main characteristics of the facility are summarised in Table 2.1. The sludge treatment wetlands were set-up in 2000 by transforming existing conventional drying beds. They are planted with *Phragmites australis*. The total surface area is 175 m^2 and the sludge loading rate around 125 kg TS m^{-2} yr^{-1}, much higher than the recommended value of 50–60 kgTS m^{-2} yr^{-1} (Burgoon, Kirkbride, Henderson, & Landon, 1997; Edwards, Gray, Cooper, Biddlestone, & Willoughby, 2001; Nielsen, 2003a). Other details on the design and operation of the wetlands may be found in Uggetti et al. (2009).

The first operating cycle lasted about 5 years; the sludge was then removed and the process re-started between 2004 and 2005. The second operating cycle was finished by the end of 2008 in two of the wetlands (named 1 and 2). After a resting

Table 2.1 Characteristics of the WWTP in Seva (Catalonia, Spain)

Treated population equivalent	1500
Type of treatment	Contact-stabilisation
Wastewater flow rate (m^3d^{-1})	180 (summer)/400 (winter)
Sludge production (kg TS d^{-1})	60
Number of treatment wetlands (beds)	7
Total surface area (m^2)	175
Nominal height for sludge accumulation (m)	~0.8
Sludge loading rate (kg TS m^{-2} yr^{-1})	125

period of 4 months, the wetlands were emptied with a power shovel and the final product was thereafter transported to a composting plant.

2.2.2 Sludge Sampling and Characterization

When the wetlands were emptied (in 2008), the layer of dry sludge was about 60 cm high. Two composite samples were prepared by mixing subsamples from each wetland; while an integrated influent sample sludge was obtained from subsamples collected during the whole week.

As recommended in the literature (Mujeriego & Carbó, 1994; Obarska-Pempkowiak, Tuszynska, & Sobocinski, 2003; Soliva, 2001), the sludge quality was characterised in terms of: pH, Electrical Conductivity (EC), Total and Volatile Solids (TS and VS), Chemical Oxygen Demand (COD), Total Kjehldahl Nitrogen (TKN), Potassium (K), Total Phosphorous (TP), heavy metals and faecal bacteria indicators (*Salmonella* spp. and *Escherichia coli*). Additionally, the Dynamic Respiration Index (DRI) was determined according to Adani, Lozzi, and Genevini (2000) and Barrena et al. (2009).

Sludge samples were analysed following the Standard Methods (APHA-AWWA-WPCF, 2001). Samples for COD, TKN, TP, K and heavy metals' analyses were previously air dried at room temperature (until constant weight); hence the results are expressed on a dry matter basis (per kg or %TS). Air dried samples were subsequently diluted in distilled water (1:5) for pH and EC measurements.

2.3 Results and Discussion

The efficiency of the process in terms of sludge dewatering and mineralisation is usually evaluated by the increase in dry matter (TS) and the decrease in organic matter (VS and COD) contents, respectively. However, total organic matter content is not sufficient to assess the stability of the product; indeed information on the amount of readily biodegradable organic matter fraction is also needed. For instance, it can be deduced from the DRI. Besides, the concentration of nutrients, heavy metals and faecal bacteria indicators are used to determine the quality of biosolids for its application on land as organic fertilizers.

Tables 2.2 and 2.3 show the main characteristics of the influent and treated sludge from the full-scale treatment wetlands. The results are here examined and discussed. A comparison with average compost characteristics is proposed to assess the requirement of additional composting post-treatment.

2.3.1 Sludge Dryness

Secondary sludge produced by the contact-stabilization process is spread on the beds with very high water content (typically around 99%) (Table 2.2). Sludge moisture is significantly reduced down to about 75% during the treatment and after

Table 2.2 Physico-chemical properties of influent sludge and biosolids from treatment wetlands

Parameter	Influent	Wetland 1	Wetland 2
Physical properties			
pH	6.75	6.21	6.27
EC 1:5 ($dS\ m^{-1}$)	0.3	1.51	1.88
TS (%)	1.1 ± 0.0	24.2 ± 0.6	25.8 ± 2.1
Organic matter			
VS (TS%)	51.5 ± 0.8	42.9 ± 1.8	44.6 ± 3.0
COD ($g\ kgTS^{-1}$)	709 ± 11	554 ± 32	494 ± 55
$DRI_{24\,h}$ ($mgO_2\ kgTS^{-1}\ h^{-1}$)	–	1400 ± 300	1100 ± 200
Nutrients			
TKN (%TS)	9.76	4.02	4.87
TP (%TS)	2.68	0.13	0.39
K (%TS)	0.27	0.18	0.62

Note: TS, VS and COD were analysed in triplicate; DRI was analysed in duplicate

Table 2.3 Concentration of heavy metals and faecal bacteria indicators in influent sludge and biosolids from treatment wetlands. Metals concentration referred to TS, while bacteria indicators to fresh weight

Parameter	Influent	Wetland 1	Wetland 2	Limit values
Heavy metals				
Cr (ppm)	51	55	59	800
Ni (ppm)	39	30	32	200
Cu (ppm)	252	318	213	800
Zn (ppm)	719	588	641	2000
Cd (ppm)	1.7	0.8	0.8	5
Hg (ppm)	<1.5	<1.5	<1.5	5
Pb (ppm)	53	73	76	500
Faecal bacteria indicators				
Salmonella spp. (presence/absence in 25 g)	Absence	Absence	Absence	Absence in 50 g
E. coli ($MPN\ g^{-1}$)	<3	<3	<3	<500

Note: Limit values proposed in the 3rd Draft EU Working Document on Sludge (Environment DG, EU, 2000)

a resting period of 4 months. The sludge volume is consequently reduced by 25%. This is the main goal of dewatering processes.

The dryness of the final product (TS~25%) is lower than that observed in other facilities after a resting period of approximately 1 year (TS around 30–40%) (Nielsen, 2003b). In this sense, a previous study indicated a poor dewatering efficiency in Seva's system compared to other Catalan facilities (Uggetti et al., 2009). One possible reason for this is that the sludge loading rate (125 kgTS m^{-2} yr^{-1}) is over twice the recommended value (50–60 kgTS m^{-2} yr^{-1}) (Burgoon et al., 1997; Edwards et al., 2001; Nielsen, 2003a). This fact suggests that the dryness of the final product could be further increased with a better system management (i.e. reducing the sludge loading rate).

2.3.2 Organic Matter Content and Stability

Table 2.2 shows organic matter contents expressed as VS and COD; as well as the DRI value. The concentration of VS in the influent is quite low (~51%TS) as a result of the high solids retention time in the contact-stabilisation process. Final values (43–45%TS) are within the range obtained after conventional sludge stabilisation techniques, such as anaerobic digestion (Ferrer, 2008; Ferrer, Ponsá, Vázquez, & Font, 2008). Therefore, VS removal (7–8%) is remarkably lower than that usually observed in these type of processes with higher VS in the sludge influents (Ferrer, 2008; Ferrer et al., 2008). Again, the results are in accordance with a previous study on this system (Uggetti et al., 2009). COD values indicate total organic matter reduction from 700 to some 500 g $kgTS^{-1}$ (Table 2.2).

Comparisons with other systems are not straightforward, since the biodegradability of the sludge depends on a number of parameters, including its nature and composition, amongst others. For instance, VS were reduced to 20%TS in pilot scale treatment wetlands in China (Yubo, Tieheng, Lihui, Tingliang, & Liping, 2008). On the other hand, in compost samples organic contents are usually much higher, around 60% for compost of sewage sludge mixed with vegetable wastes (Bertrán, Sort, Soliva, & Trillas, 2004), due to humic-like substances produced during composting process.

Organic matter in soil amendments can improve the properties and quality of soils, which is essential to guarantee long-term soil fertility (Draeger, Pundsack, Jorgenson, & Mulloy,1999). In particular, an increase in organic matter content can improve physical properties (water retention, soil structure, water infiltration, bulk density, porosity), chemical properties (cation exchange capacity, pH) and, in some cases, biological properties (Moss, Epstein, & Logan, 2002, Andreoli, Pegorini, Fernandes, & Santos, 2007). Such a response depends on the sludge:soil ratio (Singh & Agrawal, 2008).

On the other hand, higher biological stability implies lower environmental impacts (like odour generation, biogas production, leaching and pathogen's regrowth) during land application of the product (Muller, Fricke, & Vogtmann, 1998). Lasaridi and Stentiford, (1998) define biological stability as a characteristic that determinates the extent to which readily biodegradable organic matter has been decomposed. Referred to compost, the stability is a quality parameter related to the microbial decomposition or microbial respiration activity of the composted matter (Komilis & Tziouvaras, 2009).

The DRI is based on the rate of oxygen consumption and is a useful indicator of the biological stability of a sample. In this study, the DRI_{24h} from wetland samples ranged between 1100 and 1400 mgO_2 $kgTS^{-1}$ h^{-1}. Such a stability degree is much higher than that reported in the literature for a mixture of primary and activated sludge (6680 mgO_2 $kgTS^{-1}$ h^{-1}) and for anaerobically digested sludge (3740 mgO_2 $kgTS^{-1}$ h^{-1}) (Pagans, Barrena, Font, & Sánchez, 2006).

In a recent study, Ponsá, Gea, Alerm, Cerezo, and Sánchez (2008) analysed the DRI of the organic fraction of municipal solid wastes at different stages of a mechanical biological treatment. These authors observed DRI values above 7000 mgO_2 $kgTS^{-1}$ h^{-1} for the input material, a decrease to around 1500 mgO_2 $kgTS^{-1}$ h^{-1} for digested material and near 1000 mgO_2 $kgTS^{-1}$ h^{-1} for composted material, with a value of 1000 mgO_2 $kgTS^{-1}$ h^{-1} for the output material. Similarly, Scaglia and Adani (2008) found values around 2500 mgO_2 $kgTS^{-1}$ h^{-1} for input samples around 1100 mgO_2 $kgTS^{-1}$ h^{-1} for intermediate samples and between 300–600 mgO_2 $kgTS^{-1}$ h^{-1} for the final product of the stabilisation process.

If DRI values are expressed with respect to the organic matter content (VS) of the sample, the values from wetlands' biosolids correspond to 490 and 610 mgO_2 $kgVS^{-1}$ h^{-1}. According to Scaglia, Tambone, Genevini, and Adani (2000) and Adani, Confalonieri, and Tambone (2004), stability values of 1000 mgO_2 VS^{-1} h^{-1} are representative of medium compost. This is in accordance with compost classes I and II proposed by the American Society for Testing and Materials (1996), while compost classes III and IV have higher biological stabilities (<500 mgO_2 VS^{-1} h^{-1}).

From the comparison of the obtained results with those of other systems, it seems that biosolids from the studied treatment wetlands may be considered a partially (and almost) stabilised material. Therefore it can be speculated that, with sufficient resting time, the final product may be valorised in agriculture even without further post-treatment in a composting plant. Consequently, this would result in additional reduction of sludge treatment costs.

2.3.3 Nutrients

Sewage sludge may provide essential nutrients for plant growth. Biosolids are able to restore nitrogen, phosphorus, sulphur and other nutrients in soils. The concentration of nutrients in biosolids depends on sewage composition and treatment used, and on subsequent sludge treatment processes.

Although they are essential for plant growth, nutrients (particularly nitrogen and phosphorus) can be harmful when excessively applied. Different works have proved nitrogen accumulation in soil (Walter, Cuevas, García, & Martínez, 2000, Hernández, Moreno, & Costa, 1999); as well as phosphorus increase in sludge-amended soils (Hernández et al., 1999). It is well known that over application of nitrogen can lead to nitrate contamination of groundwater; although such a risk is reduced if nutrients are applied at agronomic rates (Moss et al., 2002). The great solubility of nitrate poses a high contamination hazard to groundwater and is the main reason why biosolids application in agricultural land is usually limited by the nitrogen uptake crop capability. In this sense, the application rate must not lead to nitrogen inputs greater than the crop nitrogen requirements, in order to avoid leaching to occur (Andreoli et al., 2007).

The concentration of the main nutrients (nitrogen, phosphorus and potassium) in wetland biosolids is shown Table 2.2. The results are consistent with a previous study carried out at the same facility (Uggetti et al., 2009).

Sludge's nitrogen comes from microbial biomass present in sludge and from wastewater residues. In this study, TNK values (Table 2.2) decrease about a 50% from the influent to the biosolids (~4%TS). The values of the final sludge are therefore within the range of activated sludge (Andreoli et al., 2007). For compost of sewage sludge, Bertrán et al. (2004) give slightly lower TNK values (2.53%TS). Even lower TNK values (1.5%TS) are given for aerobically digested sewage sludge used on land (Gascó & Lobo, 2007).

Phosphorus in sludge comes from biomass formed during wastewater treatment, residues and phosphate-containing detergents and soaps. Biosolids can be seen as phosphorus source assuring a slow and continued release to plants (Andreoli et al., 2007). In this study, TP values show a clear decrease from the influent to treated sludge (0.08–0.28%TS). The values of the final product are within the range of digested sludge (Gascó & Lobo, 2007) and quite lower than in composted sewage sludge (Bertrán et al., 2004).

The concentration of potasium does not seem to vary along the treatment, with values ranging between 0.18 and 0.62%TS. These values are in accordance with sludge compost (Bertrán et al., 2004); but lower than in digested sludge (Gascó & Lobo, 2007).

In general, sludge is characterized by a considerable variability in nutrient's content, depending on the wastewater source and treatment process (Moss et al., 2002). The concentration of nutrients is needed to ensure appropriate dosages of the sludge prior to land application. The required agricultural doses are frequently dependent on the fertilizer and soil characteristics (Pomares & Canet, 2001; Andreoli et al., 2007). Since nitrogen concentration in biosolids usually meets the crop needs, application rates are generally calculated based on the nitrogen requirements of each crop, whereas phosphorus and potassium can be supplemented with chemical fertilisers (Andreoli et al., 2007).

2.3.4 Heavy Metals and Faecal Bacteria Indicators

The main hazard associated to sludge application on agricultural soils is the potential long term accumulation of toxic elements (Singh & Agrawal, 2008), which may then be uptaken by crops. Such elements include both inorganic pollutants, like heavy metals, and organic micropollutants. Currently, however, only heavy metals concentrations are regulated for land application of sewage sludge (Council of the European Union, 1986). Since treated sludge may have considerable amounts of pathogens, depending on the treatment processes used, limit values for faecal bacteria indicators have also been proposed (Environment DG, EU, 2000). According to this proposal, conventionally treated sludge has to contain <500 MNP g^{-1} *E.coli* g, and *Salmonella* spp. has to be absent in 50 g.

Table 2.3 summarises the concentration of heavy metals and faecal bacteria indicators in the treatment wetlands samples, together with the limits proposed in the3rd Draft EU Working Document on Sludge (Environment DG, EU, 2000). There are only little differences between influent sludge and the final product with regards to heavy metals, suggesting that heavy metals accumulation is negligible. Furthermore, in all cases the concentrations are clearly below the limits proposed, which are more restrictive than current legislation (Council of the European Union, 1986).

Heavy metals bioavailability in soil and plants depends on the following parameters: soil pH, plant species growth stage, biosolids source, soil condition and the chemistry of the element (Warmar & Termeer, 2005). According to these authors, it is important to monitor Cu and Zn contents of plant tissues after a few years of sludge applications to verify the tolerance levels for animals and feed, or human food.

With regards to pathogens, it can be seen that *Salmonella* spp. was not detected (Table 2.3). On the other hand, *E. coli* was present but in all cases in small quantities. Both faecal bacteria indicators are well below the limits proposed.

2.4 Conclusions

This study looked at the properties of biosolids from sludge treatment wetlands and compared them with other stabilised products, such as anaerobic digested sludge and compost. Focus was put on the quality of biosolids for its use on land as organic fertilisers and soil conditioners. From this work, the following conclusions can be drawn.

In the full-scale wetlands studied, sludge water content (and volume) is reduced by 25%, from 99 to 75%. Apparently, these results would be further improved by adjusting the sludge loading rate to recommended values, which may reduce the moisture content to 60–70%.

Organic matter biodegradation leads to VS around 45%TS and COD around 500 g $kgTS^{-1}$ in the final product, within the range of digested sludge but lower than in sludge compost. Besides, DRI values (1000–1400 $mgO_2\ kgTS^{-1}\ h^{-1}$) indicate a partly stabilised product, close to a high stabilisation degree corresponding to the final product of a composting process. This suggests that composting post-treatments would not be needed, if sufficient resting time would be left at the end of the wetland's cycle. Monitoring the stabilisation degree during the final resting period would be advisable to minimise the duration of such a period.

The concentration of nutrients, heavy metals and faecal bacteria indicators suggest that the final product would be suitable as organic fertiliser and soil conditioner.

On the whole, the studied system demonstrates the efficiency of sludge treatment wetlands for sludge dewatering and stabilisation, with low treatment costs; and leading to a final product which could be used on land without further post-treatment, reducing sludge management costs. Characterization of similar systems would be advisable to corroborate these results.

Acknowledgements This work was financed by the Catalan Water Agency (ACA) and the Spanish Ministry of Environment (MMARM, Projects A335/2007 and 087/PC08). Technicians of Depuradores d'Osona S.L. are greatly acknowledged for their support to this study. Enrica Uggetti kindly acknowledges the Technical University of Catalonia and Esther Llorens the Juan de la Cierva Programme of the Spanish Ministry of Education and Science (MEC). Joan García is grateful to the School of Civil Engineering of Barcelona (ETSECCPB).

References

Adani, F., Confalonieri, R., & Tambone, F. (2004). Dynamic Respiration Index as descriptor of the biological stability of organic wastes. *Journal of Environmental Quality, 33*, 1866–1876.

Adani, F., Lozzi, P., & Genevini, P. L. (2000). Determination of biological stability by oxygen uptake on municipal solid waste and derived products. *Compost Science and Utilization 9*, 163–178.

Agència Catalana de l'Aigua (2007). La gestió de fangs de depuració d'aigües residuals municipals a Catalunya. Evolució històrica i estratègia de futur (nou programa de fangs). *III Jornades Tècniques de Gestió de Sistemes de Sanejament d'Aigües Residuals*, Barcelona, Spain. URL: http://www.mediambient.gencat.net/aca/ca//agencia/agenda/jornades_tecniques003/ponencies/inici.jsp

American Society for Testing and Materials. (1996). Standard test method for determining the stability of compost by measuring oxygen consumption. D 5975–96. ASTM Int, West Conshohocken, PA.

Andreoli, C. V, Pegorini, E. S., Fernandes, F., & Santos H. F. (2007). Land application of sewage sludge. In M. Von Sperling, C. V. Andreoli & F. Fernandes (Eds)., *Sludge treatment and disposal* (pp. 162–206). London: IWA Publishing.

APHA-AWWA-WPCF (2001). *Standard methods for the examination of water and wastewater* (20th ed.). Washington, DC: American Public Health Association.

Barrena, R., D'Imporzano, G., Ponsá, S., Gea, T., Artola, A., Vázquez, F., et al. (2009). In search of a reliable technique for the determination of the biological stability of the organic matter in the mechanical-biological treated waste. *Journal of Hazardous Materials, 162*(2–3), 1065–1072.

Bertrán, E., Sort, X., Soliva, M., & Trillas, I. (2004). Composting winery waste: sludges and grape stalks. *Bioresource Technology, 95*, 203–208.

Burgoon, P.S., Kirkbride, K.F., Henderson, M., & Landon, E. (1997). Reed beds for biosolids drying in the arid Northwestern Unitated States. *Water Science and Technology, 35*(5), 287–292.

Caselles-Osorio, A., Puigagut, J., Segú, E., Vaello, N., Granés, F., García, D., et al. (2007). Solids accumulation in five full-scale subsurface flow constructed wetlands. *Water Research, 41*, 1388–1398.

Cole, S. (1998). The emergence of treatment wetlands. *Environmental Science and Technology A-Pages, 32*, 218A–223A.

Consejo de Ministros (2001). Resolución de 14 de junio de 2001, de la Secretaría General de Medio Ambiente, por la que se Dispone la Publicación del Acuerdo de Consejo de Ministros, de 1 de junio de 2001, por el que se Aprueba el Plan Nacional de Lodos de Depuradoras de Aguas Residuales *2001–2006. Boletín Oficial del Estado* 166, de 12 de julio de 2001.

Council of the European Union (1986). Council Directive 86/278/EEC of 12 June 1986 on the Protection of the Environment, and in Particular of the Soil, when Sewage Sludge is Used in Agriculture. *Official Journal of the European Union* L 181, 04/07/1986, 6–12.

Council of the European Union (1991). Council Directive 91/271/EEC of 21 May 1991 Concerning Urban Wastewater Treatment. *Official Journal of the European Union* L 135, 30/05/1991, 40–52.

Council of the European Union (1991). Directive 2000/60/EC of the European Parliament and of the Council of 23 October 2000 Establishing a Framework for the Community action in the Field of Water Policy. *Official Journal of the European Union* L 327, 30/05/1991, 1–73.

Council of the European Union (2000). Council Directive 2000/60/EC of 23 October 2000 establishing a framework for Community action in the field of water policy. *Official Journal of the European Union, L 327,* 1–72, 22/12/2000.

Draeger, K., Pundsack, J., Jorgenson, M., & Mulloy, W.E. (1999). *Watershed Effects and biosolids land application: Literature review.* Water Environmental Research Foundation. Project 96-REM-2.

Edwards, J. K., Gray, K. R., Cooper, D. J., Biddlestone, A. J., & Willoughby, N. (2001). Reed bed dewatering of agricultural sludges and slurries. *Water Science and Technology, 44,* (11–12), 551–558.

Environment DG, EU. (2000). *Working Document on Sludge 3rd Draft.* URL: http://ec.europa.eu/environment/waste/sludge/pdf/sludge_en.pdf. Accessed December 2009.

Environment DG, EU. (2001). *Working Document on Biological Treatment of Biowaste 2nd Draft.* URL: http://ec.europa.eu/environment/waste/sludge/pdf/sludge_en.pdf.

Ferrer, I. (2008) *Study of the effect of process parameters on the thermophilic anaerobic digestion of sewage sludge, evaluation of a thermal Sludge pretreatment and overall energetic assessment.* Dissertation, Universitat Autònoma de Barcelona, Spain.

Ferrer I., Ponsá S., Vázquez F., & Font, X. (2008). Increasing biogas production by thermal (70°C) sludge pre-treatment prior to thermophilic anaerobic digestion. *Biochemical Engineering Journal, 42,* 186–192.

Fytili, D., & Zabaniotou A. (2008). Utilization of sludge in EU application of old and new methods – A review. *Renewable & Sustainable Energy Reviews, 12,* 116–140.

Gascó, G., & Lobo, M.C. (2007). Composition of a Spanish sewage sludge and effects on treated soil and olive trees. *Waste Management, 27,* 1494–1500.

Hernández, T., Moreno, J. I., & Costa, F. (1999). Influence of sewage sludge application on croop yields and heavy metals avaylability. *Soil Science & Plant Nutrition, 37,* 201–210.

Komilis, D. P., & Tziouvaras I. S. (2209). A statistical analysis to assess the madurity and stability of six composts. *Waste Management, 29,* 1504–1513.

Lasaridi, K. E., & Stentiford, E. I. (1998). A simple respirometric technique for assessing compost stability. *Water Research, 32,* 3717–3723.

Moss, L. H., Epstein, E., & Logan, T. (2002). *Evaluating risks and benefits of soil amendments used in agriculture.* Alexandria, VA: International Water Association and Water Environmental Research Foundation.

Mujeriego, R., & Carbó, M. (1994). *Sludge reuse in agriculture* (In Catalan). Technical Report. Consorci de la Costa Brava. Girona (Spain).

Muller, W., Fricke, K., & Vogtmann, H. (1998). Biodegradation of organic matter during mechanical biological treatment of MSW. *Compost Science and Utilization, 6,* 42–52.

Nielsen, S. (2003a). Sludge drying reed beds. *Water Science and Technology, 48*(5), 101–109.

Nielsen, S. (2003b). Sludge treatment in wetland systems. In V. Dias & J. Vymazal (Eds.), *The use of aquatic macrophytes for wastewater treatment in constructed wetlands.* (pp. 151–183). Lisbon, Portugal: ICN and INAG.

Nielsen, S., & Willoughby, N. (2005). Sludge treatment and treatment wetlands systems in Denmark. *Water and Environmental Journal, 19,* 296–305.

Obarska-Pempkowiak, H., Tuszynska, A., & Sobocinski, Z. (2003). Polish experience with sewage sludge dewatering in reed systems. *Water Science & Technology, 48*(5), 111–117.

Oleszkiewicz, J. A., & Mavinic, D. S. (2002). Wastewater biosolids: An overview of processing, treatment and management. *Journal of Environmental Engineering Science, 1*, 75–88.

Pagans, E., Barrena, R., Font, X., & Sánchez, A. (2006). Ammonia emissions from the composting of different organic wastes. Dependency on process temperature. *Chemosphere, 62*, 1534–1542.

Pomares, F., & Canet, R. (2001). Organic waste utilization in agriculture: Origin, composition and characterization. In J. Boixadera & M.R. Teira (Ed.), *Agricultural utilization of organic waste* (pp. 1–15). Lleida (Spain). (In Spanish).

Ponsá, S., Gea, T, Alerm, L., Cerezo, J., & Sánchez, A. (2008). Comparision of aerobic and anaerobic stability indices throught a MSW biological treatment process. *Waste Management, 28*, 2735–2742.

Reed, S. C., Crites, R. W., & Middlebrooks, E. J. (1988). *Natural systems for waste management and treatment*. New York: McGraw Hill.

Scaglia, B., & Adani, F. (2008). An index for quantifying the aerobic reactivity of municipal solid wastes and derived waste products. *Science of the Total Environment, 394*, 183–191.

Scaglia B., Tambone, F., Genevini, P. L., & Adani F. (2000). Respiration index determination: A dynamic and static approach. *Compost Science and Utilization, 8*, 90–98.

Singh, R. P., & Agrawal, M. (2008). Potencial enefits and risks of land application of sewage sludge. *Waste Management, 28*, 347–358.

Soliva, M. (2001). *Composting and management of organic residues*. Estudis i Monografies 21. Diputació de Barcelona. Barcelona (Spain), (in Catalan).

Uggetti, E., Ferrer, I., Llorens, E., & García, J. (2010). Sludge treatment wetlands: a review on the state of the art. *Bioresource Technology, 101*, 2905–2912.

Uggetti, E., Llorens, E., Pedescoll, A., Ferrer, I., Castellnou, R., & García, J. (2009). Sludge dewatering and stabilization in drying reed beds: characterization of three full-scale systems in Catalonia, Spain. *Bioresource Technology, 100*, 3882–3890.

Walter, I., Cuevas, G., García, S., & Martínez, F. (2000). Biosolid effects on soil and native plant production in a degraded semiarid ecosystem in central Spain. *Waste Management & Research, 18*, 259–263.

Warmar, P. R., & Termeer, W. C. (2005). Evaluation of sewage sludge, septic waste and sludge compost applications to corn and forage: Ca, Mg, S, Fe, Mn, Cu, Zn and B content of crops and soils. *Bioresouce Technology, 96*, 1029–1038.

Yubo, C., Tieheng, S., Lihui, Z., Tingliang, J., & Liping, Z. (2008). Performance of wastewater sludge ecological stabilization. *Journal of Environmental Sciences, 20*, 385–389.

Zwara, W., & Obarska-Pempkowiak, H. (2000). Polish experience with sewage sludge utilization in reed beds. *Water Science and Technology, 41*(1), 65–68.

Chapter 3
Process Based Models for Subsurface Flow Constructed Wetlands

Günter Langergraber

Abstract Constructed wetlands (CWs) are engineered systems designed to optimize the treatment conditions found in natural environments. A large number of physical, chemical, and biological processes are active in parallel and mutually influence each other; consequently, CWs are complex systems difficult to understand. Therefore, for a long time CWs have been often considered as "black boxes", and only little effort has been made to understand the main processes leading to wastewater purification. Numerical models that describe the transformation and elimination processes in CWs gained increasing interest during the last years and are a promising tool to get a better understanding of these processes. This chapter presents an overview of available models and recent developments in the field of CW modelling.

Keywords Constructed wetland · CWM1 · Numerical modeling · Subsurface flow

3.1 Introduction

Constructed wetlands (CWs) provide a natural way for simple, inexpensive, and robust wastewater treatment. A large number of physical, chemical, and biological processes contribute to water treatment. Because these processes occur in parallel and influence each other, detailed understanding of CW functioning is difficult (e.g. Brix, 1997; Haberl et al., 2003; Kadlec & Wallace, 2009; Kadlec et al., 2000; Tanner, 2001). Because of their complexity, CWs have long been seen as "black boxes" where wastewater enters and treated water leaves the system. Numerical models that describe the transformation and elimination processes in CWs are a promising tool to get a better understanding of the processes in CWs.

G. Langergraber (✉)
Institute of Sanitary Engineering and Water Pollution Control, University of Natural Resources and Applied Life Sciences, Vienna (BOKU), Muthgasse 18, A-1190, Vienna, Austria
e-mail: guenter.langergraber@boku.ac.at

J. Vymazal (ed.), *Water and Nutrient Management in Natural and Constructed Wetlands*, DOI 10.1007/978-90-481-9585-5_3,

During the last few years models with different complexities have been developed for describing processes in subsurface flow CWs. In subsurface flow CWs no free water level is visible, and water flows either horizontally or vertically through the porous filter media. Only few models have been published that are able to model biochemical transformation and degradation processes that occur in subsurface flow CWs. Horizontal flow (HF) systems can be simulated when only water flow saturated conditions are considered. A series or network of continuously-stirred tank reactors is most frequently used to describe the hydraulics, and reactions are modelled with various complexities. For modelling vertical flow (VF) CWs with intermittent loading, transient variably-saturated flow models are required. Due to the intermittent loading, these systems are highly dynamic, adding to the complexity of the overall system. These models use either the Richards equation to describe variably-saturated flow, or simplified approaches.

The content of the review is based on two recently published papers by Langergraber (2008) and Langergraber et al. (2009a) and gives an overview on available process based reactive transport models and includes recent developments. After a brief summary on models used to simulate the hydraulic behaviour and single-solute transport (tracer experiments) the two main types of models describing biochemical processes are discussed: Firstly, reactive transport models for saturated flow conditions that are suitable to model HF systems only and, secondly, reactive transport models for variably saturated conditions that are suitable to model both HF and VF systems. Finally, the Constructed Wetland Model N°1 (CWM1, Langergraber, Rousseau, García, & Mena, 2009b), a general model to describe biochemical transformation and degradation processes for organic matter, nitrogen and sulphur in subsurface flow CWs will be presented.

3.2 Models Describing the Hydraulic Behavior and Single-Solute Transport in Constructed Wetlands

In this section a short overview on models that only consider water flow and single-solute transport is presented. These models have been used to describe the hydraulic behaviour of the systems and to model tracer experiments. Tracer experiments are also used for calibrating the hydraulic behaviour of the system, i.e. the flow model of the more complex models described below.

Different methods have been applied: dispersion plug flow models (Chazarenc, Merlin, & Gonthier, 2003), stirred tanks-in-series models (Chazarenc et al., 2003; García et al., 2004), a multi flow dispersion model (Małoszewski, Wachniew, & Czupryński, 2006) and a model using an infinite number of small stirred tanks distributed along a set of main plug flow channels (Werner & Kadlec, 2000). All of these approaches (except the latter) assume that the flow field can be somehow simplified to one-dimensional behaviour, and the effects of dispersion, heterogeneity and dead-zones are lumped together. A more detailed discussion is provided in Langergraber (2008).

Schwager and Boller (1997) simulated tracer experiments and oxygen transport in intermittent sand filters using an older version of HYDRUS-1D (Šimůnek, Šejna, & van Genuchten, 1998), and MOFAT (Katyal, Kaluarachchi, & Parker, 1991), respectively. However, Schwager and Boller (1997) did not consider reactive transport. The effect of biomass accumulation on solute breakthrough was assessed experimentally by tracer studies. HYDRUS was used to extrapolate to different hydraulic conditions. MOFAT was used simultaneously to include the effects of water and air flow. The results strengthened the assumption that oxygen mainly enters the pores in the intervals between intermittent loadings via diffusion.

Forquet, Wanko, Mosé, and Sadowski (2009) used a diphasic numerical model to describe the movement of air and water during a loading sequence in a vertical flow filter. The authors concluded that the daily hydraulic loading does not affect the quantity of oxygen entering by convection. Increase of the daily hydraulic load increases the oxygen input. However, this phenomenon seems to be limited as it tends towards an asymptotic value. The number of flushes per day has a contradictory impact: it reduces the input of oxygen per flush but has a positive impact over a day.

3.3 Reactive Transport Models for Saturated Flow Conditions

To simulate HF CWs, several authors have been coupling reactive transport models with models describing saturated water flow. Some models are only applicable for constant flow rates (Mashauri & Kayombo, 2002; Mayo & Bigambo, 2005; Wang et al., 2009) whereas most models in this category apply a tanks-in-series approach (Chen, Wang, & Xue, 1999; Marsili-Libelli & Checchi, 2005; Rousseau, 2005; Wynn & Liehr, 2001) and can handle variable flow rates. A recently developed reactive transport model (Brovelli, Baechler, Rossi, Langergraber, & Barry, 2007) has been coupled to the groundwater flow model MODFLOW (MacDonald & Harbaugh, 1988).

Three of these seven models consider only carbon transformation processes (Chen et al., 1999; Mashauri & Kayombo, 2002; Marsili-Libelli & Checchi, 2005), two only nitrogen transformations (Mayo & Bigambo, 2005; Wang et al., 2009), and the remaining three both carbon and nitrogen transformations (Brovelli et al., 2007; Rousseau, 2005; Wynn & Liehr, 2001). The most advanced of these models are the one developed by Rousseau (2005) and Brovelli et al. (2007), who implemented a reaction model that is presented in matrix notation based on the mathematical formulation of the Activated Sludge Models (ASMs; Henze, Gujer, Mino, & van Loosdrecht, 2000). Being the most advanced models these two are described in more detail. A description of the other models (except Wang et al., 2009) can be found in Langergraber (2008).

3.3.1 Rousseau (2005)

Rousseau (2005) developed a reaction model that is coupled with a network of completely-stirred-tank-reactors (CSTRs) for describing water flow. The CSTR approach assumes a vertical uniform distribution of substrates, intermediates, products and bacteria, which may not be the case for HF CWs. Vertical mixing between the CSTRs was therefore introduced to model vertical gradients in the filter bed. For modelling the microbial conversions, the ASM approach (Henze et al., 2000) is used. The developed model considers microbiological and plant-related processes affecting COD and nitrogen in HF CWs. Phosphorus removal is not considered, and it is therefore assumed that P concentrations are non-limiting for microbial and plant growth. Bacterial groups considered include aerobic heterotrophs, fermenters, sulphate reducers, methanogens, nitrifiers and sulphide oxidisers. Aerobic and anoxic microbial carbon and nitrogen conversion processes are based on ASM processes. Anaerobic microbial processes are taken into account whereby the competition between sulphate reducing and methanogenic bacteria is modelled as described by Kalyuzhnyi and Fedorovich (1998). To avoid microbial inhibition due to sulphide accumulation, an inverse pathway was foreseen by adding sulphide oxidizing bacteria to the model. This model has been applied to evaluate porosity decreases in experimental HF CWs submitted to different pre-treatments (García et al., 2007). Figure 3.1 shows simulated and measured COD and ammonium concentrations from this study. Rousseau, Shi, Stein, and Hook (2008) presented results from lab-scale batch experiments have been modelled with Rousseau's formulation of the biokinetic model. Simulated concentrations of COD, NH_4–N and SO_4–S agree well with measured data.

3.3.2 Brovelli et al. (2007)

Brovelli et al. (2007) implemented a set of biological and geochemical reactions into the three-dimensional numerical simulator PHWAT (Mao et al., 2006). PHWAT has

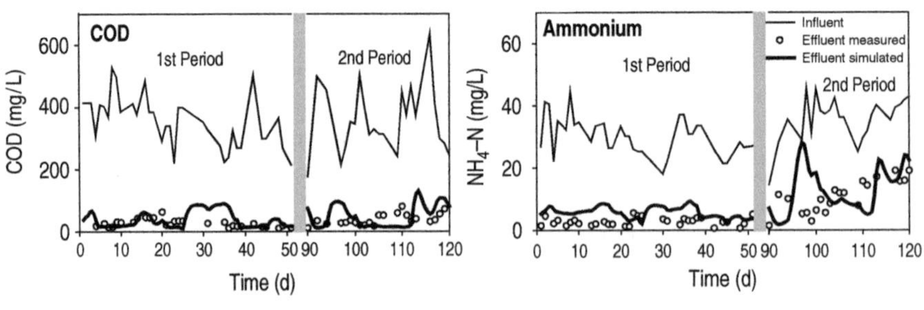

Fig. 3.1 Simulated and measured COD and ammonium concentrations in a pilot-scale HF bed fed with settled wastewater. No data on influent and effluent COD between days 50 and 90 (based on García et al., 2007)

been developed for modelling reactive transport in porous media and consists of three different modules: The flow module is based on MODFLOW (MacDonald & Harbaugh, 1988), which uses a finite difference scheme to solve the saturated water flow equation within the domain, the transport module is based on MT3DMS (Zheng & Wang, 1999), and the biogeochemistry module is based on PHREEQC-2 (Parkhurst & Appelo, 1999), a general purpose model for aqueous geochemistry. The biogeochemical reaction network is similar based on the ASM structure (Henze et al., 2000). Kinetic oxidation of carbon sources, organic matter hydrolysis, nutrient transformations and assimilation (nitrogen and phosphorous primarily) are modelled via Monod-type kinetic equations. While the ASM model consists of kinetic equations only, full water chemistry and sediment-water interactions can be modelled with PHREEQC and thus PHWAT. Some preliminary results by Brovelli et al. (2007) show reasonably good agreement with experimental data. Latela the model was updated (i) by introducing the Variably Saturated Flow (VSF) Process for MODFLOW (Thoms, Johnson, & Healy, 2006) into PHWAT thus enabling also modelling variably-saturated flow and VF CWs, respectively (Brovelli, Baechler, Rossi, & Barry, 2009a), (ii) by extending the biogeochemical reactions with anaerobic processes based on the model formulation by Maurer and Rittmann (2004) (Brovelli et al., 2009a), and (iii) by including a model able to simulate bioclogging, i.e. the effect of biomass growth on the hydraulic properties of saturated porous media (Brovelli, Malaguerra, & Barry, 2009b; Brovelli, Rossi, & Barry, 2009c).

3.4 Reactive Transport Models for Variably Saturated Conditions

The models discussed in this category are able to describe variably-saturated flow and are therefore applicable for modelling VF systems with intermittent loading. These models use either the Richards equation to describe variably-saturated flow (CW2D by Langergraber, 2001; Wanko, Mose, Carrayrou, & Sadowski, 2006; Ojeda, Caldentey, Saaltink, & García, 2008; Maier, DeBiase, Baeder-Bederski, & Bayer, 2009a), or simplified approaches such as different horizontal layers (McGechan, Moir, Castle, & Smit, 2005; FITOVERT by Giraldi, de'Michieli Vitturi, & Iannelli, 2010) or a combination of CSTRs and dead-zones (Freire, Davies, Vacas, Novais, & Martins-Dias, 2009), respectively.

The most advanced reaction models are implemented in CW2D (Langergraber, 2001) and FITOVERT (Giraldi et al., 2010) both based on the mathematical formulation of the ASMs (Henze et al., 2000), and in the model developed by Ojeda et al. (2008) that considers processes affecting solids, organic matter, nitrogen and sulphur. Ojeda's model was developed primarily for HF CWs, but because of the underlying flow model, it is also capable of simulating VF CWs. The model by Maier et al. (2009a) was developed to describe the processes in CWs for the remediation of contaminated groundwater and therefore models the degradation of benzene, ammonium and MTBE.

The model developed by McGechan et al. (2005) considers pools of organic matter, ammonium and nitrate, as well as oxygen; microbiologically controlled transformations are defined between these pools. The model developed by Wanko et al. (2006) considers organic matter removal and oxygen transport, whereas the model of Freire et al. (2009) only describes the removal of the dye AO7.

In the following, the most sophisticated models, CW2D, FITOVERT, Ojeda and Maier, are described in more detail below; descriptions of the other models can be found in Langergraber (2008).

3.4.1 CW2D (Langergraber, 2001; Langergraber and Šimůnek, 2005)

The multi-component reactive transport module CW2D (Langergraber, 2001) was developed to describe the biochemical transformation and degradation processes in SSF CWs. CW2D was incorporated into the HYDRUS variably-saturated water flow and solute transport program (Langergraber & Šimůnek, 2006; Šimůnek, Šejna, & van Genuchten, 2006). The HYDRUS program numerically solves the Richards equation for saturated/unsaturated water flow and the convection-dispersion equation for heat and solute transport. The flow equation incorporates a sink term to account for water uptake by plant roots. The solute transport equations consider convective-dispersive transport in the liquid phase, diffusion in the gaseous phase, as well as non-linear non-equilibrium reactions between the solid and liquid phases (Šimůnek et al., 2006). The CW2D module considers 12 components and 9 processes. The components include dissolved oxygen, organic matter (three fractions of different degradability, i.e. readily- and slowly-biodegradable, and inert), ammonium, nitrite, nitrate, and nitrogen gas, inorganic phosphorus, and heterotrophic and two species of autotrophic micro-organisms. Organic nitrogen and organic phosphorus are modeled as nutrient contents of the organic matter, i.e. they are calculated as a percentage of COD. The biochemical elimination and transformation processes are based on Monod-type expressions used to describe the process rates. All process rates and diffusion coefficients are temperature dependent. The processes considered are hydrolysis, mineralization of organic matter, nitrification (modeled as a two-step process), denitrification, and a lysis process (as the sum of all decay and loss processes) for the micro-organisms. CW2D assumes a constant concentration of micro-organisms (and other compounds) in each finite element. The thickness of the biofilm is not considered. The mathematical formulation of CW2D is based on the mathematical formulation of the ASMs (Henze et al., 2000). Langergraber (2005) investigated the plant uptake models provided by HYDRUS that describe nutrient uptake coupled with water uptake, and concluded that it was possible to simulate plant uptake in high loaded systems, e.g. systems treating mechanically pre-treated municipal wastewater. For low-strength wastewater, the simulation results indicate that potential nutrient uptake is overestimated by using these models. Oxygen release via roots can be modeled in a way similar to nutrient uptake (Toscano, Langergraber, Consoli, & Cirelli, 2009). For a

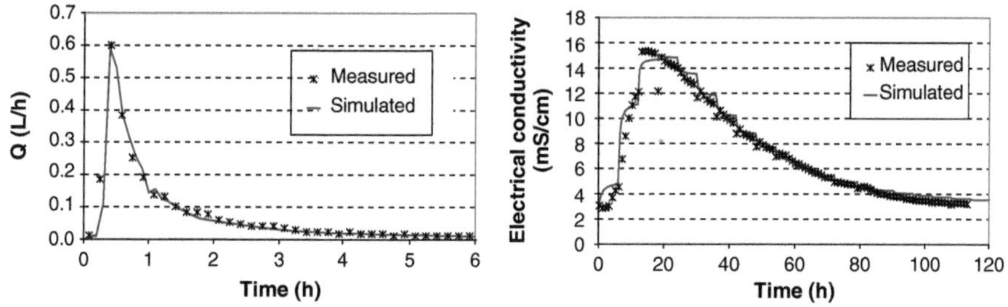

Fig. 3.2 Measured and simulated water flow (*left*) and EC data of a tracer experiment (*right*); based on Langergraber et al. (2009a)

detailed discussion of the CW2D module see Langergraber and Šimůnek (2005). Langergraber and Šimůnek (2009) recently presented an extension by including a model for describing the transport of particles.

The following figures show examples of results obtained with CW2D: Fig. 3.2 shows measured and simulated water flow for a single loading (left) and measured and simulated electrical conductivity (EC) data from a tracer experiment (right) at column experiments representing VF beds loaded intermittently with wastewater. Figure 3.3 shows measured and simulated COD and NH_4–N effluent concentrations for different temperatures for an outdoor VF bed (left, according to Langergraber, 2007) and calculated and simulated microbial biomass COD in different depths of the main layer of an indoor VF bed (right, according to Langergraber, Tietz, & Haberl, 2007).

Several groups have been using CW2D for other applications besides treating domestic wastewater, e.g. for CWs polishing the effluent of a wastewater treatment plant for reuse purposes (Toscano et al., 2009) and treatment of combined sewer overflow (Dittmer, Meyer, & Langergraber, 2005; Henrichs, Langergraber, & Uhl, 2007).

In general, results using CW2D showed that water flow, single-solute transport and reactive transport simulations were in good agreements with measured

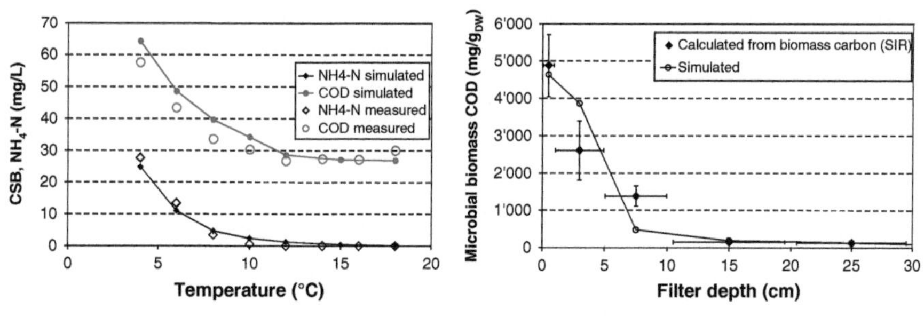

Fig. 3.3 Measured and simulated COD and NH_4–N effluent concentrations (*left*, based on Langergraber, 2007) and calculated and simulated microbial biomass COD in different depths of the main layer of a VF bed (*right*, based on Langergraber et al., 2007)

data for HF and VF CWs. The practical applications have shown that simulation results match the measured data when the hydraulic behaviour of the system can be described well. A good match of experimental data to reactive transport simulations can then be obtained for CWs treating municipal wastewater using the values for the CW2D model parameters as given by Langergraber and Šimůnek (2005). Therefore it is advisable to measure at least the porosity and saturated hydraulic conductivity of the filter material to obtain reasonable simulation results for water flow (Langergraber, 2008).

3.4.2 FITOVERT (Giraldi et al., 2010)

FITOVERT was developed specifically to simulate VF CWs and has a mono-dimensional vertical scheme discretised by a variable number of stacked horizontal layers: the wastewater flux runs through the system from the top to the bottom. The dynamic formulation of the model allows simulating the typical non stationary behaviour of VF CWs. Boundary conditions are provided to simulate both variable inflows and ponding at the surface and free drainage, drainage with a pressure head and no drainage at the bottom of the system. The hydraulic flow for unsaturated condition is described by means of the Richards equation, while the constitutive relationships among pressure head, hydraulic conductivity and water content are described using the van Genuchten parameterization. Evaporation at the CW surface and transpiration through plants are also considered. FITOVERT describes biochemical processes of organic matter and nitrogen by means of the standard ASM (Henze et al., 2000). Advective and diffusive transport of contaminants in the liquid phase is implemented according to a mass conservation equation with a limited change to the dispersion term. Gas transport and transfer to the liquid phase are described for the oxygen by a mass balance with advection and dispersion. Settling and filtration of particulate matter is included thus allowing porosity reduction due e.g. to bacteria growth and to deposition of particulate components, so that the clogging process could be also simulated; the effect of pore size reduction on the saturated hydraulic conductivity is also considered. The numerical implementation of the model is based on the Galerkin method with finite-elements discretization. FITOVERT is implemented in Matlab® with a user-friendly interface for quick access to the main input parameters and for graphic display of the simulation results. Up to now only preliminary results have been presented for FITOVERT. Figure 3.4 shows measured and simulated results of tracer experiments for a VF bed for complete saturation (left) and free drainage (right), respectively. Iannelli, Bianchi, Salvato, and Borin (2009) presented the applicability of FITOVERT for showing the effects of different macrophytes on nitrogen removal efficiency.

3.4.3 Ojeda et al. (2008)

The model by Ojeda et al. (2008) is based on the two-dimensional finite element code RetrasoCodeBright (RCB), which has been applied in hydro geological studies and simulates reactive transport of dissolved and gaseous species for

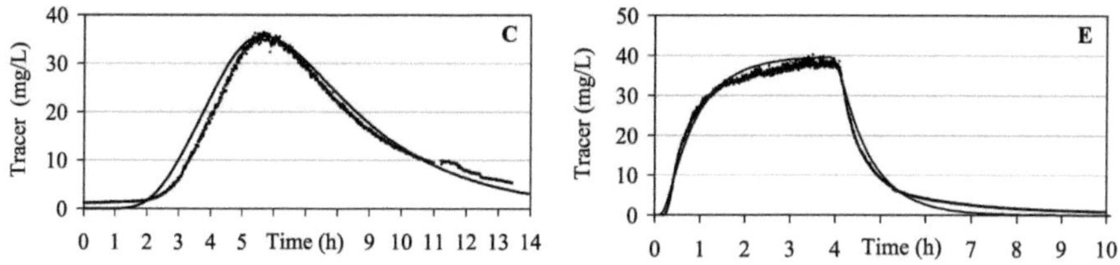

Fig. 3.4 Experimental tracer breakthrough curves (dots) and numerical simulations (*lines*) for complete saturation (Test C, *left*) and free drainage (Test E, *right*); based on Giraldi, de'Michieli Vitturi, Zaramella, Marion, and Iannelli (2009)

non-isothermal saturated or unsaturated flow domains (Rezaei et al., 2005; Saaltink, Ayora, Stuyfzand, & Timmer, 2003). In RCB, a first module calculates the flow properties, and passes it to a second module for the calculation of reactive transport. Ojeda et al. (2008) modified RCB to include the most significant biochemical pathways for organic matter removal in HF CWs. The reactive transport model is based on Van Cappellen and Gaillard (1996), and basically consists of 6 microbial kinetic reactions: hydrolysis, aerobic respiration, nitrification, denitrification, sulphate reduction and methanogenesis. The degradation rate associated with each of the 6 kinetic equations, except for hydrolysis, is described by multiplicative Monod-type expressions. The hydrolysis process is modelled by an exponential function describing the TSS removal that occurs near the inlet of the bed. Phosphorus transformations, biofilm growth and oxygen leaking from macrophytes have not been considered. Until now, the model has been used for HF CWs only, especially to evaluate the importance of hydraulic and organic loading rates on removal efficiencies of experimental HF CWs (Ojeda et al., 2008). Figure 3.5 shows as example simulated COD and NH_4–N concentrations along a HF bed for different hydraulic loading rates.

3.4.4 Maier et al. (2009a, 2009b)

Maier et al. (2009a, 2009b) developed a model to describe the degradation of benzene, ammonium and MTBE in CWs treating contaminated groundwater. The

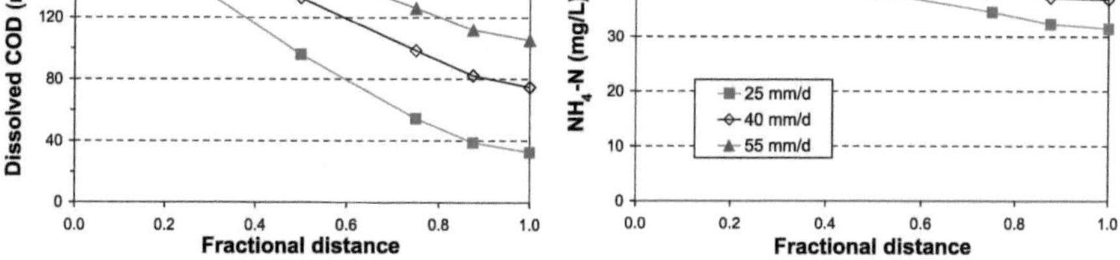

Fig. 3.5 Simulated COD (*left*) and NH_4–N (*right*) concentrations along a HF bed at 0.25 m above the bottom for different hydraulic loading rates (based on Ojeda et al., 2008)

simulation of variably saturated water flow using the numerical code MIN3P (Mayer, Frind, & Blowes, 2002) is described by the Richards equation. A simplified equilibrium formulation for preferential flow is implemented in the model using the approach of Mohanty, Van Genuchten, Bowman, and Hendrickx (1997), the formulation used in this study is described in Gérard, Tinsley, and Mayer (2004). Simulation results for two soil filters using the same sandy substrate for the main layer showed a good match between observed and simulated bottom discharges of the filters Maier et al. (2009a). Also the degradation model has been tested against experimental data. The simulations showed that volatilization is a non-negligible process during the treatment of (volatile) groundwater contaminants in VF CWs, and that the efficiency of such wetlands systems might be limited by the need to avoid high volatilization fluxes. (De Biase et al., 2009).

3.5 The Constructed Wetland Model N°1 (CWM1)

Recently, a general model to describe biochemical transformation and degradation processes for organic matter, nitrogen and sulphur in SSF CWs, the Constructed Wetland Model N°1 (CWM1), has been presented (Langergraber et al., 2009b). CWM1 describes all relevant aerobic, anoxic and anaerobic biokinetic processes occurring in HF and VF CWs required predicting effluent concentrations of organic matter, nitrogen and sulphur. 17 processes and 16 components (8 solute and 8 particulate components) are considered. According to Langergraber et al. (2009b) the main objective of CWM1 is, similar as to the ASMs, to provide a widely accepted model formulation for biochemical transformation and degradation processes in constructed wetlands that can then be implemented in various simulation tools.

The first results using CWM1 have been presented at the WETPOL 2009 symposium that was held from 20 to 24 September 2009 in Barcelona, Spain. Rousseau, Gaviano, Stein, Hook, and Lens (2009) simulated carbon, nitrogen and sulphur cycles in batch-operated, lab-scale treatment wetlands using only the CWM1 biokinetic model without coupling to any flow model. Llorens, Saaltink, and García (2009) presented results for a HF CW using the CWM1 implementation in the RETRASO code. Simulation results were promising and showed good agreement to measured data. Currently further CWM1 implementation work is ongoing for the PHWAT code (Brovelli et al., 2009c) and CW2D/HYDRUS code (Langergraber & Šimůnek, 2005), respectively.

3.6 Summary and Conclusions

In the previous sections, numerical models describing processes in subsurface flow constructed wetlands have been reviewed. Several authors aimed for modelling the hydraulic behaviour of the system including tracer experiments only. There are few published models that are able to model the biochemical transformation and degradation processes that occur in the complex system subsurface flow constructed wetland. Models describing reactive transport under water flow saturated conditions can only be used to model horizontal flow constructed wetlands. The reviewed models either use either a series or a network of ideal reactors or are only applicable for constant flow rates to model water flow. One of the models is using MODFLOW for describing water flow. This model was recently extended with the unsaturated flow extension for MOFLOW and will therefore in future be able to simulated also VF systems.

For modelling vertical flow constructed wetlands with intermittent loading, transient variably-saturated flow models are required as these systems are highly dynamic, which adds to the complexity of the overall system. Models of differing complexities have been published that consider variably-saturated flow and reaction models for constructed wetlands. Five of these models are in the rather early stages of development. For the multi-component reactive transport module CW2D, several applications have been published that include treatment of domestic wastewater with vertical flow constructed wetlands, treatment of combined sewer overflow and polishing of treated wastewater (Langergraber, 2008).

According to Langergraber et al. (2009a) current developments in numerical modelling of CWs show two well-differentiated but related objectives. Firstly, mechanistic models aimed at gaining insight into wetlands dynamics and functioning, and secondly, simplified but robust and reliable models for design purposes. Meyer, Dittmer, and Schmitt (2008) developed the RSF_Sim module based on experiences from detailed simulations with CW2D. RSF_Sim is a simplified reaction model for CWs treating combined sewer overflow that is planned to be used as design tool.

The development of simplified design models based on these numerical simulations could be a way toward bringing more reliability into the design rules and to make modelling tools more acceptable for design purposes in practice. These simplified design models should comprise only the most relevant processes for the specific application and have therefore only few parameters. This can be of great importance for a better acceptance of models in the design of CWs. Even if a user-friendly operating environment is provided for mechanistic models making their use rather simple there are a lot of parameters to consider and it takes quite some time to gain experience with the using these models (Langergraber et al., 2009a).

References

Brix, H. (1997). Do macrophytes play a role in constructed treatment wetlands? *Water Science and Technology*, *35*(5), 11–17.

Brovelli, A, Baechler, S, Rossi, L, Langergraber, G., & Barry, D. A. (2007). Coupled flow and hydro-geochemical modelling for design and optimization of horizontal flow constructed wetlands. In Ü. Mander M. Kóiv & C. Vohla (Eds.), *Second International Symposium on "Wetland Pollutant Dynamics and Control WETPOL 2007" – Extended Abstracts II* (pp.393–395). Tartu, Estonia: University of Tartu.

Brovelli, A., Baechler, S., Rossi, L., & Barry, D. A. (2009a). Comprehensive process-based modelling of sand filters and subsurface flow constructed wetlands. In J. M., Bayona, García, J. (Eds.) *3rd International Symposium on "Wetland Pollutant Dynamics and Control – WETPOL 2009" – Book of Abstracts, 20–24 September 2009* (pp. 293–294). Barcelona, Spain.

Brovelli, A., Malaguerra, F., & Barry, D. A. (2009b). Bioclogging in porous media: Model development and sensitivity to initial conditions. *Environmental Modelling and Software, 24*, 511–525.

Brovelli, A., Rossi, L., & Barry D. A. (2009c). Mechanistic understanding and prediction of bioclogging in sand filters and subsurface flow constructed wetlands. In J. M. Bayona & J. García (Eds.), *Book of Abstracts of the 3rd International Symposium on Wetland Pollutant Dynamics and Control – WETPOL 2009* (pp. 133–134).Barcelona, Spain.

Chazarenc, F., Merlin, G. & Gonthier, Y. (2003). Hydrodynamics of horizontal subsurface flow constructed wetlands. *Ecological Engineering, 21*, 165–173.

Chen, S., Wang, G. T., & Xue, S. K. (1999). Modeling BOD removal in constructed wetlands with mixing cell method. *Journal of Environmental Engineering, 125*, 64–71.

De Biase, C., Maier, U., Mosig, P., van Afferden, M., Reger, D., Oswald, S., & Thullner, M. (2009). Modeling microbial degradation of volatile organic compounds in vertical flow constructed wetlands. In J. M. Bayona & J. García (Eds.), *Book of Abstracts of the 3rd International Symposium on Wetland Pollutant Dynamics and Control – WETPOL 2009* (pp. 291–292). Barcelona, Spain.

Dittmer, U., Meyer, D., & Langergraber, G. (2005). Simulation of a subsurface vertical flow constructed wetland for CSO treatment. *Water Science and Technology, 51*(9), 225–232.

Forquet, N., Wanko, A., Mosé, R., & Sadowski, A.-G. (2009). Diphasic modelling of vertical flow filter. *Ecological Engineering, 35*(1), 47–56.

Freire, F. G., Davies, L. C., Vacas, A. M., Novais, J. M., & Martins-Dias, S. (2009). Influence of operating conditions on the degradation kinetics of an azo-dye in a vertical flow constructed wetland using a simple mechanistic model. *Ecological Engineering, 35*, 1379–1386.

García, J., Chiva, J., Aguirre, P., Alvarez, E., Sierra, J. P., & Mujeriego, R. (2004). Hydraulic behaviour of horizontal subsurface flow constructed wetlands with different aspect ratio and granular medium size. *Ecological Engineering, 23*, 177–187.

García, J., Rousseau, D., Caselles-Osorio, A., Story, A., De Pauw, N., &Vanrolleghem, P. (2007). Impact of prior physico-chemical treatment on the clogging process of subsurface flow constructed wetlands: model-based evaluation. *Water Air and Soil Poll*ution, *185*, 101–109.

Gérard, F., Tinsley, M., & Mayer, K. U. (2004). Preferential flow revealed by hydrologic modeling based on predicted hydraulic properties. *Soil Science Society of America Journal, 68*, 1526–1538.

Giraldi, D., de'Michieli Vitturi, M. & Iannelli, R. (2010). FITOVERT: a dynamic numerical model of subsurface vertical flow constructed wetlands. *Environmental Modelling and Software 25*, 633–640.

Giraldi D., de'Michieli Vitturi M., Zaramella M., Marion A., & Iannelli R. (2009). Hydrodynamics of vertical subsurface flow constructed wetlands: Tracer tests with rhodamine WT and numerical modelling. *Ecological Engineering, 35*, 265–273.

Haberl, R., Grego, S., Langergraber, G., Kadlec, R. H., Cicalini, A. R., Martins Dias, S., et al. (2003). Constructed wetlands for the treatment of organic pollutants. *JSS – J Soils & Sediments, 3*, 109–124.

Henrichs, M., Langergraber, G., & Uhl, M. (2007). Modelling of organic matter degradation in constructed wetlands for treatment of combined sewer overflow. *Science of the Total Environment, 380*, 196–209.

Henze, M., Gujer, W., Mino, T., & van Loosdrecht, M. C. M. (2000). Activated sludge models ASM1, ASM2, ASM2D and ASM3. *IWA Scientific and Technical Report No.9.* , London: IWA Publishing.

Iannelli, R., Bianchi, V., Salvato, M., & Borin, M. (2009). Modelling assessment of the nitrogen abatement in pilot subsurface flow constructed wetlands with different macrophytes. In J. M. Bayona & J. García (Eds.), *Book of Abstracts of the 3rd International Symposium on Wetland Pollutant Dynamics and Control – WETPOL 2009* (pp. 73–74). Barcelona, Spain.

Kadlec, R. H., Knight, R. L., Vymazal, J., Brix, H., Cooper, P. F., & Haberl, R. (2000). *Constructed wetlands for pollution control – Processes, performance, design and operation.* IWA Scientific and Technical Report No.8. London: IWA Publishing,

Kadlec, R. H., Wallace, S. (2009). *Treatment wetlands* (2nd ed.). Boca Raton, FL: CRC.

Kalyuzhnyi, S. V., & Fedorovich, V. V. (1998). Mathematical modelling of competition between sulphate reduction and mehtanogenesis in anaerobic reactors. *Bioresource Technology, 65*, 227–242.

Katyal, A. K., Kaluarachchi, J. J., & Parker, J. C. (1991). MOFAT: A two-dimensional finite element program for multiphase flow and multi-component transport – program documentation and user's guide. Center for Environmental and Hazardous Material Studies, Virginia Polytechnic Institute and State University, Blacksburg, VA.

Langergraber, G. (2001). Development of a simulation tool for subsurface flow constructed wetlands. Vienna, Austria: *Wiener Mitteilungen* 169.

Langergraber, G. (2005). The role of plant uptake on the removal of organic matter and nutrients in subsurface flow constructed wetlands – A simulation study. *Water Science and Technology, 51*(9), 213–223.

Langergraber, G. (2007). Simulation of the treatment performance of outdoor subsurface flow constructed wetlands in temperate climates. *Science of the Total Environment, 380*, 210–219.

Langergraber, G. (2008). Modeling of processes in subsurface flow constructed wetlands – A review. *Vadoze Zone Journal, 7*, 830–842.

Langergraber, G., & Šimůnek, J. (2005). Modeling variably saturated water flow and multicomponent reactive transport in constructed wetlands. *Vadoze Zone Journal, 4*, 924–938.

Langergraber, G., & Šimůnek, J. (2006): The multi-component reactive transport module CW2D for constructed wetlands for the HYDRUS Software Package. *Hydrus Software Series 2*, Department of Environmental Sciences, University of California Riverside, Riverside, CA.

Langergraber, G., & Šimůnek, J. (2009). Simulating particle transport in subsurface flow constructed wetlands with CW2D/HYDRUS. In J. M. Bayona & J. García (Eds.), *Book of Abstracts of the 3rd International Symposium on Wetland Pollutant Dynamics and Control – WETPOL 2009* (pp. 135–136). Barcelona, Spain.

Langergraber, G., Tietz, A., & Haberl, R. (2007): Comparison of measured and simulated distribution of microbial biomass in subsurface vertical flow constructed wetlands. *Water Science and Technology, 56*, 233–240.

Langergraber, G., Giraldi, D., Mena, J., Meyer, D., Peña, M., Toscano, A., et al. (2009a). Recent developments in numerical modelling of subsurface flow constructed wetlands. *Sciene of theTotal Environment, 407*, 3931–3943.

Langergraber, G., Rousseau, D., García, J., & Mena, J. (2009b). CWM1 – A general model to describe biokinetic processes in subsurface flow constructed wetlands. *Water Science and Technology, 59*(9), 1687–1697.

Llorens, E., Saaltink, M., & García, J. (2009): CMW1-RETRASO: a dynamic numerical model of subsurface horizontal flow constructed wetlands. In J. M. Bayona & J. García (Eds.), *Book of Abstracts of the 3rd International Symposium on Wetland Pollutant Dynamics and Control – WETPOL 2009* (pp. 77–78). Barcelona, Spain.

MacDonald, M. G., & Harbaugh, A. W. (1988). A modular three-dimensional finite-difference ground-water flow model. *U.S. Geological Survey Techniques of Water-resources Investigations, report 99–4259.* U. S. Geological Survey, Reston, VA.

Małoszewski, P., Wachniew, P., & Czupryński, P. (2006). Study of hydraulic parameters in heterogeneous gravel beds: Constructed wetland in Nowa Słupia (Poland). *Journal of Hydrology, 331*, 630–642.

Maier U., DeBiase C., Baeder-Bederski O., & Bayer P. (2009a). Calibration of hydraulic parameters for large-scale vertical flow constructed wetlands. *Journal of Hydrology, 369*, 260–273.

Maier, U., Oswald, S. Thullner, M., & Grathwohl, P. (2009b). Biogeochemical modelling of constructed wetlands for groundwater remediation. In J. M. Bayona & J. García (Eds.), *Book of*

Abstracts of the 3rd International Symposium on Wetland Pollutant Dynamics and Control – WETPOL 2009 (pp. 297–298). Barcelona, Spain.

Mao, X., Prommer, H., Barry, D. A., Langevin, C. D., Panteleit, B., & Li, L. (2006). Three Dimensional Model for Multi-component Reactive Transport with Variable Density Groundwater Flow. *Environmental Modelling and Software, 21*, 615–628.

Marsili-Libelli, S., & Checchi, N. (2005). Identification of dynamic models for horizontal subsurface constructed wetlands. *Ecological Modelling, 187*, 201–218.

Mashauri, D. A., & Kayombo. S. (2002). Application of the two coupled models for water quality management: facultative pond cum constructed wetland models. *Physics and Chemistry of the Earth, 27*, 773–781.

Maurer, M., & Rittmann, B. E. (2004). Formulation of the CBC-model for modelling the contaminants and footprints in natural attenuation of BTEX. *Biodegradation, 15*, 419–434.

Mayer, K. U., Frind, E. O., & Blowes, D. W. (2002). Multicomponent reactive transport modeling in variably saturated porous media using a generalized formulation for kinetically controlled reactions. *Water Resources Res* 38(9), 1174–1195.

Mayo, A. W., & Bigambo. T. (2005). Nitrogen transformation in horizontal subsurface flow constructed wetlands I: Model development. *Physics and Chemistry of the Earth, 30*, 658–667.

McGechan, M. B., Moir, S. E., Castle, K., & Smit, I. P. J. (2005). Modelling oxygen transport in a reedbed-constructed wetland purification system for dilute effluents. *Biosystems Engineering, 91*, 191–200.

Meyer, D., Dittmer, U., & Schmitt, T. G. (2008). Modelling CWs for CSO treatment-reasonable balancing between detailed description and practicable handling. In IWA (Ed.), *Proceedings of the 11th IWA Specialized Group Conference on Wetland Systems for Water Pollution Control –* (Vol. 2, pp. 851–857). Indore, India: Vikram University and IWA.

Mohanty, B. P., Van Genuchten, M. T., Bowman, R. S., & Hendrickx, J. M. H. (1997). New piecewise-continuous hydraulic functions for modeling preferential flow in an intermittent-flood-irrigated field. *Water Resources Research, 33*, 2049–2063.

Ojeda E., Caldentey, J., Saaltink, M. W., & García, J. (2008). Evaluation of relative importance of different microbial reactions on organic matter removal in horizontal subsurface-flow constructed wetlands using a 2D simulation model. *Ecological Engineering, 34*, 65–75.

Parkhurst, D. L., & Appelo, C. A. J. (1999). User's guide to PHREEQC – A computer program for speciation, reaction-path, 1D-transport, and inverse geochemical calculations. *U.S. Geological Survey Technical Reports 99-4529*. U.S. Geological Survey, Reston, VA.

Rezaei, M., Sanz, E., Raeisi, E., Ayora, C., Vázquez-Suñé, E., & Carrera, J. (2005). Reactive transport modeling of calcite dissolution in the fresh-salt water mixing zone. *Journal of Hydrology, 311*, 282–298.

Rousseau, D. P. L. (2005). *Performance of constructed treatment wetlands: model-based evaluation and impact of operation and maintenance*. Dissertation, Ghent University, Belgium.

Rousseau, D. P. L., Gaviano, A., Stein, O. R., Hook, P. B., & Lens, P. N. L. (2009): Simulation of carbon, nitrogen and sulphur cycles in batchoperated, lab-scale treatment wetlands: an application of CWM1. In J. M. Bayona & J. García (Eds.), *Book of Abstracts of the 3rd International Symposium on Wetland Pollutant Dynamics and Control – WETPOL 2009* (pp. 75–76). Barcelona, Spain.

Rousseau, D. P. L., Shi, W., Stein, O. R., & Hook, P. B. (2008). Simulation of COD removal in lab-scale constructed wetlands: Influence of electron acceptors. In IWA (Ed.), *Proceedings of the 11th IWA Specialized Group Conference on Wetland Systems for Water Pollution Control –* (Vol. 1, pp 526–534). Indore, India: Vikram University and IWA.

Saaltink, M. W., Ayora, C., Stuyfzand, P. J., & Timmer, H. (2003). Analysis of a deep well recharge experiment by calibrating a reactive transport model with field data. *Journal of Contaminant Hydrology, 65*, 1–18.

Schwager, A., & Boller, M. (1997). Transport phenomena in intermittent filters. *Water Science and Technology, 35*(6), 13–20.

Šimůnek J., Šejna, M., & van Genuchten, M. T. (1998). The HYDRUS-1D software package for simulating the one-dimensional movement of water, heat, and multiple solutes in variably-saturated media. Version 2.0. U.S. Salinity Laboratory, USDA, ARS, Riverside, CA.

Šimůnek J., Šejna, M., & van Genuchten, M. T. (2006): The HYDRUS software package for simulating the two- and three-dimensional movement of water, heat, and multiple solutes in variably-saturated media – technical manual, version 1.0. PC-Progress, Prague, Czech Republic.

Tanner, C. C. (2001). Plants as ecosystem engineers in subsurface-flow treatment wetlands. *Water Science and Technology, 44*(11–12), 9–17.

Thoms, R. B., Johnson, R. L., & Healy, R. W. (2006). User's guide to the Variably Saturated Flow (VSF) Process for MODFLOW. *U.S. Geological Survey Techniques and Methods 6-A18*, U.S. Geological Survey, Reston, VA, USA, 58p. (http://pubs.water.usgs.gov/tm6a18/), accessed 15 July 2010.

Toscano, A., Langergraber, G., Consoli, S., & Cirelli, G. L. (2009). Modelling pollutant removal in a pilot-scale two-stage subsurface flow constructed wetlands. *Ecological Engineering, 35*, 281–289.

Van Cappellen, P., & Gaillard, J.-F. (1996). Biogeochemical dynamics in aquatic sediments. In P. C. Lichtner, C. I. Steefel & E. H. Oelkers (Eds.), *Reviews in mineralogy and geochemistry*, (Vol. 34, pp. 335–376.). Washington, DC: Mineralogical Society of America.

Wang, Y., Zhang, J., Kong, H., Inamori, Y., Xu, K., Inamori, R., & Kondo, T. (2009). A simulation model of nitrogen transformation in reed constructed wetlands. *Desalination 235*, 93–101.

Wanko, A., Mose, R., Carrayrou, J., & Sadowski, A. G. (2006). Simulation of Biodegradation in infiltration seepage – model development and hydrodynamic calibration. *Water Air and Soil Pollution, 177*, 19–43.

Werner, T. M., & Kadlec, R. H. (2000). Wetland residence time modeling. *Ecological Engineering, 15*, 77–90.

Wynn, M. T., & Liehr. S. K. (2001). Development of a constructed subsurface-flow wetland simulation model. *Ecological Engineering, 16*, 519–536.

Zheng, C., & Wang, P. P. (1999). A modular three-dimensional multi-species transport model for simulation of advection, dispersion and chemical reactions of contaminants in groundwater systems; Documentation and user's guide. *U.S. Army Engineer Research and Development Center Contract Report SERDP-99-1*, Vicksburg, MI.

Chapter 4
Application of Vertical Flow Constructed Wetlands for Highly Contaminated Wastewater Treatment: Preliminary Results

Hanna Obarska-Pemkowiak, Magdalena Gajewska, and Ewa Wojciechowska

Abstract Based on quantitative and qualitative characteristics of reject waters (RWC) generated during dewatering of digested sewage sludge on centrifuges in conventional WWTP and municipal landfill leachate (LL), the pilot constructed wetlands for treatment of both types wastewater were designed and built. In the paper the conception, design and assumed treatment efficiencies of the pilot plants are presented. The water balance of the pilot treatment wetlands is presented and potential implications of water losses through evapotranspiration in case of treating wastewater with high concentrations of pollutants are discussed. Preliminary treatment results, obtained during the start-up period, showed good treatment effectiveness of BOD (74% for RWC and 84.5% for LL), while the effectiveness of ammonia nitrogen removal was below 20% for LL and from 48 to 59% for RWC.

Keywords Hydrophytes · Landfill leachate · Reject water from sewage sludge dewatering · Treatment wetlands

4.1 Introduction

The treatment of landfill leachate and reject water from the dewatering of digested sludge in WWTPs has become an important problem. The treatment of both types of wastewater is in many cases very costly and difficult due to quality and quantity fluctuations in time as well as high concentrations of specific pollutants.

Both types of wastewater (landfill leachate and reject water) are characterised by very high concentrations of COD (up to 8000 mg L^{-1}), TSS (up to 6000 mg L^{-1}) and total nitrogen (up to 1600 mg L^{-1}, mainly in the form of ammonia nitrogen). Fux, Boehler, Huber, Brunner, and Siegrist (2002); Fux, Valten, Carozzi, Solley, and Keller (2006) indicated that ammonia nitrogen and TSS concentrations in the reject

M. Gajewska (✉)
Faculty of Civil & Environmental Engineering, Gdansk University of Technology, Gdańsk, Poland
e-mail: mgaj@pg.gda.pl

J. Vymazal (ed.), *Water and Nutrient Management in Natural and Constructed Wetlands*, DOI 10.1007/978-90-481-9585-5_4,

water from the WWTP in Luggage Poit, Australia, varied from 943 to 1710 mg L^{-1} and from 95 to 6132 mg L^{-1}, respectively. In the reject water from WWTP in Minworth, UK, NH_4-N and TSS concentrations varied from 450 to 750 mg L^{-1} and from 220 to 2340 mg L^{-1}, respectively (Fux et al., 2003). At two WWTPs in Switzerland, the reject waters from the sludge fermentation process were similar in quality and contained 657 ±56 mg L^{-1} and 619 ± 21 mg L^{-1} ammonia nitrogen and 344 ± 112 mg L^{-1} and 384 ± 137 mg L^{-1} TSS (Hans, van der Roest, & van der Roest, 1997; Jeavons, Stokes, Upton, & Bingley, 1998).

According to Tatsi and Zoubolis (2002), landfill leachate from municipal landfills contains 3000–10000 mg L^{-1} of $N-NH_4^+$ and up to 60,000 mg L^{-1} COD. The leachate can also contain xenobiotics (BTEX, PAH, PCB) and heavy metals, as well as high concentrations of iron and chlorides (Paxeus, 2000; Christensen et al., 2001; Marttinen, Kettunen, & Rintala, 2003; Wojciechowska & Obarska-Pempkowiak, 2008).

Leachates from municipal landfills are typically characterised by the BOD/COD ratio. This parameter brings information about leachate biodegradability and decreases with the landfill age as the decomposition processes is progressing. The typical BOD/COD ratio for young landfills (3–5 years) is about 0.7, it decreases to 0.5–0.3 for mature landfills (5–10 years) and finally reaches values as low as 0.1 for old landfills (over 10 years) (Lo, 1996; Klimiuk, Kulikowska, & Koc-Jurczyk, 2007). The BOD/COD ratio for reject water from digested sludge dewatering varies from 0.14 to 0.2, which indicates that organic matter is present mainly in the form of hardly biodegradable compounds (Gajewska & Obarska-Pempkowiak, 2008; Fux et al., 2006)

The return flows of reject waters from sewage sludge dewatering alter the activated sludge process in the conventional WWTP. In WWTPs with sewage sludge digestion, 15–20% of the nitrogen load is usually redirected with the reject waters (Fux et al., 2006). While the remaining COD after anaerobic digestion is generally quite low and poorly biodegradable, a separate treatment of the high nitrogen content in this stream can considerably reduce the total nitrogen concentration in the final effluent from WWTPs (Wett & Alex, 2003; Laurich & Gunner, 2003). For the same reasons, the on-site treatment of landfill leachates is recommended rather than transport to the WWTP (Robinson, 2005).

Constructed wetlands (CWs) are applied for landfill leachate treatment in the USA and Europe, including the temperate and sub-polar climate regions (Maehlum, 1995; Peverly, Surface, & Wang, 1995; Bulc, Vrhovšek, & Kukanja, 1997; Martin, Johnson, & Moshiri, 1999; Johansson & Westholm, 2003; Kadlec, 2003; Randerson & Slater, 2005; Bulc, 2006; Rustige & Nolde, 2006; Kinsley, Crolla, Kuyucak, Zimmer, & Laflèche, 2006; Wojciechowska & Obarska-Pempkowiak, 2008). Both surface and subsurface flow (usually horizontal) CWs as well as plants which consist of several stages with changing flow conditions are applied.

Constructed wetlands create the environment for hydrophyte growth, thanks to which both aerobic and anaerobic decomposition processes are enhanced. These processes, supported by sorption, sedimentation and assimilation, are responsible for pollutant removal. It was proved that in CWs inhabited by *Phragmites australis*, the redox potential fluctuates from +200 to – 300 mV, which means that during the decay processes NO_3^-, SO_4^{2-} and other ions can act as electron acceptors while organic compounds are electrons donors (Reddy & D'Angelo, 1996; Vymazal, 2001; Obarska-Pempkowiak, 2002). Such fluctuating conditions together with a long retention time favour the degradation of many even toxic compounds such as THM, detergents or PAH. Especially suitable environment for the mineralization of organic compounds and oxidation of nitrogen compounds develops in CWs with the vertical flow of wastewater. Such beds are intermittently loaded with wastewater and this feeding mode enables better aeration of the filtration bed. During the resting periods the accumulated organic matter is decomposed, providing the protection of thee beds against clogging (Kayser, Kunst, Fehr, & Voermanek, 2001; Gajewska, Tuszynska, & Obarska-Pempkowiak, 2004).

In France, vertical flow beds are applied for the treatment of raw sewage (without mechanical pre-treatment) (Boutin, Lienard, & Esser, 1997; Molle, Lienard, Boutin, Merlin, & Iwema, 2004). According to Molle et al. (2004) two vertical flow beds (VF-CW) working in hydraulic batch, provide very effective sewage treatment. The unit area of the first VF-CW bed should be equal to 1.5 m^2 CPE^{-1} (calculated person equivalent), while the unit area of the second bed is only 1.0 m^2 CPE^{-1}. This configuration of VF-CW beds provides reduction of COD, TSS and TKN concentrations to 60, 15 and 8.0 mg L^{-1}. Molle et al. (2004) recommended that the hydraulic loading of the beds working in batches should be below 600 mm d^{-1}. Under such operation conditions, the beds provide good and stable pollutant removal efficiencies. According to Molle et al. (2004) the average pollutant removal efficiencies for the first bed were as follows: 82% COD, 89% TSS and 60% Kjeldahl nitrogen. For the second bed, the following efficiencies were reported: 60% COD, 72% TSS and 78% Kjeldahl nitrogen. According to Molle et al. (2004), the layer of sediments deposited on the surface of the first bed does not interfere with the treatment process, but it even enhances the treatment process.

The aim of the chater is to consider the application of sequentially operated VF beds for the treatment of highly polluted wastewater: reject water from sewage sludge dewatering as well as municipal landfill leachate. The quantitative and qualitative characteristics of both types of wastewater are presented in the chapter. On the basis of qualitative and quantitative analyses of wastewater, two pilot wetlands for the treatment of reject waters (RWC) and landfill leachate (LL) were designed and constructed. Both treatment wetlands consisted of two intermittently loaded beds with a vertical flow of wastewater. The pilot treatment wetlands were constructed under the research project PL 0085 "New Methods of Emission Reduction of Selected Pollutants and Application of By-products from Sewage Treatment Plants", financed by EEA Financial Mechanism and Ministry of Science and Higher Education in Poland. Preliminary treatment results are discussed and compared to the assumed treatment capacities.

4.2 Methods

4.2.1 *Study Facilities*

Two pilot treatment wetlands have been constructed. One of them treats landfill leachate from Municipal Landfill in Chlewnica near Slupsk (Pomerania Province), and the other one receives reject waters from the dewatering of digested sludge at the municipal WWTP in Gdansk.

Municipal Landfill in Chlewnica was put to operation in 2003. The facility receives municipal wastes from the town of Potegowo and surrounding villages. Initially, the landfill leachate was collected and transported to the WWTP in Potegowo. In 2007, however, the WWTP refused to treat the leachate due to high loads of COD and ammonia nitrogen. At present the leachate is being collected and recirculated to the landfill.

At the municipal WWTP in Gdansk, two types of sewage sludge are generated: primary sludge (after primary sedimentation tanks) and secondary excess sludge (separated in secondary sedimentation tanks). Both types of sludge are mixed in a 50 m^3 chamber and thickened by means of the Klein presses to approx. 6% dry matter. The next stage of sludge processing is mesophillic stabilization in two fermentation chambers with the volume of 7000 m^3 each. The digested sludge is dewatered in centrifuges to approx. 22% dry matter. The reject waters from sludge thickening and centrifugation are recirculated to the treatment process (before primary sedimentation tanks).

4.2.2 *The Landfill Leachate and Reject Waters Analyses*

The design of pilot treatment wetlands was preceded by laboratory analyses of landfill leachate (LL) and reject water (RWC) composition. The LL analyses were performed from June 2007 to May 2008 (11 samples), while the RWC analyses have been carried out for 3 years, from 2006 to 2008 (36 samples). In the samples of LL and RWC, the concentrations of organics (BOD_5 and COD), TSS (also organic suspended solids on the basis of losses on ignition), total Kiejdahl nitrogen, ammonia nitrogen and nitrates were analysed. Additionally, COD was analysed in LL and RWC after filtration (COD_f) on a 0.45 μm filter paper. The analytical procedure recommended by Hach Chemical Company and Dr Lange GmbH was used. The anlyses were performed according to the Polish norms and guidelines given in Polish Environmental Ministry Regulations of 8th July 2004 and 24th July 2006.

4.2.3 *Qualitative and Quantitative Analyses of LL and RWC*

The average concentrations of pollutants in the analysed LL and RWC were compared with the average pollutant concentrations in raw sewage discharged to the WWTP in Gdansk (Table 4.1).

Table 4.1 Comparison of average pollutant concentrations in the landfill leachate (LL, n=11), reject waters (RWC, n=36) and raw sewage (N=36) discharged to the municipal WWTP in Gdansk

Parameter	Unit	LL	RWC	Raw sewage
Flow	$m^3\ d^{-1}$		616	83764
TSS	$mg\ L^{-1}$	382.0	279.0	294.7
Organic SS	$mg\ L^{-1}$	68.0	193.5	254.7
N_{tot}	$mg\ L^{-1}$	542.0	900.0	61.9
NH_4^+-N	$mg\ L^{-1}$	217.5	734.0	29.6
N_{org}	$mg\ L^{-1}$	322.0	125.0	30.9
NO_3^-N	$mg\ L^{-1}$	2.6	3.0	1.4
COD	$mg\ L^{-1}$	2185.0	813.0	892.5
COD_f	$mg\ L^{-1}$	1259.5	353.0	318.4
BOD	$mg\ L^{-1}$	300.0	187.0	350.7
pH	–	7.9	7.6	6.8

The TSS concentration in RWC was twice higher than the corresponding concentration in LL. The share of organic SS in the TSS equalled 43.4% for RWC, 17.8% for LL and 86.4% for raw sewage. The BOD_5/COD ratios were 0.137, 0.23 and 0.4 for LL, RWC and raw sewage, respectively. The BOD_5/COD ratio of LL was surprisingly low in comparison to the values for young landfills reported in literature (Lo, 1996; Klimiuk et al., 2007). The BOD_5/COD_f ratios were significantly higher: 0.238 in LL, 0.52 in RWC and 1.1 in raw sewage. Low values of the BOD_5/COD ratio and high COD concentrations in the filtered samples indicated that organics were mostly in the form of hardly decomposable compounds.

The total nitrogen concentration in RWC was twice higher than in LL and ten times higher than in raw sewage. The share of two dominating forms of nitrogen (ammonia-N and organic-N) in the total nitrogen concentration was 40.0% of N-NH_4^+ and 59.4% of N_{org} in LL, and 81.6% of N–NH_4^+ and 13.8% of N_{org} in raw sewage.

The high variability of pollutant concentrations in time is characteristic for both LL and RWC (Gajewska & Obarska-Pempkowiak, 2008). Maximum concentrations are often more than 10 times higher than mean values, which substantially affects the median and standard deviations. Thus, further calculations of both pollutant removal effectiveness and the dimensioning of pilot treatment wetlands were based on medians. The analyses of LL and RWC composition indicate that the basic pollutant in both types of wastewater is nitrogen, present mainly in the form of Kjeldahl nitrogen (ammonia + organic). Therefore, the assumption for the design of pilot treatment wetlands was that ammonia and organic nitrogen should be effectively removed.

4.3 Dimensioning of Pilot LL and RWC Treatment Wetlands

It was assumed that the pilot wetlands for LL and RWC treatment would consist of two 1 m^3 chambers working in series, where wastewater would be collected and averaged. Additionally, the first of the chambers would be periodically aerated in order to reduce the ammonia nitrogen concentration. According to technological laboratory analyses, aeration is assumed to decrease the ammonia nitrogen concentration by approximately 40%. In order to calculate pilot VF-beds operating in a batch mode, it was assumed that the beds would treat the load of wastewater corresponding to 5 pe. The unit area of 2.5 m^2 pe^{-1} and the daily pollutant loads of 120 g COD pe^{-1} d^{-1}, 60 g TSS pe^{-1} d^{-1} and 12 g N_{tot} pe^{-1} d^{-1} were assumed. The VF-CW beds at both stages were divided into two compartments (sections), working alternately.

The total area of the first and the second stage beds is equal to:

$$F = 2.5\ \mathrm{m}^2\ \mathrm{pe}^{-1} \times 5\ \mathrm{pe} = 12.5\ \mathrm{m}^2$$

The area of the first stage bed corresponds to 60% of the total bed area:

$$F_{\mathrm{I}} = 12.5\ \mathrm{m}^2 \times 0.6 = 7.5\ \mathrm{m}^2$$

The area of the second stage bed corresponds to 40% of the total bed area:

$$F_{\mathrm{II}} = 12.5\ \mathrm{m}^2 \times 0.4 = 5\ \mathrm{m}^2$$

The daily loads of pollutants from 5 pe are as follows:

$$\mathrm{COD} = 120\,\mathrm{g\ COD\ pe}^{-1}\ \mathrm{d}^{-1} \times 5\ \mathrm{pe} = 600\ \mathrm{g\ COD\ d}^{-1}$$

$$\mathrm{TSS} = 60\ \mathrm{g\ TSS\ pe}^{-1}\ \mathrm{d}^{-1} \times 5\ \mathrm{pe} = 300\ \mathrm{g\ TSS\ d}^{-1}$$

$$\mathrm{N_{tot}} = 12\ \mathrm{g\ N_{tot}\ pe}^{-1}\ \mathrm{d}^{-1} \times 5\ \mathrm{pe} = 60\ \mathrm{g\ N_{tot.}\ d}^{-1}$$

Since nitrification which has so far been considered the main process responsible for ammonia nitrogen transformations is the most sensitive process, the calculations

Table 4.2 The operation conditions of pilot VF-CW wetlands treating RWC and LL

Type of wastewater	Hydraulic loading (mm d^{-1})	Organics loading (g COD m^{-2} d^{-1})	Kiejdahl N loading (g m^{-2} d^{-1})	Unit dose of wastewater (L d^{-1})
VF – CW I				
RWC	15.0	90.02	60.0	111.1
LL	24.6	402.7	60.0	184.3
VF-CW II				
RWC	22.0	15.9	24.0	111.1
LL	36.8	71.6	23.6	184.3

of the pilot beds were based on the daily load of total nitrogen from 5 pe equal to 60 g N_{tot} d^{-1}(Table 4.2). The assumed unit surface load of nitrogen is equal to:

$$\mathrm{N_{Tot}} = \frac{60\ \mathrm{g\,d^{-1}}}{12.5\ \mathrm{m^2}} = 4.8\ \mathrm{g\,m^{-2}d^{-1}}$$

Assuming the median concentration of total nitrogen (900 mg L^{-1} for RWC and 542 mg L^{-1} for LL) and the decrease of pollutants in the averaging chambers ($0.6 \times 900.0 = 540$ mg L^{-1} and $0.6 \times 542.0 = 325.2$ mg L^{-1}), the one-time dose per each section of the bed is equal to:

$$V_{\mathrm{RWC}} = \frac{60\ \mathrm{g\ N_{og}d^{-1}}}{540.0\ \mathrm{g\ N_{og}m^{-3}}} = 0.1111\ \mathrm{m^3d^{-1}} \cong 111.1\ \mathrm{Ld^{-1}}$$

$$V_{\mathrm{LL}} = \frac{60\ \mathrm{g\ N_{og}d^{-1}}}{325.2\ \mathrm{g\ N_{og.}m^{-3}}} = 0.1843\ \mathrm{m^3d^{-1}} \cong 184.3\ \mathrm{Ld^{-1}}$$

The one-time hydraulic loadings for both beds (transpiration process was not taken into account) are as follows:

– for the first stage bed of the area $F_{\mathrm{I}} = 7.5\ \mathrm{m}^2$

$$\mathrm{HL_{IRWC}} = \frac{V}{F_I} = \frac{0.1111\ \mathrm{m}^3 d^{-1}}{7.5\ \mathrm{m^2}} = 0.0148\ \mathrm{md^{-1}} = 15\ \mathrm{mm\,d^{-1}}$$

$$\mathrm{HL_{ILL}} = \frac{V}{F_{\mathrm{I}}} = \frac{0.1843\ \mathrm{m^3d^{-1}}}{7.5\ \mathrm{m^2}} = 0.02457\ \mathrm{m\,d^{-1}} = 24.6\ \mathrm{mm\ d^{-1}}$$

– the second stage bed of the area $F_{\mathrm{I}} = 5\ \mathrm{m}^2$

$$\mathrm{HL_{IIRWC}} = \frac{V}{F_{\mathrm{II}}} = \frac{0.111\ \mathrm{m^3\ d^{-1}}}{5\ \mathrm{m^2}} = 0.022\ \mathrm{m\,d^{-1}} = 22\ \mathrm{mm\,d^{-1}}$$

$$\mathrm{HL_{IILL}} = \frac{V}{F_{\mathrm{II}}} = \frac{0.1843\ \mathrm{m^3d^{-1}}}{5\ \mathrm{m^2}} = 0.03686\ \mathrm{m\,d^{-1}} = 36.8\ \mathrm{mm\,d^{-1}}$$

The calculated hydraulic loadings are substantially lower than 600 mm d^{-1} recommended by Molle et al. (2004). Platzer and Mauch (1996) indicated that effective nitrification and nitrogen removal at VF-CW beds take place when hydraulic loading is below 300 mm d^{-1}.

The expected removal of pollutants presented in Table 4.3, were calculated on the basis of the average treatment efficiencies reported by Molle et al. (2004) and the median pollutant concentrations in LL and RWC.

Table 4.3 Calculated influent and effluent concentrations (mg L^{-1}) in LL and RWC treatment in the pilot constructed wetlands

	COD		TSS		Kjeldahl nitrogen	
	Influent	Effluent	Influent	Effluent	Influent	Effluent
VF-CW I						
RWC	813.0	143.3	279.0	30.0	540	216.0
LL	2185.0	388.7	382.0	42.1	325.4	129.5
VF-CW II						
RWC	143.3	57.3	30.0	6.8	216.0	47.5
LL	388.7	155.5	84.9	6.3	129.5	28.5
Concentrations in municipal landfill leachate in Chlewnica						
LL	1000		24		75	
Concentrations in raw sewage discharged to the WWTP in Gdansk						
RWC	892.5		294.7		60.5	

4.4 Selection of Plant Species

Due to the high variability of LL and RWC quantities and composition, the habitat conditions of plants in the constructed wetlands for LL and RWC treatment are very difficult. Fluctuations of the water level (periodical flooding or drought), a high loading of organics, oxygen deficits, high concentrations of ammonia nitrogen, the improper nitrogen/phosphorus ratio, high salinity as well as the presence of heavy metals and xenobiotics create unfavourable conditions that only some plant species can withstand. Plants species selected for the pilot wetlands for LL and RWC treatment not only needed to survive in such conditions, but they also were supposed to support the removal and transformation of pollutants. Moreover, selected plant species should have a long vegetation period and their seedlings should be easy to plant. The capability of heavy metal retention in plant biomass was desirable. Also, fast growth and potential to produce energy from biomass might be considered as a strong advantage. In Table 4.4. hydrophyte species which could be used in constructed wetlands for LL and RWC treatment are shown.

Common reed was selected for the pilot wetland for LL and RWC treatment due to the good tolerance to elevated concentrations of chlorides and iron.

4.5 Evaluation of Evapotransportation

Water loss to atmosphere may be quite high, especially in the periods with warm and windy weather (Szymkiewicz, 1990) and the water loss is very much dependent on the plant density. In most cases when treatment wetlands are applied, the decrease of outflow is considered to be an additional benefit. However, in the case of LL and

Table 4.4 Characteristics and potential use of hydrophyte species for LL and RWC treatment

Species	Characteristics	Potential use for LL and RWC treatment
Willow (*Salix* spp.)	Fast growth (2–3 m yr^{-1}), high biomass (2.2–13.5 g m^{-2}), potential production of energy from biomass, good seedling availability, easy planting, long vegetation period, high transpiration	YES Randerson & Slater (2005) Riddell-Black, Alker, Mainstone, and Smith (2000)
Poplar *Populus* (spp.)	Fast growth (2–3 m yr^{-1}), high biomass production, potential production of energy from biomass	potentially YES
Common reed (*Phragmites australis*)	Toleration to salinity and pH fluctuations, toleration to periodical flooding and droughts (due to deep root system), ability of oxygen transfer to root zone, capability of heavy metal retention in biomass, good seedling availability	YES Urbanc-Berčič (1994), Maehlum (1995), Peverly et al.(1995), Bulc et al. (1997), Johnson, Martin, Moshiri, and McCrory (1999), Kinsley et al. (2006)
Reed canarygrass (*Phalaris arundinacea*)	Long vegetation period, fast growth, deep root system, toleration to salinity and pH fluctuations, potential production of energy from biomass, good seedling availability	potentially YES
Reed mace (*Typha latifolia*)	Fast growth, ability of oxygen transfer to root zone, capability of heavy metal retention in biomass, good seedling availability	YES Kinsley et al. (2007) Nivala, Hoos, Cross, Wallace, and Parkin (2007)
giant Chinese silver grass (Miscanthus giganteus)	Fast growth, high biomass production, potential production of energy from biomass, toleration to low temperatures, (although biomass production is dependent on temperature, insulation, water content in soil, etc). Toleration to pH in the range from 5.5 to 7.5. Well developed root system. Relatively high transpiration.	potentially YES Nixon (2001)
Giant cane (*Arundo donax*)	Fast growth, potential production of energy from biomass	no reports

Table 4.5 Water balance for pilot wetlands treating LL and RWC

Type of sewage	Inflow ($L\ d^{-1}$)	Hydraulic loading ($mm\ d^{-1}$)	Rainfall ($mm\ d^{-1}$)[a]	Evaporation ($mm\ d^{-1}$)	Transpiration ($mm\ d^{-1}$)	Outflow ($L\ d^{-1}$)
Vegetation season						
RWC	111.1	15.0	1.9	2.0	4.0	67.4
LL	184.3	24.6	1.9	2.0	4.0	77.0
Non-vegetation season						
RWC	111.1	15.0	1.4	0.6	0.0	55.2
LL	184.3	24.6	1.4	0.6	0.0	64.8

[a]Based on data from 2008 acquired from the Meteorological Station of Civil and Environmental Engineering Faculty, Gdansk University of Technology in Gdansk

RWC treatment, evapotranspiration could turn out to be disadvantageous since water losses lead to the growth of pollutant concentrations, which are already very high (for instance ammonia nitrogen, COD or chlorides). This may cause damage to the plants and lead to the decrease of treatment effectiveness in the consequence (Clarke & Baldwin, 2002). Thus, the analysis of evaporation and transpiration processes seems to be important. In Table 4.5, the average transpiration values of the pilot treatment wetlands for vegetation and non-vegetation periods were calculated. The data on rainfall in 2008 were used for calculations.

The water balance calculations showed that evapotranspiration should not have a significant impact on LL and RWC treatment processes. Rainfall values in the vegetation season are substantially higher than the transpiration capacity of reed (20 mm d^{-1} according to Headley, Davison, Huet, & Müller, 2009), thus water losses due to evapotranspiration will be compensated by rainfall.

4.6 Preliminary Results

In July 2008, the pilot treatment wetlands were planted with common reed and filled with water. The water level was held approx. 10 cm above the surface of the beds for weed control. In the case of the pilot wetland for RWC treatment, treated sewage was used instead of water. In September 2008, the discharge of LL and RWC started. Initially, the reject waters (RWC) and landfill leachate (LL) were diluted with water (volumetric ratio of wastewater/water 1:3) to avoid stress for the plants. This start-up period lasted until November 2008. The results of two preliminary investigation series, performed during this period, are presented in Figs. 4.1 and 4.2.

The effectiveness of organics removal in the preliminary treatment series was very high: 74% for RWC and 84.5% for LL. The TSS removal effectiveness for LL varied from 44 to 68%, whereas in the case of the RWC treating wetland, the TSS concentration increased after subsequent treatment stages. The analysis of TSS composition indicated that the mineral fraction is dominating. Fine-grained dust fractions were probably washed out from the beds. The effectiveness of ammonia

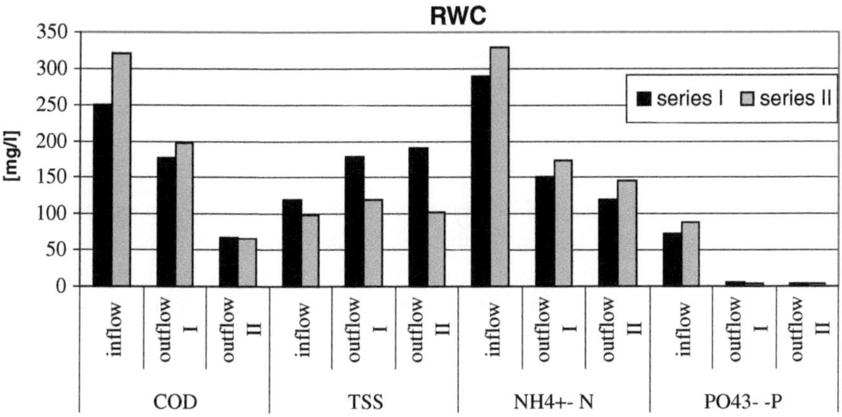

Fig. 4.1 The concentrations of pollutants in RWC after subsequent treatment stages

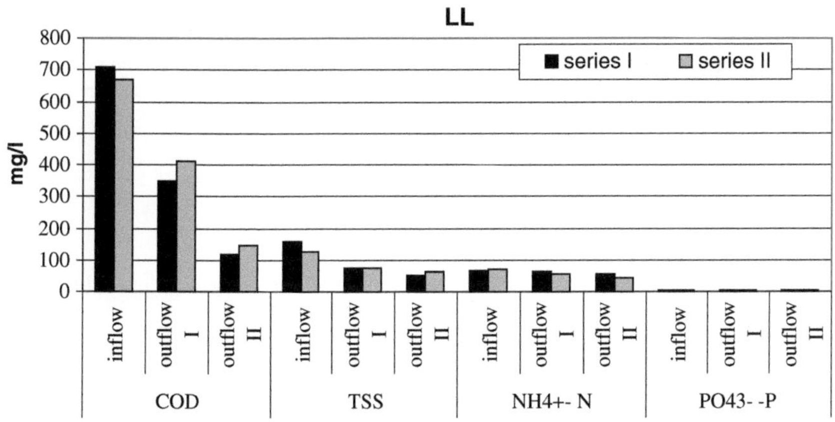

Fig. 4.2 The concentrations of pollutants in LL after subsequent treatment stages

nitrogen removal was below 20% for the LL wetland and from 48 to 59% for the RWC wetland. The significant difference in nitrogen removal effectiveness probably resulted from the fact that polycoagulants were added to the treatment process at WWTP "Wschod" for the precipitation of phosphorus compounds, which also affects the nitrogen removal (Dong & Tieheng, 2007).

4.7 Conclusions

The landfill leachate and reject waters from the dewatering of digested sewage sludge are characterized by high concentrations of pollutants, especially organics and ammonia nitrogen. The values of BOD_5/COD and BOD_5/COD_f ratios in LL and RWC indicate the high share of suspended organics, resistant to biodegradation.

High concentrations of nitrogen, mostly in the form of ammonia nitrogen in LL and RWC, make both types of sewage difficult to treat by means of conventional biological methods with activated sludge.

The evaluation of water balance in the pilot wetlands for LL and RWC treatment shows that water losses due to evapotranspiration do not cause a significant increase of pollutant concentrations.

If the assumed treatment effectiveness is reached after the start-up period, the Kjeldahl nitrogen concentrations in the treated RWC and LL will be similar to the Kjeldahl N concentrations in raw sewage discharged to the WWTP in Gdansk. Then, the treated RWC and LL could be discharged to a conventional WWTP.

The removal effectiveness in the start-up period was as follows: organics 74% for RWC and 84.5% for LL, and ammonia nitrogen – below 20% for LL and from 48 to 59% for RWC. The outflow concentrations of COD were equal to 285 mg L^{-1} for RWC and 132.5 mg L^{-1} for LL. The NH_4^+-N concentrations for RWC and LL were equal to 132.5 mg L^{-1} and 49 mg L^{-1}, respectively.

Acknowledgments Funding support from the EEA Financial Mechanism (PL 0085) and the Ministry of Science and Higher Education in Poland (E007/P01/2007/01) is gratefully acknowledged. The authors are indebted to SNG Saur Neptun Gdansk and ELWOZ for the permission to build pilot LL and RWC treating wetlands, and for allowing them to do research in WWTP in Gdansk and Municipal Landfill in Chlewnica.

References

Boutin, C., Lienard, A., & Esser, D. (1997). Development of a new generation of reed-beds filters in France: First results. *Water Science &. Technology, 35*(5), 315–322.

Bulc, T. G. (2006). Long term performance of a constructed wetland for landfill leachate treatment. *Ecological Engineering, 26*, 365–374.

Bulc, T., Vrhovšek, D., & Kukanja, V. (1997). The use of constructed wetland for landfill leachate. *Water Treatment and Technology, 35*, 301–306.

Christensen, T. H., Kjeldsen, P., Bjerg, P. L, Jensen, D. L., Christensen, J. B., Baun, A., et al. (2001). Biogeochemistry of landfill leachate plumes – Review. *Applied Geochemistry, 16*, 659–718.

Clarke E., & Baldwin A.H. (2002). Responses of wetland plants to ammonia and water level. *Ecologcal Engineering, 18*, 257–264.

Dong Z., & Tieheng S. (2007). A potential new process for improving nitrogen removal in constructed wetlands- Promoting coexistence of partial-nitrification and ANAMMOX. *Ecological Engineering, 3*, 69–78.

Fux, C., Boehler, M., Huber, P., Brunner, I., & Siegrist, H. (2002). Biological treatment of ammonium-rich wastewater by partial nitritation and subsequent anaerobic ammonium oxidation (anammox) in a pilot plant. *Journal of Biotechnology, 99*, 293–306.

Fux, C., Lange, K., Faesseler, A., Huber, P., Grueniger, B., & Siegrist, H. (2003). Nitrogen removal from digester supernatant via nitrite-SBR or SHARON? *Water Science & Technology, 48*, (8), 9–18.

Fux, C., Valten, S., Carozzi, V., Solley, D., & Keller, J. (2006). Efficient and stable nitrification and denitrification of ammonium-rich sludge dewatering liquor using SBR with continuous loading. *Water Research, 40*, 2765–2775.

Gajewska, M., & Obarska-Pempkowiak, H. (2008). The influence of return flow of reject water from sludge dewatering on WWTP operation. *Chemical Industry, 5*, 448–452 (in Polish).

Gajewska, M., Tuszynska, A., & Obarska-Pempkowiak, H. (2004). Influence of configuration of the beds on contamination removal in hybrid constructed wetlands. *Polish Journal of Environmental Studies, 13*, 149–153.

Hans, M., van der Roest, J., & van der Roest, H. (1997). Don't reject the idea of treating reject water. *Water Science & Technology, 35* (10), 27–34.

Headley T., Davison L., Huet D., & Müller R. (2009). Evapotranspiration from pilot-scale horizontal subsurface flo Phragmites australis wetlands in sub-tropical and temperate climate. In L. Kröpfelová & J. Vymazal (Eds.), *Proceedings of the 7th International Workshop Nutrient Cycling and Retention in Natural and Constructed Wetlands* (pp. 33–35). Treboň, Czech Republic: ENKI.

Jeavons, J., Stokes, L., Upton, J., & Bingley, M. (1998). Successful side stream nitrification of digested sludge liquors. *Water Science & Technology, 38*(3), 111–118.

Johansson, S., & Westholm, L. (2003). *Leachate treatment with use of SBR-technology combined with a constructed wetland system at the Isätra landfill site*, Sweden. Proceedings of the 9th international waste management and landfill symposium (pp. 75–81). Cagliari, Italy: S. Margherita di Pula,.

Johnson, K. D., Martin, C. D., Moshiri, G. A., & McCrory, W. C. (1999). Performance of constructed wetland leachate treatment system at the Chunchula landfill, Mobile County, Alabama. In G. Mulamoottil, E. A. McBean & F. Rovers (Eds.), *Constructed wetlands for the treatment of landfill leachate* (pp. 57–70). Boca Raton, FL: CRC.

Kadlec, R. H. (2003). Integrated natural systems for landfill leachate treatment. In J. Vymazal (Ed.), *Wetlands – nutrients, metals and mass cycling* (pp. 1–33). Leiden, The Netherlands: Backhuys Publishers.

Kayser, K., Kunst, S., Fehr, G., & Voermanek, H. (2001). Nitrification in reed beds-capacity and potential control methods. *Water Science and Technology, 46* (6–7), 363–370.

Kinsley, C. B., Crolla, A. M., Kuyucak, N., Zimmer, M., & Lafléche, A. (2006). Nitrogen dynamics in a constructed wetland system treating landfill leachate. In V. Dias & J. Vymazal (Eds.), *Proceedings of the 10th International Conference on Wetland Systems for Water Pollution Control* (pp. 295–305). Lisbon, Portugal: MAOTDR.

Klimiuk, E., Kulikowska, D., & Koc-Jurczyk, J. (2007). Biological removal of organics and nitrogen from landfill leachates – A review. In M. Pawłowska & L. Pawłowski (Eds.), *Management of pollutant emission from landfills and sludge* (pp. 187–204). London: Taylor & Francis.

Laurich, F., & Gunner, C. (2003). The store and treatment process for sludge liquor management. *Water Science and Technology, 47*, 269–275.

Lo, I. M. C. (1996). Characteristics and treatment of leachates from domestic landfills. *Environment International, 22*, 433–442.

Maehlum, T. (1995). Treatment of landfill leachate in on-site lagoons and constructed wetlands, *Water Science & Technology, 32*(3), 129–135.

Martin, C. D., Johnson, K. D., & Moshiri, G. A. (1999). Performance of constructed wetland leachate treatment system in Chunchula landfill, Mobile County, Alabama. *Water Science and Technology, 40*(3), 67–74.

Marttinen, S. K., Kettunen, R. H. & Rintala, J. A. (2003). Occurrence and removal of organic pollutants in sewages and landfill leachates. *Science of the Total Environment, 301*, 1–12,

Molle, P., Lienard, A., Boutin, C., Merlin, G., & Iwema, A. (2004). How to treat raw sewage with constructed wetlands: An overview of the French systems. In A. Lienard (Ed.), *Proceedings of the 9th International Conference on Wetland System for Water Pollution Control* (pp. 11–20). Lyon, France: ASTEE and Cemagref.

Nivala, J., Hoos, M. B., Cross, C., Wallace, S., & Parkin, G. (2007). Treatment of landfill leachate using an aerated, horizontal subsurface-flow constructed wetland. *Science of the Total Environment, 380*, 19–27.

Nixon, P. M. (2001). Effects of landfill leachate on the biomass production of Miscanthus. *Aspects of Applied Biology, 65*, 123–130.

Obarska-Pempkowiak, H. (2002). *Oczyszczalnie hydrofitowe* (Constructed wetlands). Gdansk, Poland: Wydawnictwo Politechniki Gdanskiej.

Paxeus, N. (2000). Organic compounds in municipal landfill leachates. *Water Science & Technology, 41*(7–8), 323.

Peverly, J. H., Surface, J. M., & Wang, T. (1995). Growth and trace metals absorption by *Phragmites australis* in wetlands constructed for landfill leachate treatment. *Ecological Engineering, 5*, 21–35.

Platzer, C., & Mauch, K. (1996). Evaluations concerning soil clogging in vertical flow reed beds-mechanisms, parameters, consequences and solutions? In IWGA – Institute for Water and Provision, Water Ecology and Wast Management, Universitaet fuer Bodenkulyur Wien, *Proceedings of the 5th International Conference on Wetland System for Water Pollution Control* (chapter IV/2-11). Vienna, Austria: Universität für Bodenkultur Wien and International Association on Water Quality.

Randerson, P. F., & Slater, F. M. (2005). The role of willow plants in the treatment of iron-rich landfill leachate. In *Proceedings of the 6th International of Conference on Environmental Engineering* (pp. 420–424). Vilnius, Lithuania: Vilnius Gediminas Technical University.

Reddy, K. R., & D'Angelo, E. M. (1996). Biochemical indicator to evaluate pollutant removal efficiency in constructed wetlands. In IWGA – Institute for Water and Provision, Water Ecology and Wast Management, Universitaet fuer Bodenkulyur Wien, *Proceedings of the 5th International Conference on Wetland Systems for Water Pollution Control* (Keynote address I, 1–21). Vienna, Austria: Universität für Bodenkultur and IWA.

Riddell-Black, D., Alker, G., Mainstone, C. P., & Smith S. R. (1996). Economically viable buffer zones – the case for short rotation forest plantations. In N. Haycock (Ed.), *Buffer zones* (pp. 103–108). Heythrop, UK: Harpenden.

Robinson, A. H. (2005). Landfill leachate treatment. *Membrane Technology, 6*, 6–12.

Rustige, H., & Nolde, E. (2006). Nitrogen elimination from landfill leachates using an extra carbon source in subsurface flow constructed wetlands. In V. Dias & J. Vymazal (Eds.), *Proceedings of the 10th International Conference on Wetland Systems for Water Pollution Control* (pp. 229–239). Lisbon, Portugal: MAOTDR.

Szymkiewicz, R. (1990). *Hydrology*. Gdank, Poland: Gdansk University of Technology (in Polish).

Tatsi, A. A., & Zoubolis, A. I. (2002). A field investigation of the quantity and quality of leachate from a municipal waste landfill in a Mediterranean climate (Thessaloniki, Greece). *Advances in Environmental Research, 6*, 207–219.

Urbanc-Berčič, O. (1994). Investigation into the use of constructed reed beds for municipal waste dump leachate treatment. *Water Science and Technology, 29*, 289–294.

Vymazal, J. (Ed.) (2001) *Transformation of nutrients in natural and constructed wetlands*. Leiden, The Netherlands: Backhuys Publishers.

Wett, B., & Alex, J. (2003). Impact of separate reject water treatment on the overall plant performance. *Water Science & Technology, 48*(4), 139–14.

Wojciechowska, E., & Obarska-Pempkowiak, H. (2008). Leachate treatment at a pilot plant using hydrophyte systems. In M. Pawłowska & L. Pawłowski (Eds.), *Management of pollutant emission from landfills and sludge* (pp. 205–210). London: Taylor & Francis.

Chapter 5
Comparison of Performance, Ageing and Eco Balance Data from Four Types of Constructed Wetlands Treating Raw Sewage

Georges Reeb and Stephanie Liey

Abstract Seven filtration systems treating raw wastewater planted with helophytes and designed by Atelier Reeb between 1996 and 2003, have been closely monitored by the appropriate authorities. The most basic system, not including sequential feeding, and the most sophisticated system involving pumps, are compared with regard to their filtering capacities (the first being less efficient), their ageing (in general, vegetated systems have a long life and they age well) and their eco balance (the more basic systems are of interest here). The filters are grouped into four types of system according to feeding method, flow type within the filter and the number of stages of filtration. The three batch-fed systems have performed well with respect to French standards. The most basic system, with gravitational feeding, requires more careful monitoring and an additional treatment phase with a wooded buffer zone in order to obtain satisfactory results. Its very low costs of operation nevertheless make it a viable system in many cases, on condition that the design has been properly adapted.

Keywords Ageing · Constructed wetlands · Helophytes · Life Cycle Assessment · Raw wastewater · Vegetated filtration systems

5.1 Introduction

In France, based on some initial findings by K. Seidel, various types of constructed wetland systems for the treatment of raw municipal wastewater have been successfully put into operation (Lienard, Boutin, & Esser, 1990; Molle, Lienard, Boutin, Merlin, & Iwena, 2005) and the Life Cycle Assessment of these systems has recently been reported by Comby, Reeb, Werckmann, and Quaranta (2008). The purpose of this study is to compare treatment performance of four combinations of hybrid constructed wetlands.

G. Reeb (✉)
Atelier Reeb, 13 Quai des Bateliers, 67000, Strasbourg, France
e-mail: georges.reeb@atelier-reeb.fr

J. Vymazal (ed.), *Water and Nutrient Management in Natural and Constructed Wetlands*, DOI 10.1007/978-90-481-9585-5_5,

5.2 Materials and Methods

5.2.1 Overview of the Treatment Plants

The study compares four types of hybrid constructed wetlands (CWs) which were designed and initiated by Atelier Reeb between 1996 and 2003 in various areas of France (Table 5.1).

The filtration material in the 1st vertical flow (VF) stages consists of layers of 15–25 cm of fine gravel (4–8 mm), gravel (10–20 mm) and coarse gravel (30–80 mm). The upper layer (20–30 cm thick) of the 2nd VF stages (in case it is present) consists of sand (0–4 mm). Horizontal flow (HF) beds are filled with a 50–60 cm layer of 10–20 mm gravel. In all VF beds, the 1st stage is planted with Common reed (*Phragmites australis*), the 2nd and 3rd stages can be planted with *Phragmites australis* but also with *Phalaris arundinacea* (Reed canarygrass), *Typha latifolia* (Broadleaf cattail), *Mentha aquatica* (Water mint), *Scirpus lacustris* (Bullrush), *Juncus effusus* (Needlerush) or *Acorus calamus* (Sweet flag).

Four types of hybrid constructed wetlands differ in stage arrangements. The first type was built at Saint Matre with continuous gravity (g) feeding and comprises two successive stages of vertical flow and a final stage of horizontal flow (g-VF-VF-HF). Two other types of hybrid constructed wetlands are fed intermittently. The CWs of the first type were built at Orbey, Cazillac and Caurel and they comprise a raw wastewater flushing device, two stages of vertical flow and one stage of horizontal flow (f-VF-VF-HF). The second intermittently fed system was built at Viviers/Artaut and is equipped with a raw wastewater flushing device followed by a stage with vertical flow, the second flushing device and another stage of vertical flow (f-VF-f-VF). The fourth CW system involves a pumping system (p) that feeds one vertical flow stage, after which it is recirculated (p-VFR-R). Constructed wetlands of this type were built at Virloup and Plufur (Table 5.1).

Four systems (Saint Matre, Caurel, Viviers/Artaut and Plufur) are complemented with a vegetated buffer zone generally planted with trees adapted to waterlogged or wet soil conditions such as *Salix* sp. (Willow), *Alnus* sp. (Alder) or *Betula* sp. (Birch) and helophytes in order to (1) improve the effluent quality, (2) protect the surrounding area (sewer overflows are directed to these buffer zones and not directly to the recipients) and (3) create biotopes that encourage regeneration and connection of the green corridors. According to Carbiener (1970), the ecological factor of forested habitats in flood areas is “the highly effective interaction between soil and vegetation for filtering flood water seeping towards the water table”, especially in the mature forest stage.

To make the comparison broader, data from constructed wetlands not designed by Atelier Reeb are also presented. For the g-VF-VF-HF systems, the data were obtained from hybrid constructed wetlands at Saint Bohaire designed by K. Seidel and Pannessière (Mettetal, Rouault, & Valero, 1990) while for the f-VF-f-VF system data are taken from the results reported by Molle et al. (2005).

The monitored treatment plants have been in operation for at least 2 years. The comparisons are based on a collection of samples over a 24-h period from six of

Table 5.1 Characteristics of the compared constructed wetlands. VF = vertical flow constructed wetland, HF = horizontal flow constructed wetland, g = continuous feeding by gravity, f = feeding by intermittent flushing, p = feeding by pumping, R = recirculation

	gVF-VF-HF	fVF-VF-HF			f-VF-f-VF	pVF-R	
System no.		1	2	3		1	2
Location	Saint Matre	Orbey	Cazillac	Caurel	Viviers/Artaut	Virloup	Plufur
Designed PE	120	100	50	350	200	40	290
Commissioning	1998	1996	1998	2000	2003	2003	2003
24 h monitoring	-	11/1999 04/2000 07/2004	06/1998	06/2008	10/2007	06/2007 10/2007	03/2007
Feeding system	Gravity	Flushing	Flushing	Flushing	Flushing/Flushing	Pump	Pump
Total CW area (m^2)	316	95	130	380	405	124	523
Specific area (m^2 PE^{-1})	2.63	0.95	2.60	1.08	2.02	3.10	1.80

the seven constructed wetlands; the gravity-fed (g-VF-VF-HF) constructed wetland could be evaluated during only one sampling campaign. The results therefore should be taken with caution. The samples were taken at the inlet and outlet points and were analyzed by authorised laboratories according to French standards.

5.3 Results

5.3.1 Performance Comparisons

The treatment performances of monitored constructed wetlands were evaluated against the regulations governing discharge limits. The discharge limits were different in 1996 and 2007 (Implementation, 1997; Regulation, 2007). The limits of BOD_5 and COD (Fig. 5.1) and TSS (Fig. 5.2) correspond to the Regulation of 2007, while the TKN standard (Fig. 5.2) refers to the Regulation of 1997. We have therefore chosen to compare the results with the more strict discharge limits.

Data shown in Table 5.2 indicate that the sewage flow in systems at Maison Oberlin, Cazillac and Caurel were at least twice lower than their nominal capacity. On the other hand the p-VF-R systems at Plufur and Viviers sur Artaut were overloaded owing to their recirculation. Three plants have lower specific area (Table 5.1) than the common French design criteria of 2 m^2 PE^{-1} recommended by Molle et al. (2005), but despite that the treatment performance is very good.

5.3.1.1 Performance of the gVF-VF-HF System Fed by Gravity

The gVF-VF-HF constructed wetland at Saint Matre is fed entirely gravitationally, without any flushing device. Effluent discharged from this system has quite poor quality: COD 294 mg L^{-1} and BOD_5 131 mg L^{-1} (Fig. 5.1). These values are about twice and five times higher than the D4 discharge limits of 125 mg L^{-1} and 25 mg L^{-1}, respectively. From this point of view, the treatment performance is weak, even after a number of modifications.

The gVF-VF-HF constructed wetland at Pannessière provides good outflow concentrations of COD (74.7 mg L^{-1}), BOD_5 (15.8 mg L^{-1}) and TSS (81.2%) which meet the D4 limits of 125 mg L^{-1} for COD, 25 mg L^{-1} for BOD_5 and 50% for TSS (Figs. 5.1 and 5.2). However, removal of 60% set by the discharge criteria were not met and removal of TKN amounted to only 54% (Fig. 5.2). In Pannessière, the average results were calculated from nine random samples (inlet/outlet) taken during the course of the year

The treatment performance of the system at Saint Bohaire (Boutin, 1987) was very high for TSS with outflow concentrations in the range of 8–28 mg L^{-1}. However, removal of other parameters fluctuated widely with concentrations exceeding the discharge limits: COD 92–184 mg L^{-1}, BOD_5 10–48 mg L^{-1}, NH_4-N 16.1–34.7 mg L^{-1}, NO_3-N <0.1–0.19 mg L^{-1} and TP 10.5–33.1 mg L^{-1}.

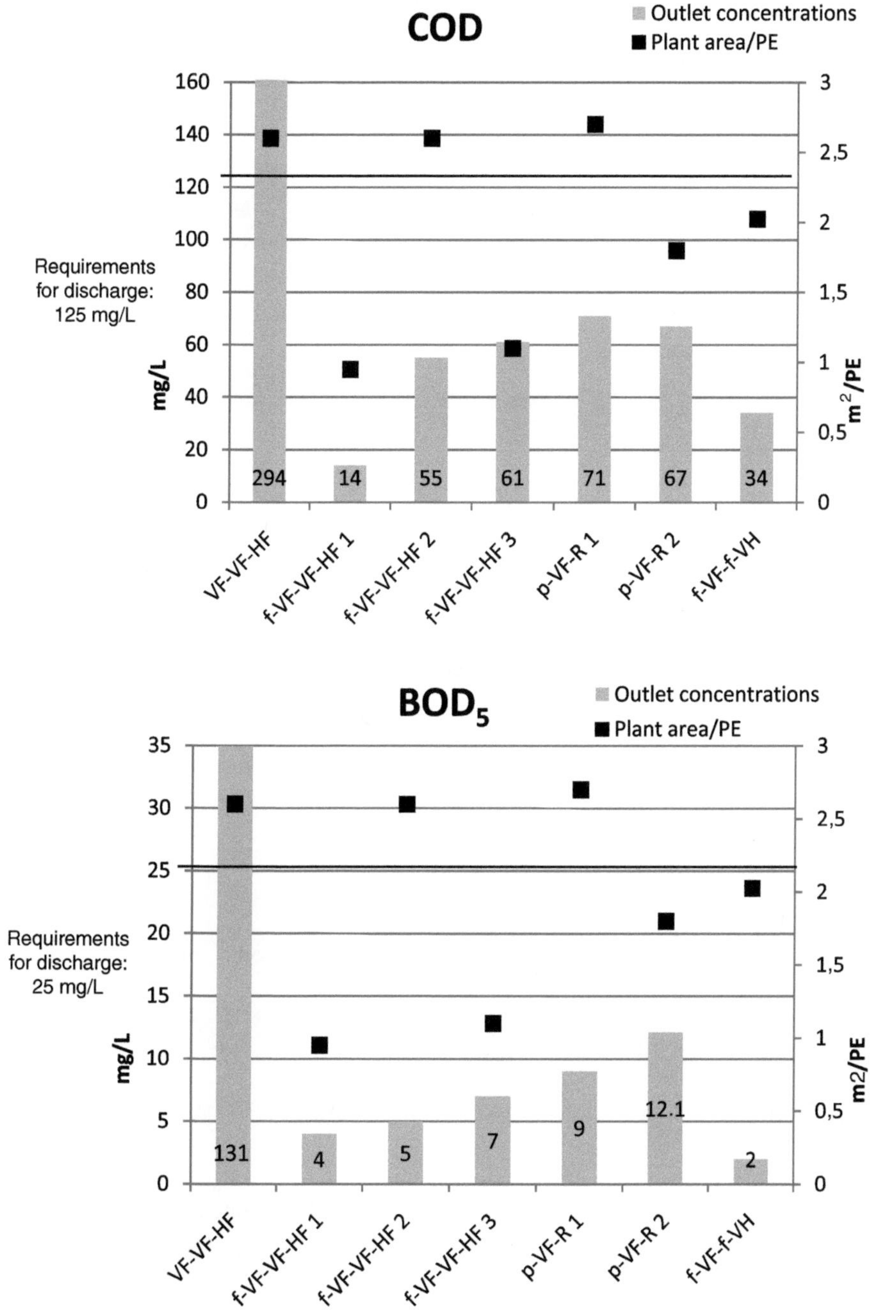

Fig. 5.1 Comparison of treatment performance for BOD_5 and COD of seven hybrid constructed wetlands in France

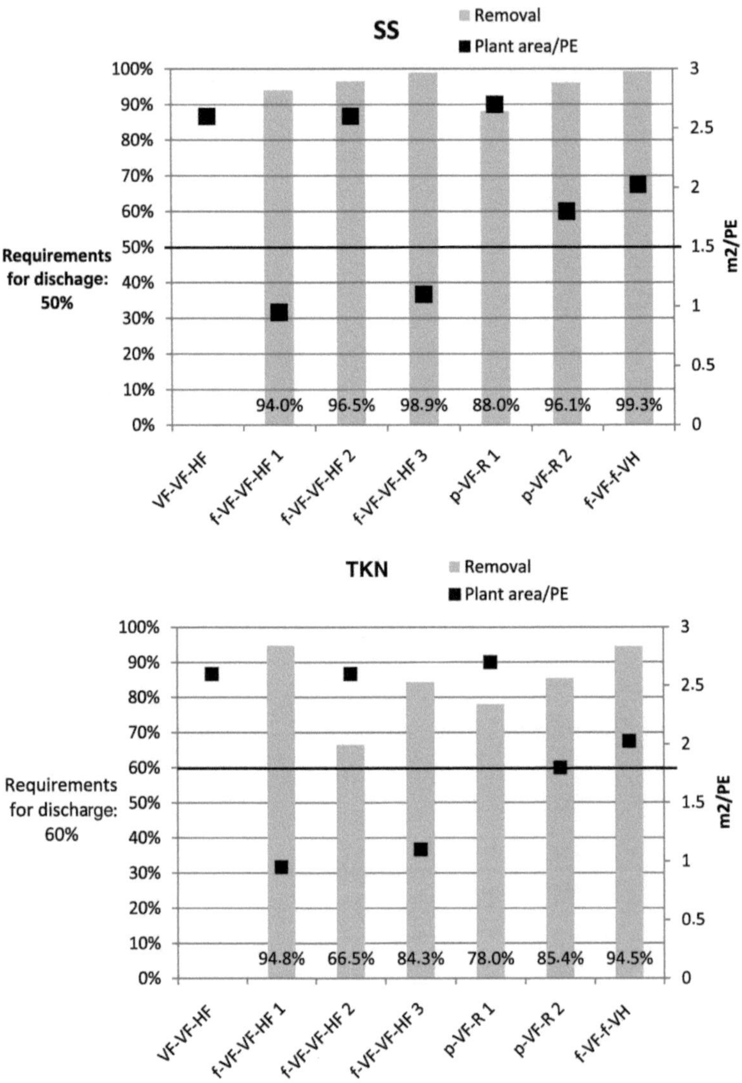

Fig. 5.2 Comparison of treatment performance for suspended solids (SS) and TKN of seven hybrid constructed wetlands in France

In general, the performance of the gVF-VF-HF treatment system does not meet the required discharge standards. Despite that, the system does have some advantages which will be discussed later in the text.

5.3.1.2 Performance of the fVF-VF-HF Systems with Intermittent Feeding

Three f-VF-VF-HF systems with intermittent feeding began to operate in 1996 (Orbey), 1998 (Cazillac) and 2000 (Caurel) (Table 5.1). For all three constructed

Table 5.2 Comparison of designed and measured wastewater flows ($m^3\ d^{-1}$) at monitored constructed wetlands

	Maison Oberlin	Cazillac	Caurel	Virloup	Plufur	Viviers/Artaut
Designed flow	15	7.5	52.5	6	43.5	30
Measured flow	7.8	2.2	11.94	8[a]	149.3[b]	12

[a] 3.4 $m^3\ d^{-1}$ without recirculation
[b] 37.3 $m^3\ d^{-1}$ without recirculation

Table 5.3 Outflow concentrations ($mg\ L^{-1}$) in Orbey constructed wetlands during sampling campaigns in 1999, 2000 and 2004

Sampling time	TSS	COD	BOD_5	TKN
November 1999	5	20	< 4	3.9
April 2000	35	105	23	22.8
July 2004	3	14	4	1.5

wetlands the treatment performance is satisfactory and discharge limits for BOD_5, COD, TKN and TSS are met (Figs. 5.1 and 5.2). In terms of TKN, the oldest constructed wetland at Orbey shows the highest removal. In Table 5.3, outflow concentrations during three sampling periods are shown. It is important to take into consideration that the systems at Orbey and Caurel provide very good discharge concentrations of pollutants using a specific area less than 1 $m^2\ PE^{-1}$ (Figs. 5.1 and 5.2).

5.3.1.3 Performance of the f-VF-f-VF System

The f-VF-f-VF constructed wetland is a very typical French system with two stages of vertical flow beds, each of them fed by a flushing device (Molle et al., 2005). We observe that this system at Viviers/Artaut provides the most satisfactory results as compared to D4 discharge limits (TSS 3 $mg\ L^{-1}$and 99.3%; COD 34 $mg\ L^{-1}$ and 96.6%; BOD_5 2 $mg\ L^{-1}$ and 99.6%, and TKN 5 $mg\ L^{-1}$ and 94.5%). However, it is difficult to make any sound conclusion based on the performance of this type of hybrid systems when only one treatment plant is evaluated. Therefore, we compared our results with those reported by Molle et al. (2005) who compared 48 constructed wetlands of this type with an average PE of 410. The constructed wetlands included in this review have been in operation for 0.2–7.2 years and the average removal efficiencies amounted to 95% for TSS, 91% for COD, and 85% for TKN. In our study, the respective removal efficiencies amounted to 99, 97 and 95%.

5.3.1.4 Performance of the pVF-R Systems

The hybrid constructed wetlands at Virloup and Plufur comprise one stage of vertical flow fed by a pump and equipped with recirculation. Both plants show very good

Table 5.4 Comparison of outflow concentrations (Out) and treatment efficiency (Eff) in p-VF-R constructed wetland at Virloup with and without recirculation

	TSS		BOD_5		COD		TKN	
	Out (mg L^{-1})	Eff (%)	Out (mg L^{-1})	Eff (%)	Out (mg L^{-1})	Eff (%)	Out (mg L^{-1})	Eff (%)
Without recirculation	55	70	21	91	160	77	52	32
With recirculation	21	88	9	98	71	93	26	78

performance in all parameters which meet the discharge limits (Figs. 5.1 and 5.2). The results in Table 5.4 compare the treatment performance in the p-VF-R constructed wetland at Virloup with and without recirculation. The disadvantage of this system is its electricity consumption which downgrades it in the Life Cycle Assessment, as will be discussed later.

5.3.1.5 Specific Area

The specific area, i.e. the area per one population equivalent, is shown in Table 5.1. When comparing the systems fVF-VF-HF and f-VF-f-VF, the former generally performs better with just half the available filtration surface area (Figs. 5.1 and 5.2).

5.3.2 Evolution of Constructed Wetlands over Time

5.3.2.1 Ageing of the gVF-VF-HF System with Continuous Feeding

A horizontal flow stage was added to the Saint Matre in 2000 to improve the plant's performance. This stage was bypassed in 2002 because it did not always work properly. Today, the results are still not satisfactory, which is why a distribution manhole is going to be installed at the outlet of the second VF stage to alternate the feeding of the horizontal flow bed and a vegetated ditch. The existing vegetated ditch, which needs to be brought back into operation, will complete the treatment without having to make any excessive modifications to the site.

At Pannessière work was carried out on the first two stages of the plant in 2006 after 15 years of operation and it was partially replanted. The development of Common reed (*Phragmites australis*) in Stage 1 is relatively slow because of the gravitational feeding. Additional plantings were done during 2009.

5.3.2.2 Ageing of f-VF-VF-HF System with Intermittent Feeding

The performance of these systems put in operation in 1996, 1998 and 2000 indicates an increase in treatment efficiency TSS (94, 96.5 and 98.9%), COD (89, 91.6 and

96.2%) and BOD_5 (94.3, 98.5 and 99.1%) while the total Kjeldahl nitrogen does not follow this tendency (94.8, 66.5and 84.3%). Thus, it is difficult to come to a single conclusion about the effects of ageing on this system.

At Orbey, the winter conditions in this region are severe (1000 m altitude, north-facing, with around 4 weeks of –20°C frosts in January/February). The components of the flushing system have to be changed regularly and a waterproof lining had to be added because the concrete walls had cracked as a result of the weather conditions. The sampling over 24-h periods, completed successively in 1997, 1999 and 2000, show that the treatment performance has improved in terms of COD. It can be seen (Table 5.3) that the plant operates at an acceptable level even in winter. The vegetation including Common reed starts to sprout in April/May at this altitude. This is later than in lowland where Common reed starts to sprout generally in March. Plants are harvested at the end of winter in order to insulate the wetland surface and protect the bacterial activity in the filtration media.

In the constructed wetland at Cazillac no major reconstruction took place with the exception of replacement of some parts of the flushing device in 2009. The hybrid constructed wetland at Caurel has been operating at half load since the start, and therefore only half of the beds are fed both in Stages 1 and 2.

5.3.2.3 Ageing of the f-VF-f-VF System

In the system at Viviers/Artaut, clogging occurred at Stage 2 some time after the start of operation but since then, everything has been working properly. The wooded buffer zone is growing and ensures the expected summer infiltration for which it has been designed.

5.3.2.4 Ageing of the p-VF-R Systems

In the system at Virloup a recirculation device was installed in 2007; it can be seen (Table 5.4) that the performance improved very rapidly because the two 24-h sampling periods were 4 months apart. The differences are very clear: the removal of BOD_5 rose from 91.2 to 97.9% and TKN from 66.5 to 84.3% which is acceptable under the required discharge limits.

At Plufur, the constructed wetland has continually met the discharge requirements since it was commissioned in 2003. The willows in the wooded buffer zone are currently reaching a height of more than 3 m. The water seeps into the bottom of the ditch well and the results of a recent sampling campaign carried out on 5 March 2009 were as follows: at the inlet to the wooded buffer zone, concentrations of NO_3- and $PO_4{}^{3-}$ were 4.5 mg L^{-1} and 6.3 mg L^{-1}, respectively while at the outlet the respective concentrations were 1.5 mg L^{-1} and 3 mg L^{-1}.

5.3.2.5 Deposits at Stage 1

The four types of hybrid constructed wetlands presented in this paper treat raw wastewater. This means that on the surface of the beds at Stage 1 the largest particles settle out and form a layer of deposits which are now recognised as being one of the best means of fine filtration and primary aerobic treatment of effluent.

Chazarenc & Merlin (2005) regard this layer as being an element of the filtration matrix as a whole and rightly draw a parallel between the time required for establishing a constructed wetland and a layer of deposits on the surface of the beds of Stage 1. On the one hand, they consider that the deposits are the cause of increased hydraulic efficiency and better distribution of effluent. In fact, without deposits, the effluent only circulates around the feeding points. The growth of the deposit layers therefore allows better flooding of the filter surfaces. Moreover, the deposits support the growth of microorganisms in aerobic conditions. This layer thus improves the treatment performance, particularly in terms of nitrogen.

The fact that deposits can offer better support for the growth of microorganisms is demonstrated by the rates of respiration within the deposits. In addition, numerous studies of biotopes confirm this hypothesis by showing more diversity in the community and the species in the older layers of the well-established CWs (Chazarenc & Merlin, 2005). Many cases presented in the literature show levels of respiration proportional to the total accumulated organic matter (e.g., Nguyen, 2001).

The information about the surface deposits comes from f-VF-VF-HF systems at Orbey and Caurel. The surface deposits in Stage 1 were collected during the summers of 2008 and 2009, after a rest period of 7 and 9 weeks, respectively. In both systems, it was found that the beds were densely colonised by vegetation. The reeds reached the height of about 2.5 m. There were a large number of insects (beetles, diptera) and epigeic earthworms such as *Lombricus enchytreides, Eisenia Andrei* and *Dendrobaena octaedra.*

The thickness of the deposit layer varied between 25 cm close to the wastewater inlet and 13 cm at the point furthest away, with an average of 20 cm across the whole surface. The deposits were composed of approximately 50% organic matter, 20% plant debris and 30% gravel. In fact, the action of the earthworms and reed rhizomes means that the layer is being constantly turned over and the gravel becomes mixed in with the organic matter.

The deposits are dark brown, both smelling and look like fresh, aerated compost with little clumps that crumble easily. Having completed their task during the pre-dredging draining phase, there are relatively few worms left, and only adults. No eggs were found.

5.3.3 Eco Balance

The initial Life Cycle Assessment (comparison of the environmental impacts) of these four systems has been evaluated by Comby et al. (2008). The study demonstrates many factors that can influence the environmental impact of the system in operation:

- soil quality at the treatment plant (underlying parent rock, water table, etc.),
- distance of the quarries for extracting the filter aggregates,

– use of energy-intensive electro-mechanical materials.
– weather conditions (risk of frost etc.),
– quality of the receiving environment (water table, surface watercourse, etc.).

Maintenance, which can be done by staff skilled to a greater or lesser degree depending on the system in place, is only influenced by the area being looked after. When complicated equipment (pumps, automatic elements, etc.) which require special maintenance staff are installed, the regular maintenance operators may lose interest in their jobs. This may lead to malfunction of the system (Comby et al., 2008). When energy consumption, and electricity in particular, is concerned, it is necessary to evaluate the long term impact on the environment, because energy is produced by various means and the actual cost varies according to the region and the source of production

Also, the question which needs to be seriously considered is under which circumstances it is necessary to use a complex treatment technology and when it is possible to use more natural systems while still guaranteeing an acceptable level of discharged pollutants. To limit the overall environmental impact in the long run must be a primary objective.

Of course, to make necessary decisions, eco balance is only one of the criteria to be taken into account – the others being, in particular, the characteristics of the site (available surface area, incline, etc.), the cost of investment and operation, as well as the performance to be achieved.

5.4 Conclusions

Overall, the fVF-VF-HF system provides good results – in the long term and with acceptable specific area as compared to other set-ups. Three intermittently fed systems give good results with regard to the standards applied in France. The most natural system, with gravitational feeding (gVF-VF-HF), needs an additional treatment stage in order to meet the required discharge limits.

Observations on four types of constructed wetlands presented here indicate that their performance may not always be consistent, and supplementary treatment in the form of vegetated buffer zone as a final polishing step is necessary. All the initial data on their contribution to constructed wetlands effluent quality improvement is very encouraging. However, the performance of wooded buffer zones needs to be carefully monitored in the long run. In general, regardless of the wastewater quality and the filtering system employed, it is obvious that the pollution of our environment is still too high. As a result the use of vegetated buffer zones will become increasingly necessary in order to achieve better protection of the environment.

This can only be successfully applied with support of vegetation and associated biotopes, thus constructed wetlands need to be designed in an increasingly "musical" fashion. We believe that vegetated systems must be considered, both by the appropriate authorities and by the whole population, more and more as a means of developing the potential benefits existing in the wastewater.

References

Boutin, C. (1987). Domestic wastewater treatment in tanks planted with rooted macrophytes: case study; description of the system; design criteria; and efficiency. *Water Science and Technology, 19*(10), 29–40.

Carbiener, R. (1970). Un exemple de type forestier exceptionnel pour l'Europe occidentale : la forêt du lit majeur du Rhin au niveau du fossé rhénan (Fraxino-Ulmetum Oberd. 53) intérêt écologique et biogéographique. Comparaison à d'autres forêts thermophiles. *Vegetatio, Acta geobotanica 20*(1–4), 97–148.

Chazarenc, F., & Merlin, G. (2005). Influence of surface layer on hydrology and biology of gravel bed vertical flow constructed wetlands. *Water Science and Technology, 51*(9), 91–97.

Comby, A., Reeb, G., Werckmann, M., & Quaranta, G. (2008). Life Cycle assessment: a method to compare different options for constructed wetlands. In S. Billore, P. Dass, & J. Vymazal (Eds.), *Proceedings of the 11th International Conference Wetland Systems Technology in Water Pollution Control* (pp. 185–191). Ujjain, India: Vikram University.

Implementation (1997). Implementation circular no. 97–31 of 17 February 1997 on municipal sewage treatment – plants with a capacity of less than 120 kg BOD/day (2000 PE). Published in the official gazette of the French Ministry of Infrastructure, Building, Tourism and Transport, (in French).

Lienard, A., Boutin, C., & Esser, D. (1990). Domestic wastewater treatment with emergent helophyte beds in France. In P. F. Cooper & B. C. Findlater (Eds.), *Constructed wetlands in water pollution control* (pp. 183–192). Oxford,: Pergamon Press.

Mettetal, J. P., Rouault, J. Y., & Valero, L. (1990). *Monitoring and evaluation of treatment systems using beds of macrophytes in Panessières (Jura).* Rhone-Mediterranean and Corsica Water Agency, Franche-Comté regional department of research and economic affaire.

Molle, P., Lienard, A., Boutin, C., Merlin, G., & Iwena, A. (2005). How to treat raw sewage with constructed wetlands: an overview of French systems. *Water Science and Technology 51*(9), 11–21.

Nguyen, L. (2001). Accumulation of organic matter fractions in a gravel-bed constructed wetland. *Water Science and Technology 44*(11–12), 281–287.

Regulation (2007). Regulation of 22 June 2007 on the collection, transport and treatment of wastewater from treatment systems as well as the monitoring of their operation and efficiency and on non-municipal treatment systems receiving a gross load of organic pollution in excess of 1.2 kg BOD_5/day (in French).

Chapter 6
Duckweed and Algae Ponds as a Post-Treatment for Metal Removal from Textile Wastewater

Diederik P.L. Rousseau, Christian B. Sekomo, Saleh A.A.E. Saleh, and Piet N.L. Lens

Abstract Untreated textile wastewater is a typical source of heavy metal pollution in aquatic ecosystems. In this study, the use of algae and duckweed ponds (AP and DP) as post-treatment for textile wastewater has been evaluated under the hypothesis that differing conditions such as pH, redox potential and dissolved oxygen would lead to different heavy metal removal efficiencies. Two lab-scale systems, each consisting of three ponds in series and seeded with algae and duckweed respectively, have been operated at a total hydraulic retention time of 7 days and under two different metal loading rates. Cr was removed at 94–98%, indifferent to the loading rate but slightly better in AP compared to DP. Zn removal proceeded well (~70%) under low loading rates, but dropped to below 40% at the higher loading rate. No effect of pond type could be demonstrated. Pb, Cd and Cu all show relatively similar patterns with removal efficiencies varying between 17 and 36%, which indicates that neither system is very suitable, or under-designed as a polishing step for removing these heavy metals. AP seems to be better suited for removing Cr whereas the DP seem to be better in removing Cd and Pb. In absolute terms however, differences between AP and DP were fairly small. A further analysis is needed to unravel whether or not the differences between both pond types are due to differences in pH, dissolved oxygen and redox potential or rather due to sorption mechanisms.

Keywords *Lemna* · Aquatic treatment systems · Textile wastewater · Heavy metals · Tracer test

6.1 Introduction

Many industrial activities discharge metal containing wastewaters, including the manufacture of textile and paper, mining and mineral processing and petrochemical industries. According to Johnson and Hallberg (2005) for example, it was estimated

D.P.L. Rousseau (✉)
Department of Environmental Resources, UNESCO-IHE Institute for Water Education, P. O. Box 3015, 2601 DA Delft, The Netherlands
e-mail: d.rousseau@unesco-ihe.org

J. Vymazal (ed.), *Water and Nutrient Management in Natural and Constructed Wetlands*, DOI 10.1007/978-90-481-9585-5_6,

that in 1989 ca. 19,300 km of streams and rivers, and ca. 72,000 ha of lakes and reservoirs worldwide had been seriously damaged by mine effluent, although the true scales of the environmental pollution caused by mine water discharge is difficult to assess accurately.

Another pollutant source for metals is textile wastewater, which is a mixture of colorants (dyes and pigments) and various organic compounds used as cleaning solvents; it also contains high concentration of heavy metals, total dissolved solids, and has a high chemical as well as biological oxygen demand. For example, its discharge has led to complete disappearance of submerged and free floating hydrophytes, and also selected marshy species in the pools and a wastewater drain in the industrial area of Sanganer town, Jaipur (Sharma et al., 2007).

Several technologies are available to remove heavy metals (HM) from wastewater such as chemical precipitation, flotation, coagulation-flocculation, ion exchange and membrane filtration; all have their advantages and limitations in application (Kurniawan, Chan, Lo, & Babel, 2006). These methods require high capital investment and they generate a sludge disposal problem (Aziz, Adlan, & Ariffin, 2008; Cohen, 2006). Thus for most developing countries, alternatives are needed that are within the economical and technological capability of the respective nations.

Aquatic treatment systems have been used as a cost-effective alternate technology to conventional wastewater treatment methods. It has been shown that they can efficiently remove HM from both domestic and industrial wastewater (Qian, Zayed, Zhu, Yu, & Terry, 1999; Rodgers & Dunn, 1992; Lakatos, Kiss, Kiss, & Juhasz, 1997; Tang, 1993; LeDuc & Terry, 2005). Plants play important roles in these systems for the removal of pollutants (Brix, 1994). They not only take up nutrients, but are also able to adsorb and accumulate metals. Water hyacinth (*Eichhornia crassipes*) and Duckweed (*Lemna sp.*) are commonly used in aquatic treatment systems. It is also well-known that these plants and also algae influence the redox and pH conditions, and therefore it may be expected that they also influence metal removal processes, as described in the paragraphs below.

6.1.1 Influence of pH on Metal Removal

Hydrogen ion activity (pH) is probably the most important factor governing metal speciation, solubility from mineral surfaces, transport, and eventual bioavailability of metals in aqueous solutions. pH affects both solubility of metal hydroxide minerals and adsorption-desorption processes. Most metal hydroxide minerals have a very low solubility under pH conditions in natural water. Because hydroxide ion activity is directly related to pH, the solubility of metal hydroxide minerals increases with decreasing pH, and more dissolved metals become potentially available for incorporation in biological processes as pH decreases. Ionic metal species also are commonly the most toxic form to aquatic organisms (Salomons, 1995).

6.1.2 Influence of Redox Potential on Metal Removal

Redox conditions affect the mobility of metals in two different ways. Firstly, there are direct changes in the oxidation states of certain metals (e.g. iron and manganese). Secondly, redox conditions may indirectly influence the mobility of metals that occur in only one oxidation state (e.g. cadmium and zinc). Such is the case when these metals are bonded with elements (sulphides, organic matter and some chelating agents) that are subject to redox changes (Stigliani, 1992). For HM, the influence of redox condition belongs to the second case, and generally, the effect mainly occurs to the HM bound to a reducible phase and an oxidisable phase.

Under anoxic conditions (i.e. low Eh) and neutral pH, sediments tend to serve as sinks for heavy metals, especially those in sulphidic and organic phases. Contrarily, Fe-Mn-oxides are unstable thermodynamically in anaerobic environments (Eqs. 6.1 and 6.2). Also, they can serve as a source of oxygen to microorganisms, whereby they are reduced and become water-soluble and lose their binding ability. This loss may result in the release of heavy metals (Stigliani, 1992).

$$2MnO_2 + CH_2O + 4H^+ \rightarrow 2Mn^{2+} + 3H_2O + CO_2, \qquad (6.1)$$

$$4Fe(OH)_3 + 3H_2O + CO_2 \rightarrow 4Fe^{2+} + 11H_2O + CO_2 \qquad (6.2)$$

If anoxic sediments are exposed to the air-atmosphere, an oxidation process will take place accompanied by an increase in Eh value. These new redox conditions will change the distribution and transformation of heavy metal binding forms in sediments. With this highest Eh, the freshly precipitated Fe-Mn-oxides act as a scavenger and cause immobilization of heavy metals. In contrast, a heavy metal-sulphide (MeS_2) is converted into sulphate (Eq. 6.3), and the initial bound heavy metals will be released (Calmano, Hong, & Förstner, 1993).

$$MeS_2 + 7/2O_2 + H_2O = Me^{2+} + SO_4^{2-} + 2H^+ \qquad (6.3)$$

Redox potential changes may occur under natural conditions as well as under artificial influence, for instance during periodical changes such as tidal currents in coastal and estuarine areas, seasonal flooding of lakes and rivers, and draining of wetlands (Calmano et al., 1993). Changes in redox potential also influence heavy metal binding forms. In general, a decrease in Eh value (anaerobic condition) would lead to dissolution of iron hydroxides and oxyhydrates, upon which the structure of the coatings will break down and the adsorbed species will be released. Compared to that, sulphate (if present) will be reduced to sulphides and the mobile heavy metals could precipitate out of solution as metal sulphides. In the opposite situation, when the Eh value increases rather than decreases, the above reaction would be reversed (Stigliani, 1992).

6.1.3 Objectives

A study was aimed at a sustainable low-cost HM treatment technology for textile wastewater based on a two-step process: (1) HM precipitation after sulphate reduction in anaerobic bioreactors and (2) a polishing step by means of algae and duckweed ponds. This chapter focuses on the second step with the objective to study the effect of HM pollutants' load on their removal efficiency and to find out how this is influenced by redox potential and other environmental conditions in duckweed (DP) and algae ponds (AP).

6.2 Materials and Methods

6.2.1 Laboratory Set-Up

The set-up (Fig. 6.1) comprised two treatment lines, one with AP and one with DP, each consisting of three glass aquaria in series (L×W×D: 50×30×30 cm; maximum water depth 23.5 cm). They were coded A1–A3 for the AP, and D1–D3 for the DP.

6.2.2 Wastewater Characteristics

Synthetic wastewater was pumped from a holding tank at equal rates to the APs and DPs. This wastewater was based on Hunter nutrient solution (Leman, 2000) and spiked with metals to a level equal to what can be expected of anaerobically pre-treated textile wastewater. The hydraulic retention time was set to 7 days; evapotranspiration losses were compensated daily by adding demineralised water.

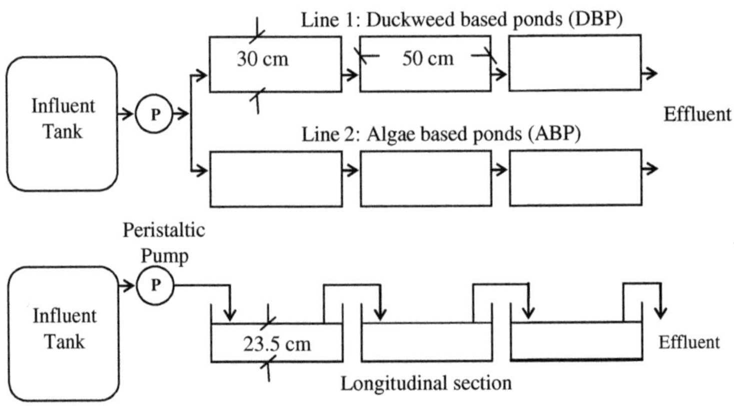

Fig. 6.1 Schematic representation of the set-up

6.2.3 Operating Conditions

The system was initially operated under 16/8 h light/dark regime with a light intensity of 125 $\mu E\ m^{-2}\ s^{-1}$ on the plant surface. DPs were started with *Lemna minor* species at a density of 600 g fresh weight m^{-2}; APs were inoculated with Delft canal water at start-up for algae development. After an initial start-up period to allow plant and algae growth, two 3-week experiments were done: (1) Zn, Cu, Pb, Cr and Cd at influent concentrations of 1.25, 0.1, 0.25, 1.5 and 0.05 mg L^{-1} respectively; and (2) double concentrations of the first run. Duckweed biomass was harvested every 5 days to restore the duckweed density to 600 g fresh weight/m^2. This density was selected to prevent overcrowding and to maintain sufficient cover to minimize the development of algae in duckweed ponds (Zimmo, 2003). Floating algae formed over the water surface were collected regularly to improve oxygen levels in the system, since the thick layer was obstructing light penetration.

6.2.4 Sampling, Preservation and Analytical Procedures

Dissolved oxygen (DO), pH, temperature and redox potential (Eh) were measured using DO-meter, pH-meter and redox-meter models HACH-LDO-HQ_{10}, HACH-LpH-HQ_{10}, and HACH-LRP-HQ_{10} respectively at 10 cm below the water surface every one hour (one pond/24 h) during the experiment's periods to produce physicochemical profiles for each pond under the different conditions of each run.

Sampling, preservation and analytical procedures further followed standard methods for water and wastewater examination (APHA, 1995). Grab samples were collected from the influent and the effluents of each pond at four instances during the last 2 weeks of each run. During the first 10 days of each run no samples were taken as it was assumed that the system was adapting to the newly applied influent conditions.

After appropriate pre-treatment of all water samples according to Standard Methods, HM analysis for Pb, Cd, Cr, Zn, and Cu was done for all water samples by using the flame and flameless atomic absorption spectrometer models Perkin-Elmer 3110 and Thermo elemental-SOLAR95/Furnace respectively according to the detection limit of each equipment. Each time, special reference standards were included for quality control purposes.

6.2.5 Tracer Test

A pond's hydraulic behaviour was studied according to a continuous tracer loading technique using Lithium chloride (LiCl). For this, the influent was spiked to a level of 4.3 mg/L lithium. Flow rate was kept at fixed value in both AP and DP. Grab samples were collected in clear plastic containers for 21 consecutive days (three times the theoretical retention time) from the effluent of each pond. Higher sampling frequency (five times per day) was used during the period when rapid changes in concentration were detected and reduced to twice per day afterwards. Lithium was analyzed on an Atomic Absorption Spectrometer model Perkin-Elmer 3110 (APHA, 1995). The initial background of tracer concentration in the pond water was measured in all sampling points immediately before the addition of the tracer (Zimmo, 2003).

Two models described by Levenspiel (1972) indicate two methods for characterization of the flow pattern. The first one is the tanks-in-series model which indicates the number of completely stirred tank reactors (CSTR) in series that would result in the same retention time distribution. The second one is the dispersion model that considers the flow to be a plug flow in which deviation of flow from plug flow to some extent is due to back mixing or intermixing. The parameters of both models are calculated from the measured mean retention time t_m and the variance, σ^2 of the retention time distribution. Variance of the time-concentration curve describes the distribution of effluent leaving the pond reactor. These parameters are calculated as follows:

$$t_m = \int t.C.\mathrm{dt}/C.\mathrm{dt}, \tag{6.4}$$

$$\sigma^2 = \int (t - t_m)^2.C.\mathrm{dt}/C.\mathrm{dt}, \tag{6.5}$$

where C = Tracer concentration (mg L^{-1}) and t = Time (hours).

For the tanks-in-series model,

$$R_n = t_m^2/\sigma^2 \tag{6.6}$$

If $R_n = 1$, the flow pattern is complete mixing and if R_n = infinity, it indicates a plug flow.

The percentage of dead volume is usually calculated as

$$D_{Vol} = 1 - (t_m/\mathrm{HRT}_{theo})\ 100, \tag{6.7}$$

where D_{Vol} = Dead volume (%).

For the dispersion model, the dispersion number which is a measure of the amount of mixing is given by

$$d = 0.5 \times (\sigma^2/t_m^2). \tag{6.8}$$

When d = zero, the flow pattern is a plug flow, and if $d > 0.01$, then the flow pattern of systems have a significant degree of deviation from the PF model (García et al., 2004).

6.2.6 Statistical Analysis

For statistical analysis of the experimental data, Microsoft Excel^{TM} was used (*t* test for the physico-chemical data and ANOVA: Two-Factor With Replication for the metal data).

6.3 Results and Discussion

6.3.1 Tracer Test

Tracer recovery was consistently 97% or higher. Results show a good agreement between theoretical and actual hydraulic retention times, indicating the absence of short-circuiting and/or dead zones (Table 6.1). Dispersion numbers and N (number of tanks according to the tanks-in-series model) indicate well-mixed conditions.

6.3.2 Physico-Chemical Parameters

Water temperature was in the range of 19–22°C, with only small daily variations. Redox potentials (ORP), O_2 concentrations and pH levels are summarized in Table 6.2. As can be expected, algae have a big influence on oxygen and pH levels through photosynthetic activity and show high day-night fluctuations. In contrast the complete cover of duckweed plants provides stable conditions below the surface.

6.3.3 Temperature

Different temperature profiles were recorded when comparing DPs to APs. This could be explained by the fact that the duckweed ponds are totally covered by plants whereas the algae ponds are uncovered. In the algae ponds, maximum and minimum temperatures were recorded respectively at 1:30 AM and 9:30 PM. This is explained

Table 6.1 Hydraulic characteristics in duckweed and algal pond reactors

Parameters	Eff_D1	Eff_D2	Eff_D3	Eff_A1	Eff_A2	Eff_A3
HRT_{theo} (days)	2.31	4.62	6.93	2.31	4.62	6.93
HRT_{act} (days)	2.29	4.81	6.90	2.27	4.88	6.92
σ^2 (hr^2)	4006	17573	13112	4039	15624	13160
R_n	0.75	0.76	2.09	0.73	0.88	2.10
Dispersion number d	0.66	0.65	0.24	0.68	0.57	0.24
Dead zone (%)	0.87	−4.17	0.43	0.87	−5.52	0.43
Recovery (%)	100.00	100.00	100.00	100.00	97.60	97.60
Dead time (hours)	21.25	44.68	69.50	21.25	44.68	69.50

Table 6.2 Minimum and maximum readings of oxidation-reduction potential (ORP), dissolved oxygen (DO) and pH for all ponds during both runs

	ORP (mV)		DO (mg L^{-1})		pH	
	Run_1	Run_2	Run_1	Run_2	Run_1	Run_2
D1	230–295	296–369	0.1–1.2	1.2–1.9	7.0–7.2	6.7–6.8
D2	201–215	219–237	3.0–3.5	1.5–2.9	6.8–6.9	6.4–6.7
D3	276–294	325–345	3.2–4.5	1.9–3.4	6.9–7.0	6.8–6.8
A1	123–151	212–340	12.6–19.4	10.3–6.9	9.1–9.7	7.9–8.9
A2	122–184	241–301	14.1–19.4	8.0–15.2	9.2–9.8	8.5–9.2
A3	149–181	229–273	9.7–13.8	7.9–12.0	9.0–9.4	8.7–9.1

by a direct heating of the water occurring in the algae ponds, using lamps in the absence of a covering layer on the ponds' surface, so it becomes easy to gain and lose the heat leading to increase or decrease of the water temperature during the day/night regime. On the other hand, in the duckweed ponds, the temperature was more or less constant due to the effect of the covering of the upper layer of the ponds by that plant, providing a good insulation layer, therefore limiting the heat loss or gain.

6.3.4 Dissolved Oxygen

High oxygen content was recorded in the algae ponds when compared to duckweed ponds. The reason for high oxygen levels in algae ponds is due to algae photosynthetic activity within the water phase. In the duckweed system, oxygen production is at the top surface of the plants (which is in the air phase) and oxygen tends to be lost to the atmosphere. Only a small proportion is transported via the roots to the water phase (Bonomo, Pastorelli, & Zambon, 1997). The duckweed mat further prevents atmospheric re-aeration and light penetration, hence limiting algae growth (Caicedo, 2005). Also the biological degradation of organic compounds contributes to the reduction of oxygen content in that system (Fig. 6.2).

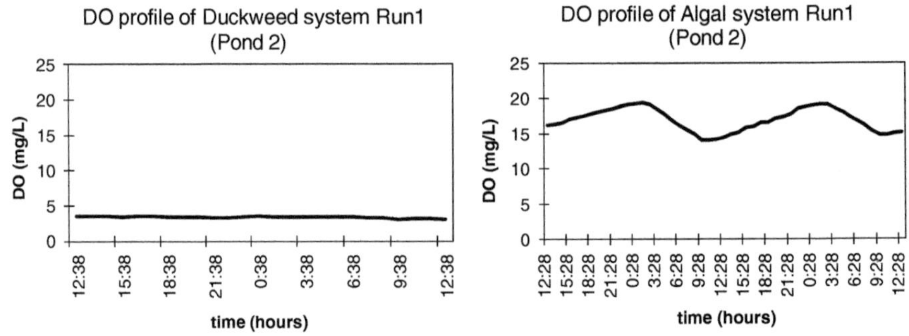

Fig. 6.2 Typical DO profiles for algae and duckweed ponds

Significant differences ($P < 0.05$) in DO values have been found for all duckweed ponds comparing to the algae ponds. Furthermore, decreases in DO values for both duckweed and algae in run 2 have been observed. This decrease in DO was most probably due to the decreasing plant activities affected by the increased pollutants load of HM.

6.3.5 pH

Lower pH was recorded in the duckweed ponds compared to the algae ponds. However, the pH values stayed almost constant in the duckweed ponds during the experimental period. For the algae ponds, daily maximum and minimum values were recorded, being caused by photosynthesis activity within the pond. High pH values were recorded in the algae ponds with the highest value of 9.8. Significant differences ($P < 0.05$) in pH values have been observed for all DP comparing to the AP. Different trends between duckweed and algal ponds have been recorded (Fig. 6.3). This variation of pH between duckweed and algal ponds is related to the decreasing plant activities affected by doubling the HM in run 2. During photosynthesis, carbon dioxide is taken from the water resulting in a pH increases.

6.3.6 Oxidation-Reduction Potential

Redox potential was positive in both systems, generally higher in duckweed ponds when compared to algal ponds. Values were almost constant for the duckweed ponds, whereas some variations were recorded in the algae ponds with daily maxima and minima (Fig. 6.4). The variation observed in the AP is due to the photosynthesis and respiration activity within the pond during the day and night regimes. Significant differences ($P < 0.05$) in ORP values were revealed for all duckweed ponds compared to the algae ponds. Moreover, an increasing trend was observed

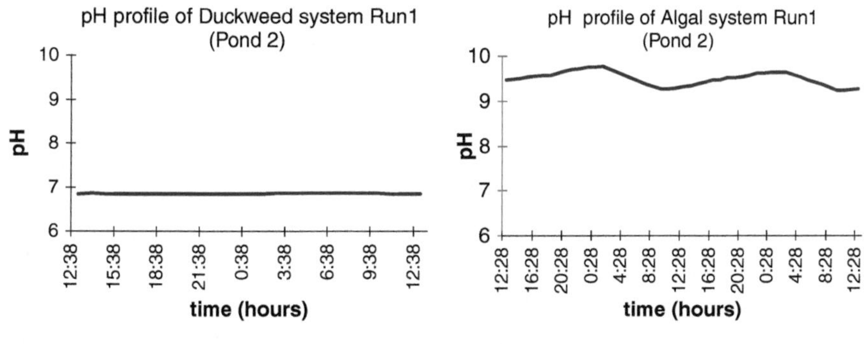

Fig. 6.3 Comparison between pH profiles for algae and duckweed ponds

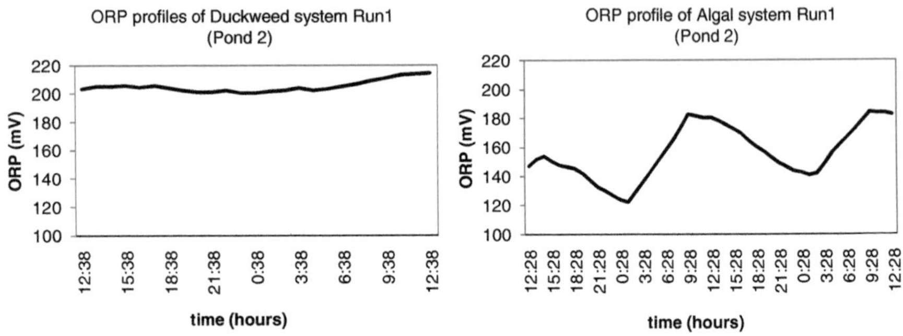

Fig. 6.4 ORP Profiles for algae and duckweed pond reactors

Table 6.3 Correlations between DO, pH and ORP for all ponds during different runs

		DO	ORP	pH
	Pearson Correlation	1	−723(**)	.972(**)
DO	Sig. (2-tailed)		0.000	0.000
	N	396	396	396
	Pearson Correlation	−.723(**)	1	−.666(**)
ORP	Sig. (2-tailed)	.000		.000
	N	396	396	396
	Pearson Correlation	0.972(**)	−.666(**)	1
pH	Sig. (2-tailed)	0.000	.000	
	N	396	396	396

** Correlation is significant at the 0.01 level (2-tailed).

both for algae and duckweed ponds in run 2. The increase in ORP is directly linked to the decreasing DO influenced by a higher metal load in the system.

A strongly significant positive correlation was observed between pH and DO, which is expected since the photosynthesis occurring in the algae pond produces more oxygen; while at the same time consuming CO_2 and thus increasing the pH (Table. 6.3). Moreover, a strongly significant negative correlation was recorded for ORP with both DO and pH as well.

6.3.7 Metal Removal

Table 6.4 gives an overview of influent and effluent concentrations based on four grab samples taken once the system had adapted to the applied influent conditions. Overall, the highest removal rates were achieved for Cr, both in the DPs and APs and both under low and high loading conditions, and they amount to 94–98%. No significant differences were found between loading rates (ANOVA, $P = 0.66$) but the type of organisms did have a highly significant effect on metal removal

Table 6.4 Influent and effluent concentrations (mg L^{-1}, average ± standard deviation; n = 4) of the various ponds

	influent		Run_1						Run_2					
	Run1	Run2	D1	D2	D3	A1	A2	A3	D1	D2	D3	A1	A2	A3
Pb	0.26 ± 0.01	0.61 ± 0.01	0.23 ± 0.01	0.19 ± 0.01	0.17 ± 0.01	0.24 ± 0.01	0.22 ± 0.01	0.21 ± 0.01	0.52 ± 0.03	0.48 ± 0.03	0.41 ± 0.01	0.54 ± 0.03	0.48 ± 0.02	0.44 ± 0.02
Cd	0.060 ± 0.010	0.130 ± 0.010	0.053 ± 0.005	0.048 ± 0.005	0.040 ± 0.000	0.055 ± 0.006	0.053 ± 0.005	0.048 ± 0.005	0.118 ± 0.005	0.108 ± 0.005	0.098 ± 0.005	0.118 ± 0.005	0.110 ± 0.000	0.108 ± 0.010
Cr	1.35 ± 0.08	2.68 ± 0.15	0.38 ± 0.04	0.18 ± 0.02	0.07 ± 0.01	0.19 ± 0.02	0.15 ± 0.01	0.05 ± 0.01	0.78 ± 0.15	0.25 ± 0.08	0.17 ± 0.02	0.53 ± 0.11	0.18 ± 0.10	0.06 ± 0.01
Zn	1.70 ± 0.05	3.14 ± 0.03	1.25 ± 0.04	0.87 ± 0.05	0.48 ± 0.05	1.29 ± 0.03	0.94 ± 0.04	0.54 ± 0.09	2.82 ± 0.09	2.48 ± 0.16	2.25 ± 0.11	2.53 ± 0.18	2.25 ± 0.20	1.95 ± 0.32
Cu	0.11 ± 0.01	0.24 ± 0.01	0.10 ± 0.00	0.10 ± 0.01	0.08 ± 0.01	0.1 ± 0.01	0.09 ± 0.01	0.08 ± 0.01	0.21 ± 0.03	0.20 ± 0.01	0.20 ± 0.01	0.21 ± 0.02	0.20 ± 0.02	0.19 ± 0.02

efficiency (ANOVA, $P < 0.001$) with algae ponds having a better performance. This effect is even more clearly visible when one compares the effluent concentrations of A1 and D1. In the subsequent ponds, the removal seems to approach a maximum level and therefore no important differences can be distinguished between the final effluents. A larger system, with longer retention, will therefore not result in a better Cr removal.

In Run 1, Zn exhibited the second best removal, reaching similar levels of around 70% in both APs and DPs (ANOVA, $P = 0.36$). Each of the three ponds in series seems to contribute in an almost equal way to the Zn removal, resulting in linearly increasing removal efficiency. Further improvement can thus be expected by enlarging the system; extrapolating the data shows that removal levels of 90% and more could be reached by just adding one more pond. For Run 2 the situation was quite different, with much lower overall removal efficiencies: 28% for DP and 38% for AP. Higher loading rates thus seem to have a negative effect on Zn removal (ANOVA, $P < 0.001$). The removal efficiency is still linearly increasing (about 10% per pond); extrapolating these data shows that one would have to triple the number of ponds to reach removal levels of 90% and more.

Pb, Cd and Cu all showed relatively similar patterns with removal efficiencies varying between min. 17% (Run 2: Cd in APs and Cu in DPs) and max. 36% (Run 1: Pb in DPs) which indicates that neither system is very suitable (or under-designed) as a polishing step for HM removal. ANOVA showed statistically significant effect of the loading rate on Cu removal ($P < 0.05$) and of pond type on Cd and Pb removal ($P < 0.01$ and $P < 0.001$ respectively).

In summary, the following orders were observed in metal removal under different conditions:

- Cr > Zn> Pb > Cd > Cu in run 1 for duckweed system,
- Cr > Zn> Cu > Cd > Pb in run 1 for algae system,
- Cr > Zn = Pb > Cd = Cu in run 2 for duckweed system,
- Cr > Zn> Pb > Cu > Cd in run 2 for algae system.

6.4 Conclusions

Overall removal efficiencies of Pb, Cd and Cu were quite low in both AP and DP and for both low and high loading rates. Cr on the contrary was removed very well under all experimental conditions. Zn removal reached intermediate levels of about 70% at low loading rates, but dropped drastically under the high loading conditions.

From the statistical results, AP seems to be better suited for removing Cr whereas the DP seem to be better in removing Cd and Pb. In absolute terms however, differences between AP and DP were fairly small. A further analysis is still needed to unravel whether or not the differences between both pond types are due to differences in pH, dissolved and redox or rather due to sorption mechanisms.

References

APHA (1995). *Standard methods for the examination of water and wastewater* (19th ed.). Washington, DC: American Public Health Association/American Water Works Assosication/Water Environment Federation.

Aziz, H. A., Adlan, M. N., & Ariffin, K. S. (2008). Heavy Metals (Cd, Pb, Zn, Ni, Cu and Cr(III)) removal from water in Malaysia: Post treatment by high quality limestone. *Bioresource Technology, 99*, 1578–1583.

Bonomo, L., Pastorelli, G., & Zambon, N. (1997). Advantages and limitations of duckweed-based wastewater treatment systems. *Water Science and Technology, 35*(5), 239–246.

Brix, H. (1994). Functions of macrophytes in constructed wetlands. *Water Science and Technology, 29*(4), 71–78.

Caicedo B. J. R. (2005). *Effect of operational variables on nitrogen transformations in duckweed stabilization ponds*. PhD thesis, IHE Delft, The Netherlands.

Calmano, W., Hong, J., & Förstner, U. (1993). Binding and mobilization of heavy metals in contaminated sediments affected by pH and redox potential. *Water Science and Technology, 28*(8–9), 223–235.

Cohen, R. R. H (2006). Use of microbes for cost reduction of metal removal from metals and mining industry waste streams, *Cleaner Production, 14*, 1146–1157.

Garcia J., Chiva J., Aguirre P., Alvarez E., Sierra J. P., & Mujeriego R. (2004). Hydraulic behaviour of horizontal subsurface flow constructed wetlands with different aspect ratio and granular medium size. *Ecological Engineering, 23*, 177–187.

Johnson D. B., & Hallberg, K. B. (2005). Acid mine drainage remediation options: a review. *Science of the Total Environment, 338*, 3–14.

Kurniawan, T. A., Chan, G. Y. S., Lo, W.-H., & Babel, S. (2006). Physico-chemical treatment techniques for wastewater laden with heavy metals. *Chemical Engineering Journal, 118*, 83–98.

Lakatos, G., Kiss, M. K., Kiss, M., & Juhasz, P. (1997). Application of constructed wetlands for wastewater treatment in Hungary. *Water Science and Technology, 35*(5), 331–336.

LeDuc, D. L., & Terry, N. (2005). Phytoremediation of toxic trace elements in soil and water. *Journal of Industrial Microbiology & Biotechnology, 32*, 514–520.

Leman M. (2000). *Zinc removal by aquatic plants: Pistia stratiotes*. MSc. Thesis, IHE Delft, The Netherlands.

Levenspiel, O. (1972). *Chemical Reaction Engineering* (2nd ed.). Wiley, New York.

Qian, J. H., Zayed, A., Zhu, Y. L., Yu, M., & Terry, N. (1999). Phytoaccumulation of trace elements by wetland plants. Part III. Uptake and accumulation of ten trace elements by twelve plant species. *Journal of Environmental Quality, 28*, 1448–1455.

Rodgers, J. H. Jr., & Dunn, A. (1992). Developing design guidelines for constructed wetlands to remove pesticides from agricultural runoff. *Ecological Engineering, 1*, 83–95.

Salomons, W. (1995). Environmental impact of metals derived from mining activities: Processes, predictions, prevention. *Journal of Geochemical Exploration, 52*, 5–23.

Sharma K. P., Sharma, S., Sharma, S., Singh, P. K., Kumar, S., Grover R., et al. (2007). A comparative study on characterization of textile wastewaters (untreated and treated) toxicity by chemical and biological tests. *Chemosphere, 69*, 48–54.

Stigliani, W. M. (1992). Overview of the chemical time bomb problem in Europe. In G. R. B. ter Meulen, W. M. Stigliani, W. Salomons, E. M. Brigdges & A. C. Imeson, (Eds.), *Proceedings of the European State-of-the-Art Conference on Delayed Effects of Chemicals in Soils and Sediments* (pp. 23–56). Hoofddorp: The Foundation for Ecodevelopment "Stichting Mondiaal Alternatief ".

Tang, S. (1993). Experimental study of a constructed wetland for treatment of acidic wastewater from an iron mine in China. *Ecological Engineering, 2*, 253–259.

Zimmo O. R. (2003). *Nitrogen transformations and removal mechanisms in Algal and Duckweed Wastewater Stabililization Ponds*. Ph.D Dissertation, and UNESCO – IHE institute for water education Wageningen University, The Netherlands.

Chapter 7
Diel Fluctuations of Redox Potential in a Horizontal Subsurface Flow Constructed Wetland for Wastewater Treatment

Jiří Dušek and Tomáš Picek

Abstract Redox potential (Eh) was measured in a treatment bed of the constructed wetland (CW) with subsurface horizontal flow for municipal wastewater treatment at Slavošovice, Czech Republic. The system for 150 person equivalents was planted with *Phragmites australis* and put into operation in 2001. The study is focused on diel fluctuation of Eh measured in situ in a reed bed of a constructed wetland. Redox potential fluctuated from –400 to +800 mV and significant changes of Eh were recorded in very short periods. Two interesting trends have been found in the diurnal course of Eh. The first trend is characterised by decreasing Eh during the light period (warm period in 0.2 m depth). The second trend is characterised by increasing Eh during the light period and is closely correlated with solar global radiation. The amplitudes of daily fluctuations of averaged Eh for warm and cold periods, respectively, were not too wide (maximally 60 mV). Some trends of Eh fluctuations were found in the studied system during the day, however, daily trends of Eh fluctuations studied in situ conditions in horizontal subsurface flow constructed wetland for wastewater treatment, cannot be easily explained by one factor.

Keywords Constructed wetlands · *Phragmites australis* · Platinum electrode · Redox potential · Redox processes · Wastewater treatment

7.1 Introduction

Wetlands are known as very dynamic and diverse ecosystems in which nutrients cycling is very intense (Mitsch & Gosselink, 1986). This important feature is often used in constructed wetlands for a treatment of different types of wastewaters such

J. Dušek (✉)
Institute of Systems Biology and Ecology, v.v.i., Academy of Sciences of the Czech Republic, Na Sádkách 7, České Budějovice, Czech Republic
e-mail: jiri.dusek@usbe.cas.cz

J. Vymazal (ed.), *Water and Nutrient Management in Natural and Constructed Wetlands*, DOI 10.1007/978-90-481-9585-5_7,

as municipal, agricultural and industrial ones (Hammer, 1989). Main processes responsible for the transformation of organic pollution and nutrients proceed in hydric soils or substrates of wetlands (Richardson & Vepraskas, 2000). In flooded soils anaerobic conditions prevail and the aerobic conditions are limited to surroundings of roots and nodes of rhizomes where radial losses of oxygen occur (Armstrong, 1979; Armstrong & Armstrong, 1991). As aerobic and anaerobic conditions may fluctuate intensely, redox conditions inside hydric soils are difficult to simply monitor. Redox conditions within wetland soils cannot be determined directly because they result from the oxidation-reduction status of many redox couples. Most of organic matter and nutrients transformations run through microbial oxidation-reduction processes. When oxygen is missing in a system, alternative acceptors must be used. Prevailing redox couples in a system can be determined by redox potential (Eh) measurement (Rowell, 1981). Eh can be therefore used for characterising current redox conditions in soil. Eh measured in a reed bed of a constructed wetland for wastewater treatment with horizontal subsurface flow may fluctuate widely from –400 to +800 mV (Dušek, Picek, & Čížková, 2008). Fluctuations of Eh were recorded in various time scales (months to days). Different factors are responsible for Eh fluctuations in various time scales, however short-time (diurnal) Eh fluctuations were attributed mainly to the plant effect (root exudation, transport of oxygen) (Wießner et al., 2005). Our chapter is focused on diel fluctuation of Eh measured in situ in a reed bed of a constructed wetland.

7.2 Methods

7.2.1 Site Description

Redox potential was measured in a treatment bed of the constructed wetland (CW) with subsurface horizontal flow treating wastewater in the village of Slavošovice, Czech Republic (Dušek et al., 2008). CW is located at an altitude of 480 m above sea level. The annual average air temperature is 7.9°C and the average annual precipitation is 634 mm. CW started operation in August 2001 and it is used for treatment of wastewater from a combined sewerage (municipal sewage and storm water runoff) from the village. The CW is designed for 150 person equivalents (PE), and at present, about 100 PE are connected to the CW. Detailed design parameters of the system are summarized in Table 7.1. After pre-treatment, the water flows through a splitter chamber, which divides the water between two parallel reed beds. The reed beds are sealed with a clay layer which occurred naturally at the site. The beds are planted with Common reed (*Phragmites australis*) in a gravel substrate. The wastewater is distributed to the treatment beds through perforated inlet and outlet pipes placed horizontally in the inflow and outflow zones at a depth of 0.5 m. The average wastewater inflow rate was 0.12±0.10 L s^{-1} (maximum of 1.0 L s^{-1} during extremely strong precipitation). Hydraulic retention time varied from 8 to 16 days (Holcová, 2007).

Table 7.1 Design parameters of the constructed wetland with subsurface horizontal flow in Slavošovice

Pretreatment units	Vegetated bed
Storm overflow	Number of beds: 2
Screens	Bed length: 17 m
Imhoff tank	Bed width: 22 m
	Slope of bed: 1%
	Substrate: gravel (size 1.0–2 cm)
	Area of each bed: 374 m^2
	Total area of beds: 748 m^2
	Number of person equivalents: 150
	Area per 1 person equivalent: 5 m^2
Water level in the beds are kept at 5 cm under the gravel surface	Beds are planted by *Phragmites australis* (Cav.) Trin. ex Steudel

7.2.2 Redox Potential Measurements

Redox potential (Eh) was measured using platinum electrodes. The electrodes were inserted into perforated plastic tubes (100 mm in diameter) and installed vertically in two depths (0.2 and 0.5 m) of the treatment bed along a transect laid from inflow to outflow. The transect was situated in the middle of the treatment bed and Eh was measured at the distances of 1 and 13 m from the inflow to the treatment bed. Ten platinum electrodes (made by EDT, Turnov, Czech Republic) were individually connected to the high terminals and the reference electrode (Ag/AgCl) was connected to the low terminals of the data-logger channels (M4216, Fiedler, Electronics for Ecology, Czech Republic). Redox potential was measured continuously from June 2002 to September 2004 every 15 min. MySQL 5.0 (MySQL AB, Sweden) database was used for data storage and was managed by the Internet browser using phpMyAdmin 3.1.1. Eh values were corrected relative to the normal H electrode by adding +200 mV (Bochove, Beauchemin, & Thériault, 2002). The Eh values were not corrected to pH 7.00, because "mixed redox potential" (more redox couples in the pore water) was measured and the pH of pore water was very close to 7.00 (pH 6.87±0.25) during the whole studied period. The electrode platinum tips were brushed in order to remove any surface coatings and cleaned in an HCl–HNO_3 solution. Poisoning of the Pt tips reported by Austin and Huddleston (1999) was not found. The platinum electrode functioning was controlled in the standard ORP solution ($K_3[Fe(CN_6)]$ +$K_4[Fe(CN)_6]$ in phosphate buffer). The inflow and outflow flows were measured with an ultrasonic probe US1000 (Fiedler, Electronics for Ecology, Czech Republic). Temperature was measured using Pt100 sensors. The measuring period (June 2002 to November 2004) was divided into the cold and warm period of particular years. The cold period is defined as the time between November to April and the warm period is defined as the time from May to October.

7.3 Results

Redox potential was continuously measured in the inflow and outflow zones of the reed bed. The inflow zone is characterized by eutrophic conditions with a relatively high concentration of total organic carbon (COD $= 70.7\ \mathrm{mg\,L^{-1}}$ in average) and nutrients ($N_{tot} = 23.1\ \mathrm{mg\,L^{-1}}$, $P_{tot} = 4.9\ \mathrm{mg\,L^{-1}}$), while the outflow zone is characterized by much lower concentrations of organics and nutrients (COD $= 41.9\ \mathrm{mg\,L^{-1}}$; $N_{tot} = 10.3\ \mathrm{mg\,L^{-1}}$; $P_{tot} = 1.7\ \mathrm{mg\,L^{-1}}$, on average). For detailed description of Eh fluctuations 1-h averages in the inflow and outflow zones were used. The Eh values were calculated only for selected days (with no precipitation) separately for warm (443 days) and cold (359 days) periods. The days with any rain were excluded as precipitation is one of the main factors significantly affecting Eh. The effect of precipitation on Eh is (a) direct through rainwater falling on the reed bed oxygenating water column, and (b) indirect by increasing flow rate of incoming wastewater from sewage system. During rainy periods the Eh values fluctuated widely from negative to positive values (data not shown).

Diurnal courses of Eh are different during the warm and cold periods and also different for the two depths. During the warm period, Eh decreased in 0.2 m depth in the inflow and outflow of the reed bed during the light period of the day and the difference between maximal and minimal Eh in these cases were about ±40 mV. However, Eh increased during the light period of the day from +50 mV to +100 mV in the 0.5 m depth in the outflow zone during the warm period. In sunny days of the warm period, the diel Eh fluctuations were closely correlated with the course of global radiation ($r = 0.77$; $p < 0.001$) and Eh was affected most probably by oxygen transfer to the reed bed by plants.

7.3.1 *Eh Fluctuations in the Inflow Zone*

Redox potential showed a very consistent diel fluctuation (Fig. 7.1) as measured in the 0.2 m and 0.5 m depth of the reed bed in the inflow zone. The fluctuations were more pronounced in the depth of 0.2 m than in the 0.5 m depth. The Eh fluctuations were also different during the warm period as compared with the cold period of the year (Fig. 7.1). During the warm period Eh increased in the early morning (at 2:00–4:00 local time) in 0.2 m depth. After 8:00 Eh decreased to minimal values which were measured in the late afternoon (at 16:00–20:00) and then slowly increased again till midnight. The fast decrease of Eh during the day (from 8:00 till 16:00) corresponded with increasing solar radiation (Fig. 7.1). The flow rate of wastewater started to increase from the early morning (3:00) with local maxima at 7:00 and at 19:00. The relation of Eh to the wastewater flow rate was weaker in comparison with its relation to solar radiation. During the cold period the amplitude of Eh fluctuations was small and the Eh decreased earlier than during the warm period. Eh was more negative during the cold period. Fluctuations of Eh in the 0.5 m depth differed between the warm and the cold periods. The Eh was significantly lower during the warm period than in the cold period (Fig. 7.1). During the warm period a slight increase in Eh during the day was recognized, when the values of Eh were

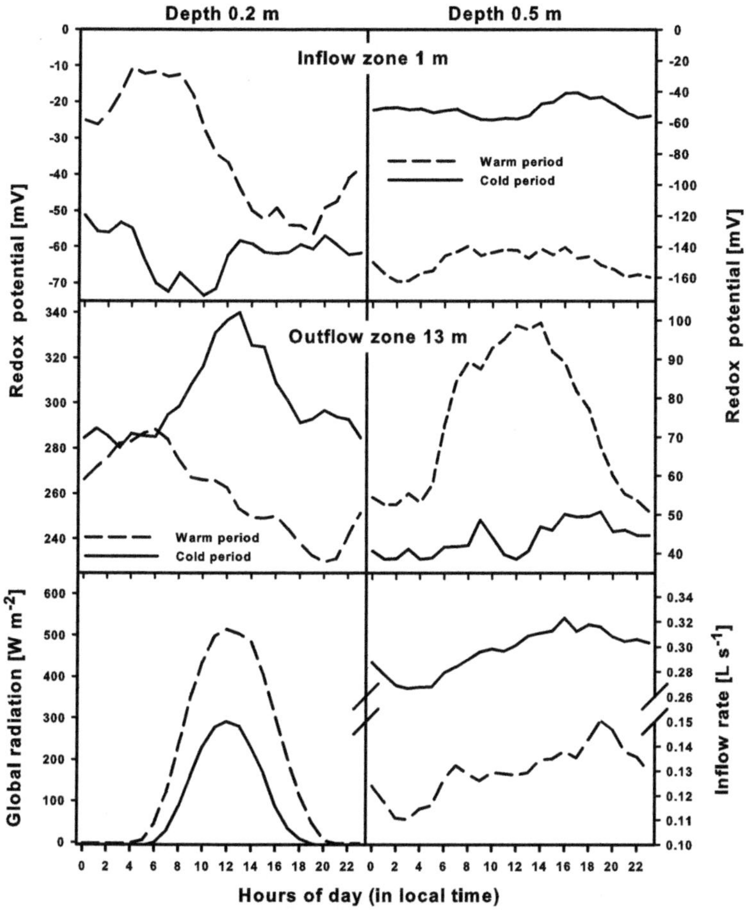

Fig. 7.1 Diel fluctuation of the redox potential (Eh) in the 0.2 and 0.5 m depths of the reed bed at 1 m and 13 m distances from the inflow of the Slavošovice constructed wetland. Presented data of the Eh values were corrected relative to the normal H electrode. Redox potential, global radiation and water flow rates are presented separately for warm (May to October) and cold (November to April) periods of the studied years. Means of Eh ($n = 40000$ measurements for each warm and cold periods) measured from June 2002 to September 2004 are shown

negative (about –150 mV), however the fluctuation was less than ± 20 mV. Similar slight fluctuations were recorded also during the cold period. In the afternoon Eh increased by 20 mV to its maximal values. Eh fluctuations in the 0.5 m depth were not statistically significant.

7.3.2 Eh Fluctuations in the Outflow Zone

Eh showed a consistent trend of diel fluctuation in the 0.2 m and 0.5 m depths in the outflow zone of the reed bed (Fig. 7.1), but the trend of Eh fluctuation was different for different periods of the year. Fluctuations were more pronounced in the 0.2 m depth as compared to 0.5 m depth. During the warm period Eh decreased in the 0.2 m depth from 4:00 in the morning local (Central European) time till 20:00 local time. The fast decrease of Eh was negatively related to the increase of water flow rate during the day (Fig. 7.1). Water flow rate increased from 3:00 till 19:00, and then decreased till 2:00. Maximal water flow rate was measured at 19:00. The relation of Eh to solar radiation in the outflow zone was not so close in comparison with the same relation found in the inflow zone. The fluctuation of Eh in the 0.5 m depth in the outflow zone was more pronounced in comparison with the Eh measured in the same depth in the inflow zone. Mainly during the warm period Eh closely correlated with global radiation. During the cold period the fluctuation of Eh was less pronounced in the 0.5 m depth as compared to the 0.2 m depth, similarly as in the inflow zone of the reed bed.

7.4 Discussion

The recorded Eh measured in wetlands soils is a result of mutual interactions between several redox systems present in the soil solutions (Schüring, Schulz, Fischer, Böttcher, & Duijnisveld, 1999: Richardson & Vepraskas, 2000). The resulting redox potential is referred to as "*mixed redox potential*" (Schüring et al., 1999). Mixed potentials can always be detected when several redox systems which are not in equilibrium exist simultaneously in the same solution (Ponnamperuma, 1972). In situ measurements of the redox potential are mainly valuable when relative changes and their fluctuations are recorded continuously in a long time scale. These trends and fluctuations may be used for description of soil redox conditions. Redox potential fluctuated from –400 to +800 mV and significant changes of Eh were recorded in very short periods (Dušek et al., 2008). The lowest Eh values may be detected in situations when anaerobic microbial processes are intense.

We found two interesting trends in the diurnal courses of Eh. The first trend is characterized by decreasing Eh during the light period (warm period in 0.2 m depth). The second trend is characterized by increasing Eh during the light period and is closely correlated with solar global radiation. Similar diel fluctuations of Eh have also been observed; for example, Wießner et al. (2005) found daily Eh fluctuation ranging from −200 mV to +200 mV in the *Juncus effusus* L. rhizosphere under strictly controlled laboratory conditions, driven by daylight.

The decrease of Eh during the light period could be attributed to the release of root exudates to the rhizosphere, which are released when plants are photosynthetically active. Root exudation is part of the rhizodeposition process, which is a major source of soil organic carbon released by plants (Hutsch, Augustin, & Merbach, 2000; Nguyen, 2003). Root exudates may consist of a vast range of substances varying from low-molecular weight compounds such as sugars, alditols, phenolics, organic acids, amines, and amides to more complex substances like exo-enzymes and mucilage (e.g., Neumann & Römheld, 2001; Uren, 2000). Root exudates are chemical substances, which are often not involved in redox processes directly, but they are involved in complicated microbial processes. Carbon-based compounds serve as substrates for heterotrophic microorganisms (Paul & Clark, 1989; Bertin, Yang, & Weston, 2003), which are present in the reed bed. Microbial processes can be promoted by the release of root exudates in systems limited with carbon (Gale, Devai, Reddy, & Graetz, 1993; Picek, Čížková, & Dušek, 2007). This promotion may result in a decrease of Eh especially in the densely rooted layer (upper layer, usually about 0.2 m depth) of the reed bed (Fig. 7.1). A decrease of Eh was also recorded in the 0.5 m depth without roots and rhizomes during the day, but it was slower than in the upper layer. This layer of the reed bed was probably less affected by the root exudation. The carbon-based root exudates were consumed (transformed) by bacteria in the upper rooted layer of the reed bed.

Diurnal cycling of Eh was documented also by Flessa (1994) and Wießner et al. (2005), who concluded that Eh was affected by the oxygen transport through the plant aerenchyma to the belowground plant parts and then by the radial loss of oxygen from the roots to the rhizosphere increases redox potential. However, we found an opposite trend of Eh in the heavily loaded inflow zone of the reed bed. The fluctuations and changes of Eh cannot be explained solely by a release of oxygen in this case. The resulting Eh involves mutual interactions between roots and their surrounding environments.

7.5 Conclusion

The amplitudes of daily fluctuations of averaged Eh for warm and cold periods, respectively, are not too wide (maximally 60 mV). Some trends of Eh fluctuations were found in the studied system during the day, however, daily trends of Eh fluctuations studied in situ conditions in horizontal subsurface flow constructed wetland for wastewater treatment, cannot be easily explained by one factor, as is known from laboratory experiments (e.g. Wießner et al., 2005).

Acknowledgements The authors gratefully acknowledge the financial support of research provided by projects 206/03/P021, 206/06/0058 and 526/09/1545 from the Grant Agency of the Czech Republic, project MSM6007665801 from Ministry of Education. We also warmly thank Slavošovice and Libín village authority and the mayor Mr. Ludvík and the constructed wetland manager Mr. Adamík for technical assistance and help with continuous measurement in the Slavošovice constructed wetland. We would also like to thank Dr. Jan Pokorný (ENKI, o.p.s., Třeboň) for providing part of the radiation data for the studied years. For the English revision of the manuscript and valuable remarks we thank to Dr. Hana Čížková.

References

Armstrong, W. (1979). Aeration in higher plants. In: H.W. Woolhouse (Ed.), *Advances in Botanical Research*, (Vol. 7, pp. 225–332). London: Academic.

Armstrong, J., & Armstrong, W. (1991). A convective through-flow of gases in *Phragmites australis* (Cav.) Trin. ex Steud. *Aquatic Botany, 39*, 75–88.

Austin, W. E., & Huddleston, J. H. (1999). Viability of permanently installed platinum redox electrodes. *Soil Science Society of America Journal, 63*, 1757–1762.

Bertin, C., Yang, X. H., & Weston, L. A. (2003). The role of root exudates and allelochemicals in the rhizosphere. *Plant and Soil, 256*, 67–83.

Bochove, E., Beauchemin, S., & Thériault, G. (2002). Continuous multiple measurement of soil redox potential using platinum microelectrodes. *Soil Science Society of America Journal, 66*, 1813–1820.

Dušek, J., Picek, T., & Čížková, H. (2008). Redox potential dynamics in a horizontal subsurface flow constructed wetland for wastewater treatment: Diel, seasonal and spatial fluctuations. *Ecological Engineering, 34*, 223–232.

Flessa, H. (1994). Plant-induced changes in the redox potential of the rhizospheres of the submerged vascular macrophytes *Myriophyllum verticillatum* L. and *Ranunculus circinatus* L. *Aquatic Botany, 47*, 119–129.

Gale, P. M., Devai, I., Reddy, K. R., & Graetz, D. A. (1993). Denitrification potential of soils from constructed and natural wetlands. *Ecological Engineering, 2*, 119–30.

Hammer, D. A. (1989). *Constructed wetlands for wastewater treatment municipal, industrial and agricultural*. Chelsea, Michigan: Lewis Publishers.

Holcová, V. (2007). *The affect of macrophytes on retention times during summer and winter periods in a constructed wetland for wastewater treatment*. M.Sc. Thesis. Faculty of Biological Sciences, University of South Bohemia, České Budějovice, Czech Republic, (in Czech).

Hutsch, B. W., Augustin, J., & Merbach, W. (2000). Plant rhizodeposition an important source for carbon turnover in soils. *Journal of Plant Nutrition and Soil Science, 165*, 397–407.

Mitsch, W., & Gosselink, J. G. (1986). *Wetlands* (1st ed.). New York: Van Nostrand Reinhold Company Inc.

Neumann, G., & Römheld, V. (2001). The release of root exudates as affected by the plant's physiological status. In R. Pinton, Z. Varanini & P. Nannipieri (Eds.), *The rhizosphere: Biochemistry and organic substances at the soil-plant interface* (pp. 44–93). New York: Marcel Dekker.

Nguyen, C. (2003). Rhizodeposition of organic C by plants: Mechanisms and controls. *Agronomie, 23*, 375–396.

Paul, E. A., & Clark, F. E. (1989). *Soil microbiology and biochemistry*. London: Academic.

Picek, T., Čížková, H., & Dušek, J. (2007). Greenhouse gas emission from a constructed wetland—plants as important source of carbon. *Ecological Engineering, 31*, 98–106.

Ponnamperuma, F. N. (1972). The chemistry of submerged soils. *Advances in Agronomy, 24*, 29–96.

Richardson, J. L., & Vepraskas, M. J. (2000). *Wetland soils: Genesis, hydrology, landscapes and classification*. Boca Raton, FL: CRC.

Rowell, D. L. (1981). Oxidation and reduction. In: D. J. Greenland & M. H. B. Hayes (Eds.), *The chemistry of soil processes* (pp. 401–462). Toronto: Wiley.

Schüring, J., Schulz, H. D., Fischer, W. R., Böttcher, J., & Duijnisveld, W. H. (Eds.). (1999). *Redox: Fundamentals, processes and applications*. Heidelberg: Springer.

Uren, N. C. (2000). Types, amounts and possible functions of compounds released into the rhizosphere by soil-grown plants. In R. Pinton, Z. Varanini, & P. Nannipierri (Eds.), *The rhizosphere, biochemistry and organic substances at the soil–plant interface* (pp. 19–40). New York: Marcel Dekker.

Wießner, A., Kappelmeyer, U., Kuschk, P., & Kästner, M. (2005). Influence of the redox condition dynamics on the removal efficiency of a laboratory-scale constructed wetland. *Water Research, 39*, 248–256.

Chapter 8
A Finite Element Approach to Modelling the Hydrological Regime in Horizontal Subsurface Flow Constructed Wetlands for Wastewater Treatment

Paul R. Knowles and Philip A. Davies

Abstract A Finite Element Analysis (FEA) model is used to explore the relationship between clogging and hydraulics that occurs in Horizontal Subsurface Flow Treatment Wetlands (HSSF TWs) in the United Kingdom (UK). Clogging is assumed to be caused by particle transport and an existing single collector efficiency model is implemented to describe this behaviour. The flow model was validated against HSSF TW survey results obtained from the literature. The model successfully simulated the influence of overland flow on hydrodynamics, and the interaction between vertical flow through the low permeability surface layer and the horizontal flow of the saturated water table. The clogging model described the development of clogging within the system but under-predicted the extent of clogging which occurred over 15 years. This is because important clogging mechanisms were not considered by the model, such as biomass growth and vegetation establishment. The model showed the usefulness of FEA for linking hydraulic and clogging phenomenon in HSSF TWs and could be extended to include treatment processes.

Keywords Clogging · Horizontal subsurface flow · Modelling · Suspended solids · Hydraulics

8.1 Introduction

Horizontal Subsurface Flow Treatment Wetlands (HSSF TWs) have proven to be highly effective for the tertiary treatment (polishing) of municipal wastewaters. Resultantly, many systems have been commissioned in the UK since the technology was introduced to the country in the mid-1980s (Cooper, Job, Green, & Shutes, 1996), with Cooper (2007) reporting that the 2006 version of the UK Constructed

P.R. Knowles (✉)
Sustainable Environment Research Group, Aston University, Birmingham, UK
e-mail: knowlesp@aston.ac.uk

J. Vymazal (ed.), *Water and Nutrient Management in Natural and Constructed Wetlands*, DOI 10.1007/978-90-481-9585-5_8,

Wetland Association database holds records for 677 such systems. One of the major attractions of the technology is the low capital and maintenance costs in comparison to alternatives. However, the longevity of these systems has proven far shorter than original predictions of 15–20 years (EC/EWPCA, 1990) because of their propensity to clog (Cooper, Griffin, & Cooper, 2008). Utility providers now expect to refurbish clogged systems as frequently as every 8 years (Griffin, Wilson, & Cooper, 2008), which markedly reduces the economic advantage proposed by HSSF TWs.

Clogging in HSSF TWs has been reported internationally (Batchelor & Loots, 1997; Caselles-Osorio et al., 2007; Kadlec & Watson, 1993; Pedescoll et al., 2009; Tanner, Sukias, & Upsdell, 1998: Wallace & Knight, 2006), although the exact nature of clogging varies between systems depending on design and operating conditions. Generally, the loss of gravel porosity to clog matter accumulation and corresponding reduction in hydraulic conductivity will result in some degree of hydraulic malfunction that warrants intervention depending on operator expectations (Kadlec & Wallace, 2008). The composition of clog matter is variable although components include those derived from wastewater, such as organic and inorganic wastewater solids and chemical precipitates; and internally generated contributions from plant roots, plant detritus and biomass growth (Kadlec, 1993; Llorens, Puigagut, & García, 2009; Nguyen, 2001; Puigagut, Caselles-Osorio, Vaello, & García, 2008; Tanner et al., 1998).

Figure 8.1 is adapted from Knowles, Griffin, and Davies (2010) and shows the typical development of clogging in the longitudinal cross-section of a UK HSSF TW. Typically, systems in the UK employ distributors that span the width of the bed at the inlet, and load the wastewater directly onto the surface. When the macrophyte root-mat develops in the upper strata the infiltration of the wastewater into the subsurface is impeded, and a low permeability surface layer begins to accumulate which is deepest towards the inlet. The percolation head-loss through the surface region results in two equilibrium water levels within the system; one associated with

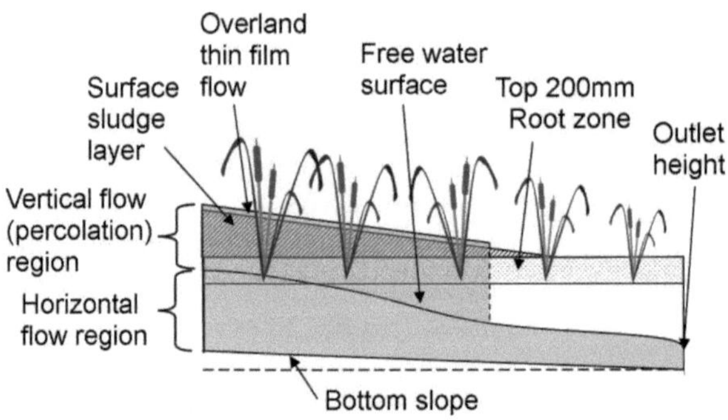

Fig. 8.1 The clogging profile typical to HSSF TWs for the tertiary treatment of municipal wastewater, as used in the UK. Diagram depicts a transverse cross-section and is not to scale (adapted from Knowles et al., 2010, with permission of Elsevier)

ponding on the surface; and a subsurface water table. The profile of the water table is non-linear due to the effects of evapotranspiration and precipitation, heterogeneous hydraulic conductivity, and variable percolation over the length of the overland flow region (Knowles et al., 2010). A combination of in situ constant head hydraulic conductivity tests, and multiple-point internal dye tracer tests demonstrated that clogging had resulted in two distinct preferential flow-paths through the system. The path with the lowest hydraulic retention time accounted for 80% of the flow and represented overland flow across the surface layer to a point approximately 7 m downstream, where the flow subsequently infiltrated into the subsurface and vertically short-circuited below the root-zone. The second flow path was heavily retarded in comparison to the first and accounted for net infiltration of all flow upstream of 7 m.

Overestimates of system longevity emphasise the inability of current design guidelines and models to account for the aforementioned clogging scenario. Despite two decades of TW research, modelling capabilities to simulate and predict clogging are still in their infancy. Previously, provision for clogging has often been made by proposing design guidelines for equilibrium hydraulic conductivity. In the UK it was assumed that a combination of coarse media (3–6 mm or 6–12 mm gravel) and the expansion of the macrophyte root network would counteract clogging caused by treatment processes and result in an equilibrium hydraulic conductivity of approximately 90 m d^{-1} (Cooper et al., 1996). However, recent in situ measurements performed at a 15 year old HSSF TW in the UK found that values of substrate hydraulic conductivity in the inlet region of the bed were two to three orders of magnitude lower than this guideline (Knowles et al., 2010).

An early attempt to model clogging in Vertical Flow (VF) TWs assumed that porosity was diminished cumulatively by the volume of influent suspended solids loaded into the system over time, such that system longevity corresponded to zero porosity (Blazejewski & Murat-Blazejewska, 1997; Langergraber, Haberl, Laber, & Pressl, 2003; Zhao, Sun, & Allen, 2004). However, the model simplifies the mechanism of clogging and does not consider additional factors such as biological growth or spatial heterogeneity in clogging development. Resultantly the extent of clogging was under-predicted when the model was applied to a pilot-scale VF TW in Austria (Langergraber et al., 2003). García et al. (2007) adapted a dynamic HSSF TW treatment process model (Rousseau, 2005) so that rules were included for particle transport, and biomass growth was linked to substrate utilisation. The model was used to explore the extent to which clogging could be mitigated by various physico-chemical pre-treatment options within two pilot-scale HSSF TWs, but was not used for comparison with the actual clogging present in those systems. In a similar fashion, Leverenz, Tchobanoglous, & Darby (2009) applied a Finite Element Analysis (FEA) model (Langergraber & Ŝimůnek, 2005) that had been developed to describe treatment performance in a VF TW to study clogging in a wastewater sand filter. The model combines physical descriptions of unsaturated flow and reactive transport with the treatment process nomenclature presented in the Activated Sludge Models (Henze, Gujer, Mino, & Van Loosedrecht, 2000), and was used by Leverenz et al. (2009) to study biomass growth under the assumption that this could be used to represent clogging. However, clogging associated with particle transport and the impact of biomass growth on hydrology could not be directly simulated using the model and was considered separately. At the time of writing there are no HSSF TW models that successfully integrate the symbiotic effects of hydrology, treatment and clogging.

Numerous attempts have been made to model the clogging of porous media associated with particle transport in other applications, mainly in the field of Deep Bed Filtration (DBF). An in-depth consideration of the available DBF models is beyond the scope of this paper although several comprehensive reviews exist that also consider suitability for applications such as biological transport and wastewater treatment (Hubbe, Chen, & Heitmann, 2009; Thullner, 2010; Tufenkji, 2007; Zamani & Maini, 2009).

One example which warrants mention is the use of a DBF model to represent clogging in a Landfill Leachate Collection Systems (Cooke & Rowe, 2008). The sophisticated FEA model, called BioClog, considers how flow reacts to clogging caused by particle transport and biomass growth. The particle transport model is based on the trajectory analysis model of Rajagopalan and Tien (1976) with formulations for attachment coefficients presented by Reddi and Bonala (1997), and allows clog matter to be transported, attached, detached and reattached from the porous media depending on surface conditions and hydrodynamic effects (Cooke, Rowe, & Rittman, 2005). Biomass growth is derived from several major wastewater constituents using Monod kinetics and biomass transport is also governed by the trajectory analysis model (Martin, Bouwer, & Hanna, 1992). Precipitation of calcium carbonate is also considered as a component of clogging . The subsequent impact on hydrology is resolved by assuming each accumulation forms a uniform coating on the surface of the rock substrate, such that the change in specific surface area of the bulk media (and hence porosity) can be calculated (Rittman, 1982).

A simpler approach to modelling DBF is to experimentally deduce the collector efficiency of the system such that the model only requires an expression for continuity (Einstein, 1968; Mays & Hunt, 2005; O'Melia & Ali, 1978). This approach requires less modelling assumptions and parameters than trajectory analysis, however, the non-physically based expression for collector efficiency is system specific and cannot be extrapolated to different systems or used to explore the consequence of parameter changes (Rege & Fogler, 1988). A DBF continuity model presented by Cui et al. (2008) corrects the previous model presented by Sakthivadivel & Einstein (1970) such that it can be applied to describe infiltration of sediment into a gravel bed. The collector efficiency was based on the experimental system of Wooster et al. (2008), which employed media and fine sediment with physical properties respectively analogous to the gravel media and wastewater solids typical to UK HSSF TWs.

The modularity of FEA offers great scope to implement models which couple clogging, hydraulics and treatment performance (Langergraber et al., 2009), especially for improved description of particle transport phenomena and surface layer formation. The aim of the paper is to create an FEA model which represents the hydrology of a clogged UK HSSF TW as described by Knowles et al. (2010). Secondly, the clogging model of Cui et al., (2008) will be coupled to the hydraulic model so that it can be used to explore clogging by particle transport. The model will incorporate parameters that allow it to be validated against a 15 year old HSSF TW in the UK, as profiled by Knowles et al. (2010).

8.2 Model Formulation

Modelling is achieved using COMSOL Multiphysics 3.5 FEA software (COMSOL A.B., Sweden) by coupling 5 different physical sub-models; three for hydrology and two for clogging. The hydraulic formulation will be considered first.

8.2.1 Hydraulic Model

The hydraulic model is represented in Fig. 8.2 and depicts two sub-domains, seven boundaries numbered according to the bold font, and various hydraulic parameters applicable to each sub-domain. A full list of the boundary conditions is given in Table 8.1, whilst the nature and source of hydraulic parameters is disclosed in Table 8.2. The sub-domain size is based on a transverse cross section of the HSSF TW surveyed by Knowles et al. (2010), with length 15 m and height 0.6 m. The sub-domains are separated by the surface of the water-table, with flow through the upper sub-domain represented using Richard's Law for variably saturated flow, and the water table flow described using Darcy's Law (Bear, 1979). The profile of each sub-domain is resolved based on modelling parameters.

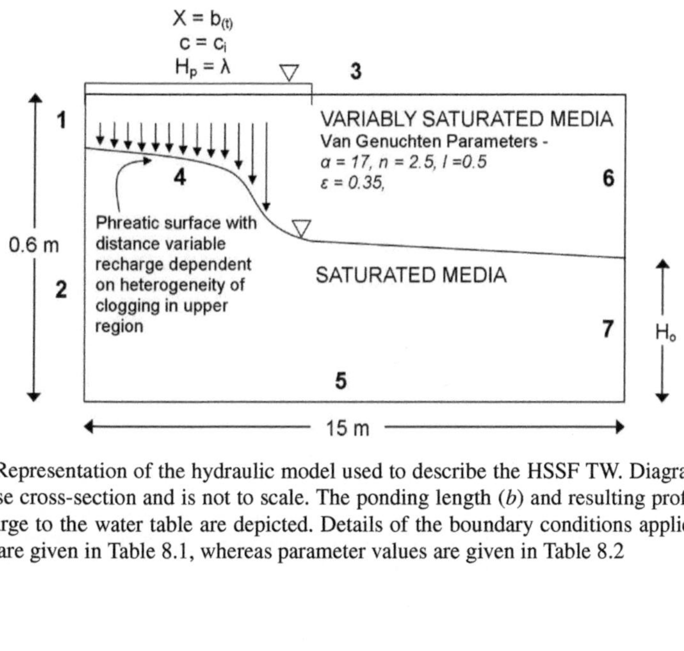

Fig. 8.2 Representation of the hydraulic model used to describe the HSSF TW. Diagram depicts a transverse cross-section and is not to scale. The ponding length (*b*) and resulting profile of vertical recharge to the water table are depicted. Details of the boundary conditions applied to each boundary are given in Table 8.1, whereas parameter values are given in Table 8.2

Table 8.1 A full list of boundary conditions prescribed in the model. The boundary numbers are consistent with the labelling in Fig. 8.2

Boundary	Condition	Description
Darcy's Law boundary conditions		
2,5	$\mathbf{n}.\left[k_s\nabla\left(H_p+D\right)\right]=0$	Neumann – zero flux
4	$\mathbf{n}.\left[k_s\nabla h\right]=-\mathbf{v}_r$	Neumann – inward flux
7	$h_{(x,t)}=H_\circ$	Dirichlet – outlet height
1,3,6		Does not apply
Richard's equation boundary conditions		
1,6	$\mathbf{n}.\ [k\nabla\ (H_P+D)]=0$	Neumann – zero flux
3	$H_{P(0\leq x\leq b)}=\lambda$	Dirichlet – film thickness pressure head
4	$H_P=0$	Dirichlet – saturated transition
2,5,7		Does not apply
Advection dispersion equation boundary conditions		
1,2,5,6	$\mathbf{n}.[-\varepsilon\, D\,\nabla c+\bar{\mathbf{u}}\, c]=0$	Neumann – zero flux
4		Neumann – Continuity
3	$c_{(0\leq x\leq b)}=c_\circ$	Dirichlet – influent solids concentration
7	$\mathbf{n}.[-\varepsilon\, D\,\nabla c]=0$	Neumann –advective flux only
Deep Bed filtration boundary conditions		
1,2,3,4,5,6,7		Neumann – Continuity
Deformed Mesh boundary conditions		
1,2,6	$dx=0$	Horizontally constrained
3,5,7	$dx=0,\quad dy=0$	Fully constrained
4	$dy=h-y$	Vertical displacement until $H_P=0$

The minimum depth of water in the system corresponds to the outlet height (H_o) and flux is only permitted through boundary 7. Wastewater is applied on boundary 3 over a region that extends from the inlet to a distance representative of the ponding length (b). The pressure head generated by ponding is calculated using the assumption that water depth is governed by gravitational water film thickness (λ) which is approximately 5 mm for water (de Gennes, Brochard-Wyart, & Quéré, 2004). Although this is a simplification it is based on the observation that open channel flow cannot be used to describe overland flow in these systems, as has been done previously (Kadlec & Wallace, 2008). The two sub-domains are hydraulically coupled across boundary 4 by assuming the profile of outward flux from the upper sub-domain equals the profile of inward flux into the upper sub-domain. All other boundaries have a zero flux condition.

The surface infiltration corresponding to λ at a point along b depends on the vertical distance between the top boundary and the water-table, and the subsurface hydraulic conductivity. However, the profile of the water-table depends on the infiltration profile, such that an iterative solver is required to calculate the equilibrium condition. This is achieved using an additional "moving-mesh" model available in COMSOL 3.5, which allows mesh movement depending on certain physical criteria. A further iterative solver calculates the value of b required to achieve the input flow-rate through the system (Q). It should be emphasised that at equilibrium a

Table 8.2 A list of the parameters included in the model, along with the reference from where the parameter was obtained and justification for use

Description	Symbol	Units	Value	Reference and notes
Median gravel diameter	d	mm	5.2	(Knowles et al., 2010) – based on conditions in modelled system
Flow Rate	Q	$m\ d^{-1}$	0.77	
Outlet Height	H_o	m	0.542	
Ponding depth	λ	mm	5	
Saturated porosity	ε	–	0.35	
Parameter in van Genuchten model	α	$1\ m^{-1}$	17	(Leverenz et al., 2009) – van Genuchten parameters for substrate with high conductivity and high specific yield substrate
Parameter in van Genuchten model	n	–	2.5	
Parameter in van Genuchten model	l	–	0.5	
Dispersivity in longitudinal direction	σ_L	cm	10	(Cooke & Rowe, 2008) – based on gravel diameter of 6 mm
Dispersivity in transverse direction	σ_T	cm	10	
Influent solids concentration	c_i	$mg\ L^{-1}$	29	(Cooper, 2007) – based on influent solids concentration data of 71 UK HSSF TWs
Clean filter efficiency	β_o	–	0.0244	(Wooster et al., 2008) – based on gravel diameter of 7.9 mm and hydrosol diameter of 0.35 mm
Clogging coefficient	Φ	–	0.8	

saturated region will exist that is bounded between *b* and the water-table, although the exact profile will depend on the model parameters for variably saturated flow.

8.2.2 Clogging Model

Initially only clogging by particle transport is considered, thus focusing on the ability of FEA to simulate this phenomenon in HSSF TWs. The effects of biomass growth, vegetation growth and sludge layer formation are ignored. It is assumed that solids are non-degradable and once they have become attached they cannot detach for transportation further downstream. The clogging model of Cui et al. (2008) uses an expression (Eq. (8.1)) which describes the ripening of the filter coefficient β over time according to filter and wastewater conditions. This includes parameters for clean filter coefficient (β_O), media porosity (ε) and a coefficient which describes the propensity of the system to clog given operating and design conditions (Φ). The variable ε_S describes the specific volume lost to clog matter accumulation.

$$\beta(t) = \beta_O \exp\left(\phi \frac{\varepsilon_s(t)/\varepsilon}{1 - \varepsilon_s(t)/\varepsilon}\right) \tag{8.1}$$

Physically this equation represents an exponential increase in filter coefficient as the filter begins to clog, thus increasing subsequent solids removal. Based on the formulation of Sakthivadivel and Einstein (1970), the change in the specific volume of accumulated clog matter is given by:

$$\frac{\partial \varepsilon_s}{\partial t} = \beta q \tag{8.2}$$

where q is the net volumetric clog matter flux through the system. The corresponding removal of solids from the wastewater is achieved using a continuity statement:

$$\frac{\partial \varepsilon_s}{\partial t} = -\nabla q \tag{8.3}$$

The parameters required in Eq. (8.1) are given in Table 8.2 and are based on gravel with mean diameter 7.9 mm being infiltrated by suspended solids with mean diameter 0.35 mm.

The clogging relationship is coupled to the hydraulic model using a reactive transport model which links the volumetric concentration of solids in the wastewater with velocity using the advection-dispersion equation; thus allowing q to be calculated anywhere in the system. The clogging and reactive transport models are applicable over both sub-domains and the clogging model requires no boundary conditions to be specified. The reactive transport model specifies the volumetric concentration of solids in the influent wastewater (c_i) across the ponding length (b), and assumes an advective flux condition exists at boundary 7. All other boundaries have a zero flux condition.

8.2.3 Model Implementation

A mesh consisting of 4416 triangular elements, with a 10 times scaling factor in the longitudinal direction, was used to represent the system (Fig 8.3). In the first instance only the hydraulic model was solved for, to allow validation against field measurements. The spatial hydraulic conductivity of the subsystem was specified to represent the most clogged longitudinal transect measured by Knowles et al. (2010) during the site survey. The nested iterative solver and moving mesh model calculate the equilibrium flow field corresponding to the inlet flow rate, outlet height and subsurface hydraulic conductivity profile. The effect of the modelled flow regime on hydrodynamics is investigated by simulating the propagation of a tracer plume through the subsurface (similar to the test performed in Knowles et al. (2010).

Secondly the model was used to explore how clogging develops within a clean version of the system. The initial hydraulic conductivity of the subsurface was calculated using the Kozeny-Carman equation for flow through porous media (Zamani &

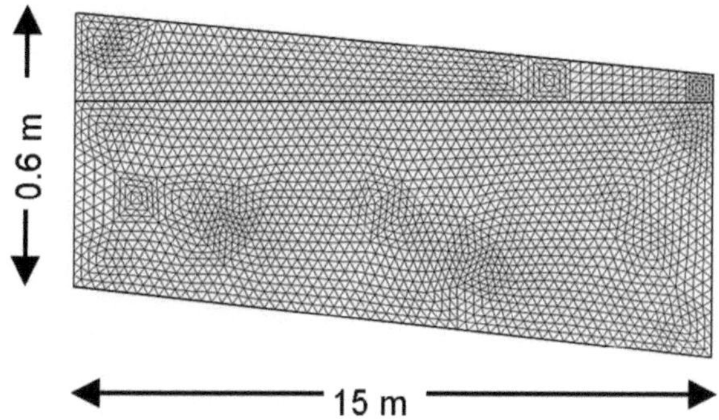

Fig. 8.3 The finite element mesh for the model solver which consists of 4416 triangular elements, with a 10 times scaling factor in the longitudinal direction. Diagram depicts a transverse cross-section

Maini, 2009), based on a median gravel diameter of 5.2 mm. The flow field calculated in the hydraulic model was used to dynamically simulate wastewater hydrodynamics. The time dependent model elapses for one day and then the flow-field is recalculated again using the Kozeny-Carman equation to relate changes in porosity to changes in hydraulic conductivity. This loop was repeated for a period representing 15 years. It would be possible to solve both the hydraulic model and clogging model simultaneously to reduce errors associated with the segregated solver, but this requires extensive computational time.

8.3 Results

Figure 8.4 shows the hydraulic conductivity profile of the system with values varying across 6 orders of magnitude. The most clogged region corresponds to the surface layer at the inlet with values on the order of 0.01 m d^{-1}, whilst lower depths towards the outlet of the bed have a hydraulic conductivity on the order of clean gravel (1000 m d^{-1}). The computed flow regime that corresponds to this hydraulic conductivity profile at a flow-rate of 2 L s^{-1} is shown in Fig. 8.5. The length of the ponding region (b) was found to be 7.72 m and two different resulting zones of flow are clear: a vertical percolation zone below the ponding length indicated by the presence of pressure head contours (upper sub-domain), and a horizontally flowing saturated water table with a profile that corresponds to percolation through the variably saturated region, as indicated by the greyscale shading from left to right (lower sub-domain). The area of the variably saturated region with no contours corresponds to zero flow through this part of the system.

Overall the hydraulic head across the horizontal flow region varied from 0.666 m at the inlet to 0.542 m at the outlet.

The model indicates that surface infiltration increases with distance downstream according to increasing hydraulic conductivity and increasing pressure heads (due to increasing vertical distance between the ponding surface and the water table). Infiltration rates at the termination of the ponding region are 72 m d^{-1} whilst rates at the inlet are approximately an order of magnitude lower (Fig. 8.6). The high flux

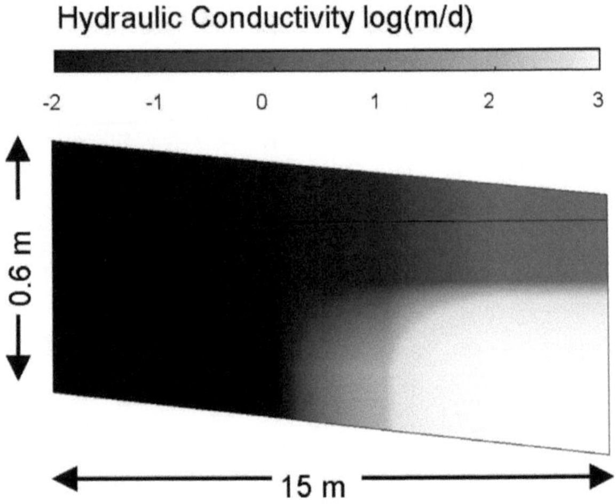

Fig. 8.4 The hydraulic conductivity profile of the most clogged transect surveyed by Knowles et al. (2010), as depicted using the model. The shading bar is logarithmic and represents six orders of magnitude variation, with darker areas being more clogged

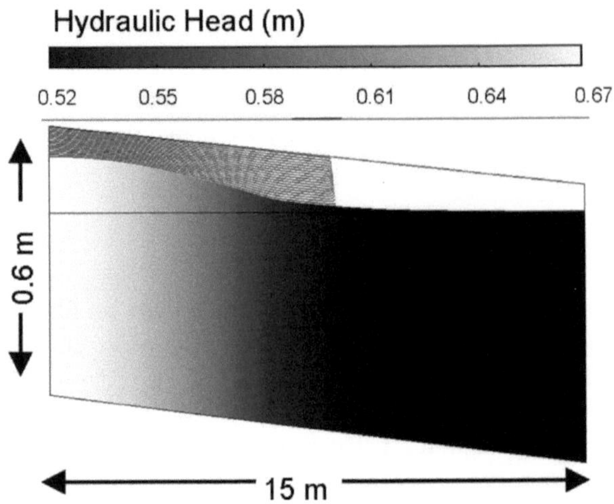

Fig. 8.5 The flow field which corresponds to the modelling parameters specified in Table 8.2 and the hydraulic conductivity profiled illustrated in Fig. 8.4. The vertical contours in the upper sub-domain represent vertical infiltration through the surface layer and the shaded profile represents the variation of hydraulic head in the horizontal water table

towards the middle of the bed results in the "S-shaped" water table profile illustrated in Fig. 8.7, which corresponded closely to that observed in the field survey.

Three snap-shots from the tracer impulse simulation are shown in Fig. 8.8a–c for times 20, 40 and 140 min respectively. The shading shows the spatial concentration profile of the tracer plume normalised against the inlet tracer concentration, where darker shades indicate greater concentration. Note that the shading bar is rescaled for each different time frame so that as the plume becomes more dispersed with time the concentration gradient is still discernible. In Fig. 8.8a the variable infiltration below the ponding region can be seen. Plume advancement near the inlet is fairly stagnant whereas faster infiltration rates downstream have led to rapid vertical spreading of the plume towards the middle of the bed. This effect is emphasised in Fig. 8.8b where the vast majority of the injected tracer is engaging the water-table

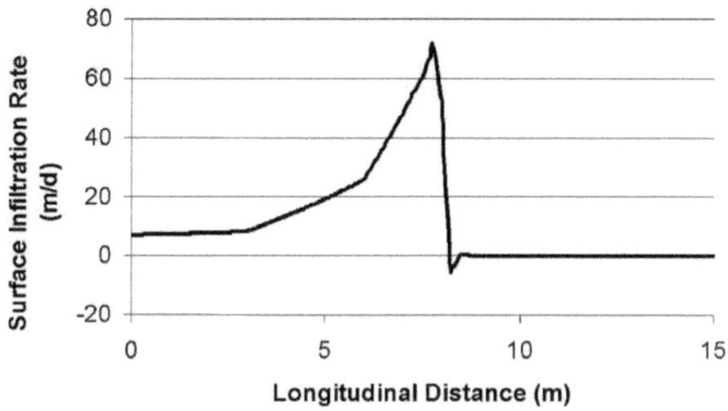

Fig. 8.6 Variation of surface infiltration rate with longitudinal distance. Infiltration corresponds to the ponding length and is greatest at the downstream edge

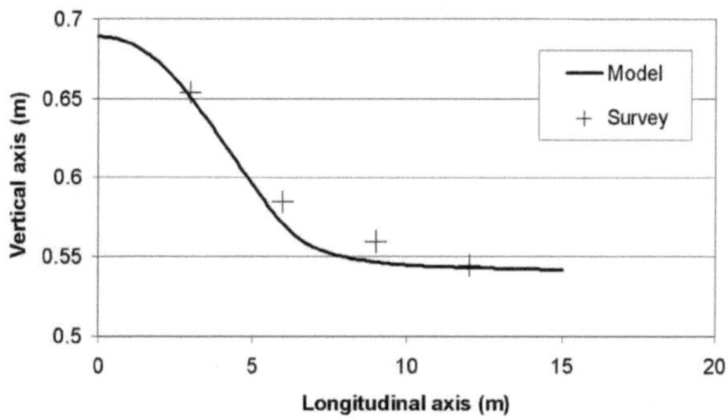

Fig. 8.7 A comparison between the modelled water table surface and that surveyed by Knowles et al. (2010), showing the model successfully describes the profile of the water table

approximately 8 m downstream of the inlet, and has quickly spread towards the bottom of the bed.

Figure 8.8c shows that the plume migrates horizontally once within the water-table, although low hydraulic conductivity within the root-zone causes the majority of the plume to vertically short-circuit through the relatively clean media along the bottom of the bed. Once within the water table, the fraction of the tracer which manages to infiltrate near the inlet moves with relatively slow speed in comparison to flow that infiltrates downstream, such that an interesting effect happens which may help to explain the results of Knowles et al. (2009). The plume breaks into two visible phases: the first phase to reach the outlet represents the majority of the flow fraction which rapidly propagates along the overland flow-path and then short-circuits below the root-zone; whilst the second phase represents flow through the clogged upstream media and consequently reaches the outlet much later.

Figure 8.9 illustrates the clogged hydraulic conductivity profile at the end of 15 years. As evident the lowest conductivities are at the surface near the inlet whilst the higher values correspond to the base of the bed near the outlet. Overall values vary by three orders of magnitude, from 0.76 to 56 m d^{-1}, which is a similar range to that measured in the field (0.11–28 m d^{-1}). Figure 8.10 shows the variation in depth-averaged hydraulic conductivity versus longitudinal distance, according to both the clogging model results and the survey of Knowles et al. (2009). This confirms that

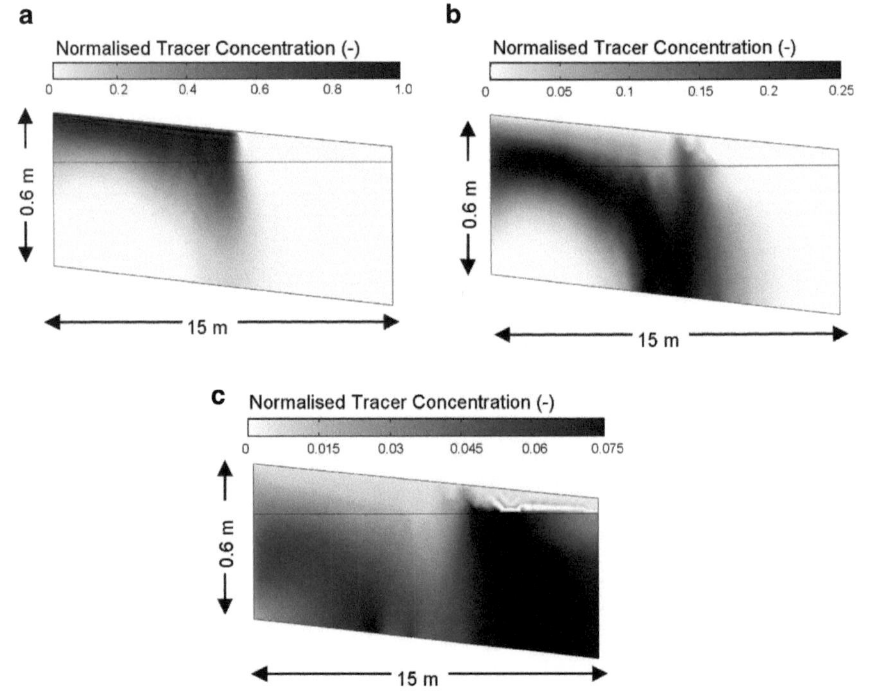

Fig. 8.8 The passage of tracer through the system illustrated in Fig. 8.5 The shading represents the spatial concentration of the tracer plume relative to the influent concentration at times (**a**) 20 min (**b**) 40 min (**c**) 140 min. Note the shading bar is rescaled in each time-frame so the plume remains discernible as it becomes dispersed

Fig. 8.9 The modelled hydraulic conductivity profile of the system after 15 years of particle transport, as predicted using the Deep Bed Filtration analysis of Cui et al. (2008). The shading bar is logarithmic and represents three orders of magnitude variation overall, with darker areas being more clogged

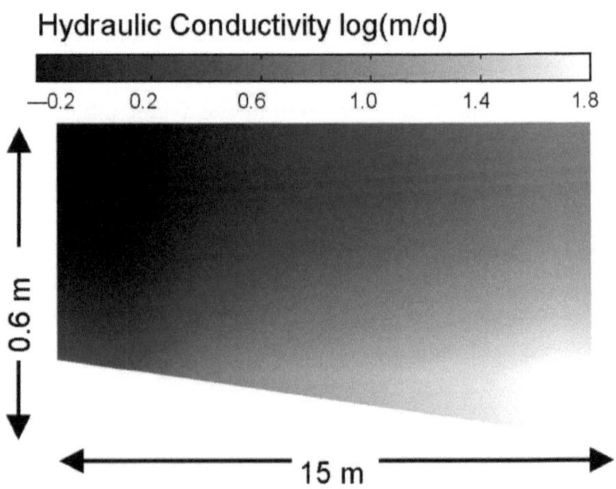

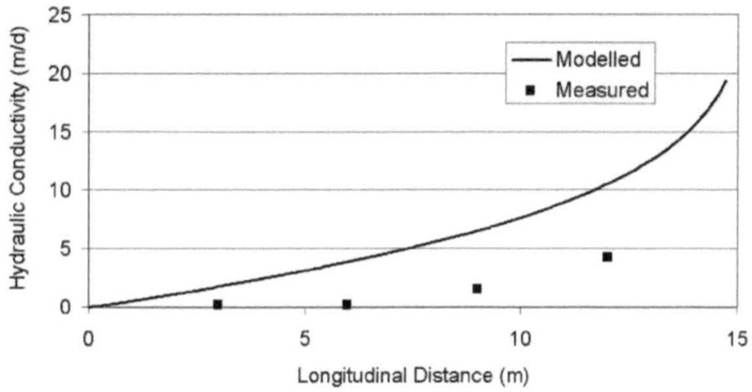

Fig. 8.10 The depth averaged hydraulic conductivity of the subsurface versus longitudinal distance, as measured by Knowles et al. (2010) and as predicted using the model. The model describes the general trend of clogging versus distance but under-estimates the extent of clogging

although the model is capable of mimicking the typical development of clogging, in this case it has underestimated the extent of clogging that has occurred.

8.4 Discussion

The hydraulic model emphasised the poor volumetric efficiency of this clogged system. Most of the flow bypasses the first half of the bed via overland flow and then short-circuits below the root-zone, representing a malfunction which could have deleterious consequences for treatment performance. It is felt the hydraulic model describes the hydrological situation of UK HSSF TWs very adequately, whereas the clogging model warrants further discussion.

The results indicate that the clogging model under-predicts the extent of clogging in the system after 15 years. This is expectable because the model only explores one facet of clogging, particle transport, and neglects potentially important mechanisms such as biomass and vegetation growth, and surface layer formation from plant detritus and cake filtration. For instance the extracellular polymeric substances secreted by biofilms are analogous to gel networks with pore diameters on the nanometric scale (deBeer & Stoodley, 1995), thus making them relatively impermeable. Furthermore, the root mat of *Phragmites australis* is reported to have a hydraulic conductivity on the order of 1–20 m d^{-1} (Baird, Surridge, & Money, 2004), similar to the range of values found in the field survey, thus suggesting root establishment could easily provide a bottleneck to flow. Surface formation by cake filtration could be implemented relatively easily using FEA (Mays & Hunt, 2005), whereas the influence of plant contributions is harder to account for. Additionally, neglecting root establishment may be responsible for the discrepancies between surveyed and modelled hydraulic conductivity profiles. What is evident in the survey results is a very sudden increase in conductivity below the root-zone, in the back half of the bed. In the model results the vertical hydraulic conductivity gradient is less pronounced. The root-zone would be very effective at filtering solids from the wastewater, which would make this sudden change in vertical hydraulic conductivity feasible, whereas this facility is not accounted for by the model, thus allowing solids to penetrate deeper into the bed.

Alternatively the model may not adequately describe the filter efficiency of the TW. The clogging model formulation suggests that removal is proportional to flux such that high velocities will lead to high solids removal. This may not be the case as low velocities will promote sedimentation (Rajagopalan & Tien, 1976) whilst high velocities will increase viscous shear and lubrication effects, thus promoting detachment and transport rather than attachment (Stevenson, 1997). Indeed, HSSF TWs are relatively low velocity systems making sedimentation one of the dominant solids removal processes (Kadlec & Wallace, 2008), and subsurface velocities are limited by the maximum available head (freeboard height). Furthermore, using the Kozeny-Carman equation to relate changes in porosity to hydraulic conductivity is inaccurate without accounting for the change in effective particle diameter due to accumulation. Instead it may be better to formulate the model using an approach similar to that used by Cooke and Rowe (2008) so that the collector efficiency reflects the physical conditions of the system more accurately.

8.5 Conclusions

The following salient conclusions are drawn:

An FEA model was devised to describe the hydrology and clogging processes which occur within UK HSSF TWs. The modularity of FEA models enables realistic representation and coupling of multi-physical systems, making FEA ideal for relating complex hydraulic, treatment and clogging interactions in TWs.

The hydraulic model is able to describe the important physical attributes which are evident in clogged UK HSSF TWs, but which have previously been overlooked. These include a ponding front with a length that is determined by the flow-rate, outlet height and subsurface hydrology; and a region of vertical percolation above a horizontal water table such that two equilibrium water-levels exist in the system.

Simulating the hydrodynamics of an existing clogged HSSF TW using the model showed that two preferential flow paths exist through the subsurface. The major path corresponds to overland flow via the ponded region and subsequent vertical short-circuiting below the root-zone.

A clogging model was developed based on conservation of mass and a single collector efficiency derived from a study which used similar media and hydrosol size distributions. The model described the general clogging behaviour observed in these systems, with hydraulic conductivity at the inlet reduced by four orders of magnitude in comparison to clean media (from 2000 m d^{-1} to approximately 1 m d^{-1}), and values increasing longitudinally from inlet to outlet.

However, the extent of clogging was underestimated. This is because the model currently simplifies the clogging process by only considering particle transport. Potentially important biological clogging mechanisms contributed by biomass, root networks and plant detritus are neglected. It is thought that a trajectory analysis model may relate clogging to physical conditions in the system more accurately.

Considering the encouraging initial results, the model will now be refined so that it can be used as a predictive tool for wetland scientists.

Acknowledgements This work was made possible thanks to joint funding from Severn Trent Water Plc. (UK) and a CASE studentship granted by the ESPRC UK (ref. CASE/CNA/06/28). The authors gratefully acknowledge the collaboration of Dr. Paul Griffin from Severn Trent Water Plc.

References

Baird, A. J., Surridge, B. W. J., & Money, R. P. (2004). An assessment of the piezometer method for measuring the hydraulic conductivity of a *Cladium mariscus-Phragmites australis* root mat in a Norfolk (UK) fen. *Hydrological Processes, 18*, 275–291.

Batchelor, A., & Loots, P. (1997). A critical evaluation of a pilot scale subsurface flow wetland: 10 years after commissioning. *Water Science and Technology, 35*(5), 337–343.

Bear, J. (1979). *Hydraulics of groundwater*. New York, London: McGraw-Hill.

Blazejewski, R., & Murat-Blazejewska, S. (1997). Soil clogging phenomena in constructed wetlands with subsurface flow. *Water Science and Technology, 35* (5), 183–188.

Caselles-Osorio, A., Puigagut, J., Segu, E., Vaello, N., Granés, F., García, D., & García, J. (2007). Solids accumulation in six full-scale subsurface flow constructed wetlands. *Water Research, 41*, 1388–1398.

Cooke, A. J., & Rowe, R. K. (2008). 2D modelling of clogging in landfill leachate collection systems. *Canadian Geotechnical Journal, 45*, 1393–1409.

Cooke, A. J., Rowe, R. K., & Rittman, B. E. (2005). Modelling species fate and porous media effects for landfill leachate flow. *Canadian Geotechnical Journal, 42*, 1116–1132.

Cooper, D., Griffin, P., & Cooper, P. F. (2008). Factors affecting the longevity of sub-surface horizontal flow systems operating as tertiary treatment for sewage effluent. In J. Vymazal (Ed.), *Wastewater treatment, plant dynamics and management in constructed and natural wetlands* (pp.191–198). Leiden, The Netherlands: Backhuys Publishers.

Cooper, P. F., Job, G. D., Green, M. B., & Shutes, R. B. E. (1996). *Reed beds and constructed wetlands for wastewater treatment*. Medmenham: Water Research Centre.

Cui, Y., Wooster, J. K., Baker, P. F., Dusterhoff, S. R., Sklar, L. S., & Dietrich, W. E. (2008). Theory of fine sediment infiltration into immobile gravel bed. *Journal of Hydraulic Engineering, 134*, 1421–1429.

De Gennes, P.-G., Brochard-Wyart, F., & Quéré, D. (2004). *Capillary and wetting phenomena-drops, bubbles, pearls, waves*. Berlin: Springer.

Debeer, D., & Stoodley, P. (1995). Relation between the structure of an aerobic biofilm and transport phenomena. *Water Science and Technology, 32*(8), 11–18.

EC/EWPCA (1990). *European design and operations guidelines for reed bed treatment systems*. Swindon, UK: WRc.

Einstein, H. A. (1968). Deposition of suspended particles in a gravel bed. *Journal of the Hydraulics Division, American Society of Civil Engineers, 95*, 1197–1205.

García, J., Rousseau, D., Caselles-Osorio, A., Story, A., De Pauw, N., & Vanrolleghem P. (2007). Impact of prior physico-chemical treatment on the clogging process of subsurface flow constructed wetlands: Model-based evaluation. *Water, air, and soil pollution, 185*, 101–109.

Griffin, P., Wilson, L., & Cooper, D. (2008). Changes in the use, operation and design of subsurface flow constructed wetlands in a major UK water utility. In S. Billore, P. Dass & J. Vymazal (Eds.), *Proceedings of 11th International Conference on Wetland Systems for Water Pollution Control* (pp. 419–426). Indore, India: Vikram University.

Henze, M., Gujer, W., Mino, T., & Van Loosedrecht, M. (2000). *Activated Sludge Models ASM1, ASM2, ASM2d and ASM3* (5th ed.). Water Intelligence Online.

Hubbe, M. A., Chen, H., & Heitmann, J. A. (2009). Permeability reduction phenomena in packed beds, fiber mats, and wet webs of paper exposed to flow of liquids and suspensions: A review. *BioResources, 4*, 405–451.

Kadlec, R. H. (1993, July 25–30). Flow patterns in constructed wetlands. In H.W. Shen, S.T. Su & F. Wen (Eds.), *Proceedings of the national conference on hydraulic engineering* (pp. 131–136). San Francisco, CA.

Kadlec, R. H., & Wallace, S. (2008). *Treatment wetlands* (2nd ed). Boca Raton, FL: CRC.

Kadlec, R. H., & Watson, J. T. (1993). Hydraulics and Solids Accumulation in a Gravel Bed Treatment Wetland. In G. A. Moshiri (Ed.), *Constructed wetlands for water quality improvement* (pp. 227–235). Boca Raton, FL: CRC /Lewis Publishers.

Knowles, P. R., Griffin, P., & Davies, P. A. (2010). Complementary methods to investigate the development of clogging within a horizontal sub-surface flow tertiary treatment wetland. *Water Research, 44*, 320–330.

Langergraber, G., Giraldi, D., Mena, J., Meyer, D., Peña, M., Toscano, A., Brovelli, A., & Korkusuz, E. A. (2009). Recent developments in numerical modelling of subsurface flow constructed wetlands. *Science of the Total Environment, 407*, 3931–3943.

Langergraber, G., Haberl, R., Laber, J., & Pressl, A. (2003). Evaluation of substrate clogging processes in vertical flow constructed wetlands. *Water Science and Technology, 48*(5), 25–34.

Langergraber, G., & Šimůnek, J. (2005). Modeling Variably Saturated Water Flow and Multicomponent Reactive Transport in Constructed Wetlands. *Vadose Zone Journal, 4*, 924–938.

Leverenz, H. L., Tchobanoglous, G., & Darby, J. L. (2009). Clogging in intermittently dosed sand filters used for wastewater treatment. *Water Research, 43*, 695–705.

Llorens, E., Puigagut J., & García, J. (2009). Distribution and biodegradability of sludge accumulated in a full-scale horizontal subsurface-flow constructed wetland. *Desalination and Water Treatment, 1*, 54–58.

Martin, R. E., Bouwer, E. J., & Hanna, L. M. (1992). Application of clean-bed filtration theory to bacterial deposition in porous media. *Environmental Science & Technology, 26*, 1053–1058.

Mays, D. C., & Hunt, R. J. (2005). Hydrodynamic aspects of particle clogging in porous media. *Environmental Science & Technology, 39*, 577–584.

Nguyen, L. (2001). Accumulation of organic matter fractions in a gravel-bed constructed wetland. *Water Science and Technology, 44*(11–12), 281–287.

O'Melia, C. R., & Ali, W. (1978). The role of retained particles in deep bed filtration. *Progress in Water Research, 10*, 167–182.

Pedescoll, A., Uggetti, E., Llorens, E., Granés, F., García, D., & García, J. (2009). Practical method based on saturated hydraulic conductivity used to assess clogging in subsurface flow constructed wetlands. *Ecological Engineering, 35*, 1216–1224.

Puigagut, J., Caselles-Osorio, A., Vaello, N., & García, J. (2008). Fractionation, biodegradability and particle-size distribution of organic matter in horizontal subsurface-flow constructed wetlands. In J. Vymazal (Ed.), *Wastewater treatment, plant dynamics and management in constructed and natural wetlands* (pp. 289–297). Leiden, The Netherlands: Backhuys Publishers.

Rajagopalan, R., & Tien, C. (1976). Trajectory analysis of deep-bed filtration with the sphere-in-cell porous media model. *AIChE Journal, 22*, 523–533.

Reddi, L. N., & Bonala, M. V. S. (1997). Critical shear stress and its relationship with cohesion for sand-kaolinite mixtures. *Canadian Geotechnical Journal, 34*, 26–33.

Rege, S. D., & Fogler, H. S. (1988). A network model for deep bed filtration of solid particles and emulsion drops. *AIChE Journal, 34*, 1761–1772.

Rittman, B. E. (1982). The effect of shear stress on biofilm loss rate. *Biotechnology and Bioengineering, 24*, 501–506.

Rousseau, D. P. L. (2005). *Performance of constructed treatment wetlands: model based evaluation and impact of operation and maintenance*. Dissertation, Ghent University, Belgium.

Sakthivadivel, R., & Einstein, H. A. (1970). Clogging of porous column of spheres by sediment. *Journal of the Hydraulics Division, American Society of Civil Engineers, 96*, 461–472.

Stevenson, D. G. (1997). Flow and filtration through granular media - the effect of grain and particle size dispersion. *Water Research, 31*, 310–322.

Tanner, C. C., Sukias, J. P. S., & Upsdell, M. P. (1998). Organic matter accumulation during maturation of gravel-bed constructed wetlands treating farm dairy wastewaters. *Water Research, 32*, 3046–3054.

Thullner, M. (2010). Comparison of bioclogging effects in saturated porous media within one- and two-dimensional flow systems. *Ecological Engineering, 36*, 176–196.

Tufenkji, N. (2007). Modeling microbial transport in porous media: Traditional approaches and recent developments. *Advances in Water Resources, 30*, 1455–1469.

Wallace, S. D., & Knight, R. L. (2006). *Small-scale constructed wetland treatment systems: Feasibility, design criteria and O & M requirements*. Virginia, Alexandria: Water Environment Research Foundation (WERF).

Wooster, J. K., Dusterhoff, S. R., Cui, Y., Sklar, L. S., Dietrich, W. E., & Malko, M. (2008). Sediment supply and relative size distribution effects on fine sediment infiltration into immobile gravels. *Water Resources Research, 44*, WO3424.

Zamani, A., & Maini, B. (2009). Flow of dispersed particles through porous media – Deep bed filtration. *Journal of Petroleum Science and Engineering, 69*, 71–88.

Zhao, Y. Q., Sun, G., & Allen, S. J. (2004). Anti-sized reed bed system for animal wastewater treatment: A comparative study. *Water Research, 38*, 2907–2917.

Chapter 9
The Evolution of Horizontal Subsurface Flow Reed Bed Design for Tertiary Treatment of Sewage Effluents in the UK

Clodagh Murphy and David Cooper

Abstract The Horizontal Subsurface Flow (HSSF) reed bed is the most widely used concept of constructed wetland. (Vymazal, 2005). Typically rectangular in design, wastewater is fed at the inlet and passes through the media below the surface in a horizontal path before discharging at the outlet. This basic classification of a HSSF wetland is based on the hydraulic flow pattern through the system. Designed to achieve maximum performance from the system, the optimal design characteristics for HSSF have evolved separately in each country, due to the unique set of conditions (i.e. extremes of temperature, high rainfall, etc), in order to maximise their removal efficiencies and keep the area to a minimum. (Akratos & Tsihrintzis, 2007). Climatic conditions are the main reason for HSSF bed design differences in different parts of the world, with each country having its own set of design guidelines. The Constructed Wetland Association Database lists 1025 reed bed sites in the UK, of which 893 are HSSF. HSSF reed beds have been used for tertiary sewage treatment in the UK since the 1980's. Over this time, the HSSF concept has been widely studied and the design criteria adapted in light of practical experience gained from the operation of numerous systems in order to achieve optimum treatment efficiency. Lessons have been learned from failing systems leading to the "designing out" of problem areas. As well as achieving maximum performance and reducing the footprint, the most important factor in designing constructed wetlands today is to prolong the life of beds by minimising clogging of systems.

Keywords Clogging · Constructed wetlands · Design criteria · Design evolution · Horizontal subsurface flow

---

C. Murphy (✉)
ARM Ltd, Rydal House, Colton Road, Rugeley, Staffordshire, WS15 3HF, UK
e-mail: clodagh.murphy@armreedbeds.co.uk

J. Vymazal (ed.), *Water and Nutrient Management in Natural and Constructed Wetlands*, DOI 10.1007/978-90-481-9585-5_9,

9.1 Introduction

9.1.1 The Origin of Constructed Wetlands

Natural wetlands have been utilised by civilisations to provide treatment of wastewater for centuries, however Vymazal (1998) suggests this is more to do with the wetlands being a convenient outlet for disposal rather than a deliberate means of providing treatment. Kadlec and Wallace (2009) suggest that the first recorded wetland to be deliberately engineered to provide wastewater treatment occurred in 1901. The US patent belonged to Cleophas Monjeau who essentially devised a vertical flow system similar in design to a vegetated trickling filter. An essay written by Nemo to the head of the Hornsby Literacy Institute in 1904 provides additional documentation of the early use of deliberately engineered wetlands for domestic wastewater treatment in New South Wales, Australia (Brix, 1994; Kadlec & Wallace, 2009).

The origins of constructed subsurface wetland technology began in Germany in the 1950s at the Max Planck Institute with the investigations of water purification using *Schoenoplectus lacustris* by Käthe Seidel, who had observed that this species had the ability to remove organic and inorganic substances from contaminated water. Further work in the 1960s by Seidel, developed a series of vertical flow and horizontal flow gravel filled basins in several cascading stages planted with emergent macropytes known as the Max Planck Institute Process (MPIP) or Krefeld System (Brix, 1994; Nuttall, Boon, & Rowell, 1997). Collaborations between Seidel and Reinhold Kickuth at the University of Göttingen resulted in Kickuth developing the concept of the Root Zone Method (RZM); a soil-based, horizontal flow single stage process. The RZM is based on the theory of the root formation of the macrophytes increasing the hydraulic conductivity of the soil matrix, and the capacity of the macrophytes to transfer atmospheric oxygen to the rhizosphere. These studies lead to the first operational system being installed in Othfresen, Germany in 1974 (Cooper & Green, 1995; Decamp & Warren, 2000; Vymazal, 2005).

Interest in the United Kingdom began in 1985 following a visit by the water industry to systems in Germany and Denmark. By late 1985, the Water Services Association formed the Reed Bed Treatment System Co-ordinating Group setting up a 5 year R&D program to set guidelines for the design of reed bed systems, with the Water Research Centre (WRc) acting as the coordinator. (Cooper & Green, 1995). The group exchanged experiences with Germany and Denmark with lead to the formation of the European Community Expert Contact Group on Emergent Hydrophye Treatment Systems (EHTS) which culminated in the organisation of the first International Conference on Constructed Wetlands in Cambridge, and the production of the European Design and Operation Guidelines for Reed Bed Treatment Systems (EC/EWPCA) in 1990. (Cooper & Green, 1995; IWA, 2000; Vymazal, 2005; Kadlec & Wallace, 2009).

By the late 1980s, there was global interest in HSSF constructed wetlands with systems being introduced to Germany, Denmark, Austria, France, Sweden, Switzerland, the Netherlands, the UK, North America and Africa. By 1990s this has been extended to China, India and Australia and other European countries including the Czech Republic, Poland, Norway and Slovenia. (Brix, 1994; IWA, 2000).

9.2 Constructed Wetland Design

Constructed Wetlands are generally classified into categories depending on how the water passes through the system, surface (also known as free water surface) and subsurface. These categories can be further subdivided into horizontal flow and vertical flow paths. The Horizontal Subsurface Flow (HSSF) reed bed is the most widely used concept of constructed wetland for the treatment of secondary and tertiary effluents (Vymazal, 2005, 2009). The primary benefit of subsurface flow systems is that the water is not exposed during the treatment process, minimising energy losses through evaporation and convection, and making these systems more suitable to cold climate applications (Wallace, Parkin, & Cross, 2001.)

Constructed Wetlands have several advantages over natural wetlands. These engineered systems utilise processes that take place in natural systems, but in a more controlled environment. They offer a significant potential for performance optimisation through the selection of system configuration which can all be can be built with a much greater control including site selection, sizing, flexibility of control over the hydraulic pathways and retention time (Vymazal, 2005; WPCF, 1990). Kadlec and Wallace (2009) suggest that the designing of constructed wetlands can be roughly divided into two categories; sizing calculations and physical specifications.

9.2.1 Sizing Calculations

The first step in designing a constructed wetland usually involves determining the size (surface area) of the system. Sun and Zhang (2008) state that the lack of understanding for pollution dynamics often leads to inaccurate design, which is typically reflected in the undersizing or over-sizing of constructed wetland surface areas, despite there being over 20 years experience of intensive research into the understanding of the hydraulics of wastewater and the degradation kinetics of individual pollutants such as first-order and Monod kinetics (Sun & Zhang, 2008).

Rousseau, Vanrolleghem, and De Pauw (2004) reviews the various applications of model-based designs for HSSF. The "Rule of Thumb" Model is the most conservative model when dealing with parameter uncertainty. These HSSF wetland designs have an added "safety factor" due to parameter uncertainty resulting in larger bed sizes requiring increased investment costs. The Black Box concept uses regression equations to focus on input and output data rather than focusing on the internal processes data. The Kickuth equation in wetland design is seen as a Black Box concept. A combination of first-order kinetics and the plug flow model, it is used to size wetlands based on organic matter removal and is considered one of the most user friendly design equations for wetland practitioners. The model, determined in the 1990s, assumes ideal plug flow and constant rate, based on regression analysis of inlet and outlet BOD values taken from existing wetlands in the UK and Denmark. The accuracy of the Kickuth equation raises some concerns as this model oversimplifies such complex systems and neglects important factors such as climate, bed dimensions and matrix material, leading to uncertainty in the design. (Rousseau et al., 2004; Sun & Zhang, 2008). Sun, Cooper, & Zhaoxiang (2008) investigated data from 78 HSSF wetlands from the CWA database to determine the K_1 value, deriving a figure of 0.0829 m d^{-1}, comparable to values previously reported for tertiary systems (Cooper et al., 1996; Cooper, 1999). If the Kickuth equation is used with this derived K_1 value to determine the wetland surface area, the probability of meeting the design target is 78% based on the performances of the 78 beds (Sun et al., 2008). Sun goes on to suggest that the high success rate of the Kickuth equation is not indicative of its accuracy, although it provides a reliable, if conservative design, with inflated capital costs as the surface area of the wetland is likely to be overestimated.

Rousseau et al. (2004) concludes that the state-of-the-art First-Order Model (k-C*) is the best available model to use, however this model also assumes constant conditions and ideal plug flow behaviour. In reality short circuiting and dead zones are a frequent phenomenon in constructed wetlands causing non ideal plug flow and the required constant conditions such as the influent concentrations, water depth, maturity of the bed and hydraulic loading rate are rarely constant in an operational system. Removal rates continue to increase with increasing loading rates, something that first-order models cannot take into account (Rousseau et al., 2004). Monod-type design models represent first-order rate reactions for relatively low concentrations, but zero-order rate reactions for higher concentrations of contaminants. Monod-type models better describe the kinetics of organic matter removal in a HSSF system with the variability of the data than a first-order model, with however, still with the assumption of plug flow (Rousseau et al., 2004; Sun & Zhang, 2008).

In practice, the design of HSSF wetlands are carried out using the Black Box concept, and 'Rule of Thumb' (Kadlec & Knight, 1996; Kadlec, 2003; Rousseau et al., 2004; Sun & Zhang, 2008; Sun et al., 2008). Important design aspects when dealing with parameter uncertainties such as organic and hydraulic loading rates, aspect ratio, media size and water depth are defined mostly from previous experience from a wide range of systems and are at best, from an engineering point of view, the fastest but most crudely designed method (García et al., 2005; Rousseau et al., 2004).

Once the surface area of the wetland is determined, there are further detailed engineering design requirements such as flow calculations, which are determined using Darcy's Law for fluid flow through a porous media, which is intended to provide flow through the gravel under average conditions (Green & Upton, 1995). This formula determines the cross sectional area of the bed and assumes that the hydraulic conductivity will stabilise at 10^{-3} ms^{-1} for a gravel media, however the European Guidelines (EC/EWPCA, 1990) advise not to assume a hydraulic conductivity greater than that of the original media. The base of the bed and the surface of the gravel would have had a 1% slope from inlet to outlet to provide a hydraulic gradient. (EC/EWPCA, 1990; Cooper et al., 1996).

9.2.2 Physical Specifications

Designed to achieve maximum performance from the system, the optimal design characteristics for HSSF have evolved separately in each country due to the unique set of conditions (i.e. extremes of temperature, high rainfall, etc), in order to maximise their removal efficiencies and keep the area to a minimum (Akratos & Tsihrintzis, 2007). Climatic conditions are the main reason for design differences in HSSF beds throughout the world. Countries with extreme temperatures such as the USA are required to factor in weather conditions to prevent freezing of the system which results in reduced performance. This factor alone has led to the fundamental difference between the US and UK designs of HSSF systems. In the UK, the distribution system is on the surface, whereas, in the US, the distribution is below the surface. The variability of size, aspect ratio, loading rates and regimes, plant species, distribution system, composition of media, and composition of the wastewater have all had an influence on the design of HSSF beds (Brix, 1994), which has lead to each country having their own design guidelines.

Hydraulic conductivity is one of the most important determinants in pollutant removal efficiency (Ellis et al., 2003). Primary mechanisms for bed-clogging predominate in the inlet region and HSSF systems will not maintain "clean bed" hydraulic conductivity once the system is in operation (Kadlec & Wallace, 2009). Bed-width is determined by the hydraulic conductivity and bed length by pollutant removal requirements, therefore a submerged bed may have an aspect ratio (width: length) of less or greater than 1:1 depending on performance treatment goals (WPCF, 1990). Originally, to overcome problems of overflowing and channelling, the Danish systems were designed with wide beds with short lengths. This design aspect has continued in many countries to minimise organic loading across the inlet. There has been much debate on the relevance of aspect ratio in the design of wetlands. The WPCF Manual of Practice FD-16 report suggests that *"most removal takes place in the vicinity of the discharge area with decreasing total assimilation per incremental distance at increasing distances from the discharge point"*. Geller (1997) backs this up by stating that most organic substances decompose within 1–2 m of the inlet zone. Experiments by García et al. (2005) confirmed that organic matter was removed by physical mechanisms mainly near the inlet of HSSF and that these mechanisms are similar regardless of the aspect ratio and water depth. Differences in performances are therefore related to the biochemical processes operating along the length of the bed (García et al., 2005). Recommendations in the various guidelines for width to length ratios vary. The Environment Agency Guidance Manual (Ellis et al., 2003) suggests an aspect ratio of 4:1 for HSSF systems, whilst the European design guidelines (EC/EWPCA, 1990) suggest one of 2:1. However the IWA (2000) advises that any aspect ratio with a good inlet distribution can be applied, based on the assertion that the expected outcome of high aspect ratios to favour a more effective (close to plug) flow have been proven to be untrue in tracer tests.

It is imperative to get an even distribution across the bed and therefore the correct design of the inlet structure is paramount. There is some trade off made between the capacity to spread the effluent evenly across the inlet of the bed (critical for the establishment of hydraulic efficiency) and the ability to resist clogging at the inlet (Davison et al., 2005). The various guidelines suggest a variety of permutations of inlet structures often providing conflicting information, as the preferred option is often selected based on experience of the performance of particular distribution structures. The inlet structures consist of either a level pipe or trough to provide a uniform gravity fed distribution, with a series of spreader devices (Fig. 9.1) such as a castellated or V notch weir plates, slotted pipes, adjustable tees or riser pipes (EC/EWPCA, 1990; Cooper et al., 1996; Ellis et al., 2003). Inlet structures are generally positioned on the surface in the UK, however in countries with extremes of temperature such as the USA, the inlet structures are below ground to prevent freezing. (EC/EWPCA, 1990; Wallace & Knight, 2006). Subsurface inlet structures however are difficult to adjust and to maintain. Surface structures also avoid problems associated with back pressure. Recent studies into hydraulic conductivity and clogging mechanisms (Knowles, Griffin, & Davies, 2010) suggest that surface and subsurface manifolds dictate the clogging pattern within the wetland and as such, have an impact on the internal clogging mechanisms (Personal correspondence with Paul Knowles and Scott Wallace).

Fig. 9.1 Various inlet distribution structures available in the UK. (**a**) Concrete or plastic trough systems with castellated (**b**) or V-notch (**c**) weirs, (**d**) fixed position riser pipe and (**e**) pipe system with adjustable tees

Keeping the inlet structure level to achieve an even distribution over a long length of weir creates some problems, especially where inevitable ground movement and settlement occurs. Compartmentalisation of constructed wetlands into several cells arranged either in series or parallel, will allow better control over gaining a uniform distribution (WPCF, 1990). Compartmentalisation has the additional benefit of taking cells "off line" to allow for maintenance and any refurbishment requirements without taking the wetland system off line completely.

Media choice for the majority of European and North American systems is gravel, with the Water Research centre recommending gravel media for sewage treatment since 1986. EC/EWPCA (1990) report recommended 3–6 mm and 5–10 mm gravel size whilst European gravels sizes vary from 3–16 (IWA, 2000; Knowles et al., 2010). Similarly to RZM, it was postulated that as the gravel bed gets blocked with solids, the roots and rhizomes would keep the beds open and maintain hydraulic conductivity however the European Guidelines advise not to assume hydraulic conductivity greater than that of the original media (EC/EWPCA, 1990; Cooper et al., 1996).

As hydraulic conductivity decreases over time, there is a propensity for systems to flood, reducing the performance of the system. The outlet structure has an important role in controlling the level of the water in the bed, typically keeping the water level about 50 mm below the surface of the gravel. Various structures exist including weirs or a pipe with a 90° swivelling elbow (EC/EWPCA, 1990; WPCF, 1990). By lowering the level control device, the hydraulic gradient needed to maintain subsurface flow can be adjusted (Griffin & Pamplin, 1998).

Over the past 50 years, multiple variations of HSSF wetland design features have been tested and gradually developed. As a result, there are a number of referenced design documents that contain outdated concepts that should no longer be used for design purposes, as listed in the 2nd edition of Treatment Wetlands. (Cooper & Green, 1995; Kadlec & Wallace, 2009). The CIRIA Report (Nuttall et al., 1997) states that systems fail due to over-optimistic design, use of unsuitable configurations, inappropriate media, invalid assumptions about their capabilities, built on the basis of inadequate design guidance or poor information regarding wastewater characteristics. Pries (1994) suggests that it is the failed systems that provide wetland designers and researchers with valuable data, experience and knowledge.

9.3 Global Adaptations to Design

The HSSF concept has been widely studied and the design criteria adapted in light of practical experience in order to achieve optimum treatment efficiency. The first major adaptation of horizontal flow design occurred in 1983, when the German design based on the RZM was introduced to Denmark. Some of the basic design characteristics of the RZM systems have been proven effective, however some have been shown to be inadequate or wrong when used in full scale systems (Cooper & Green, 1995). The RZM claim to increase hydraulic conductivity over time was not realised. The hydraulic conductivity remained stable, but often decreased resulting in systems clogging and were frequently observed to be overflowing causing channelling and scouring of the surface, and bypassing treatment areas resulting in reduced performance (IWA, 2000). The Danish reengineered the German design to get over the problem of channelling and overflowing. The Danish systems were designed to be very wide with short flow paths. Flow distribution proved to be an issue over such wide beds resulting in beds being separated into a series of cells, each being subject to separate hydraulic loading to get better control over the distribution of water (Brix, 1998; Kadlec & Wallace, 2009).

Following a visit to Danish and German systems in 1985, the UK adopted HSSF systems with one further change in design, the introduction of coarser filtration to ensure subsurface flow. From 1986 onwards, UK water industry treatment wetlands were generally constructed using gravel medium, similarly in North American systems (Brix, 1994; Kadlec & Wallace, 2009).

Freezing systems result in hydraulic failure, and as a result, North American adaptations to constructed wetlands in sub-freezing conditions introduced an insulation layer (Wallace et al., 2001). To be effective, the insulation layer must be uniform in coverage and therefore becomes an integral part of the wetland design. Additionally sub-freezing conditions have led to sub-surface inlet distribution being the structure of choice in North America, whereas surface inlet structures are generally prevalent in temperate regions.

Early beds were constructed using gabion walls, some with geotextile over the inner face of the wall at the outlet zone. This was included in the design to prevent gravel penetration into the outlet collection zone which contained large stones. This has proven ineffective as the geotextile often accumulated a biofilm growing on the surface, impeding the flow. The solution was to excavate the outlet zone, remove the geotextile and gabions, replace the collection pipe and relay large stone at a natural angle of repose. (Cooper et al., 2005).

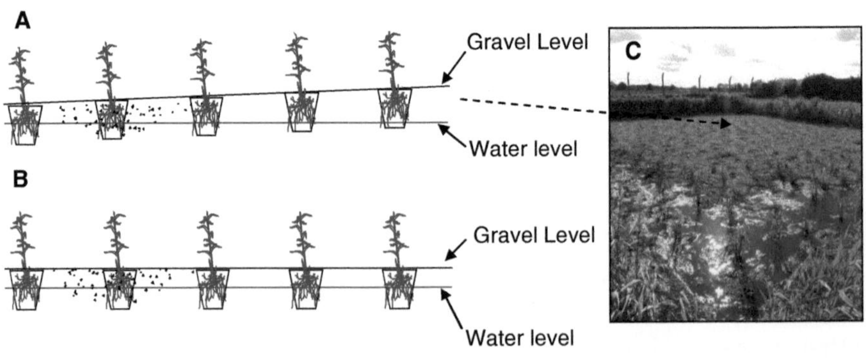

Fig. 9.2 Problems associated with sloping gravel in a newly planted system, as demonstrated in (**a**) and (**c**) where the plants at the top of the slope have very poor growth and high mortality. The water level is above the surface of the gravel at the bottom of the slope, and plant growth is good. Having a flat gravel surface (**b**) allows for uniform growth of reeds, reducing the plant mortality rate

Sizing the cross sectional area of reed beds using Darcy's Law resulted in the base of the bed and the gravel having a 1% slope from inlet to outlet in order to provide a hydraulic gradient. This however resulted in problems with plant growth. Water will find its level, consequently, reeds planted at the top of the slope will not have their roots in water resulting in poor growth and a high plant mortality rate in newly planted systems (Fig. 9.2).

9.4 Evolution of Design: A UK Perspective

The UK has been interested in wetland systems since 1985 following a visit by the water industry to wetland systems in Germany and Denmark. Twenty systems had been installed in the UK by 1986 and by 2008, the UK had over 1000 systems (CWA, 2008; Kadlec & Wallace, 2009). Brix (1994) suggests that in Europe, the systems tend to be used for providing secondary treatment for village sized communities up to the population equivalent (pe) of 1000, where as in North America, they tend to be used for tertiary treatment for much larger populations. UK water authorities wanted low cost, low maintenance systems for small villages of 50–1000 pe. Wetlands are capable of treating up to 10,000 pe but due to land constraints in the UK, they were generally limited to 1000 pe (IWA, 2000; Nuttall et al., 1997). The initial constructed wetlands in the UK were for secondary treatment, however, this rapidly shifted towards tertiary treatment for up to 2000 pe largely due to studies by a water company; Severn Trent Water. Green & Upton (1995) and Griffin & Pamplin (1998) suggest that small works (under 2000 pe) were at greatest risk of compliance failure under the EU Urban Wastewater Treatment Directive. At the time that these papers were written, 70% of Severn Trent's treatment works fell into this 'small' category, although, collectively, they served less than 3% of the population. With this in mind, Severn Trent policy established reeds beds as the preferred option for tertiary treatment at works serving populations of less than 2000 (Green & Upton, 1995). As a consequence, Severn Trent is the largest user of constructed wetlands for sewage treatment in the UK. They have been utilising Reed Bed Treatment Systems (RBTS) since 1985, and currently have reed beds on 419 sites, 334 of which are for tertiary treatment using HSSF.

ARM Ltd has over 20 years experience in designing and constructing reed beds and has a long standing relationship with Severn Trent. Data from the ARM maintenance database combined with the Constructed Wetland Association database (CWA, 2008), suggests that there are 699 sites for tertiary treatment using horizontal reed beds in the UK ranging in size from 8 m^2 to 44000 m^2. The small number of large systems (>10000 m^2) skews the average size of wetland systems in the UK to 694 m^2 (Table 9.1). However, largely influenced by the number of Severn Trent reed beds, the majority of systems are less than 1000 m^2, with an average size of 318 m^2, and the median size for all HSSF beds in the database is 313 m^2. Analysis of the CWA and ARM Maintenance databases suggest that the aspect ratio when plotted against the size of the bed (up to 1000 m^2) gives a scattered pattern (Fig. 9.3)

Table 9.1 Average size and aspect ratio for HSSF constructed wetlands in the UK for tertiary treatment (CWA and ARM Maintenance databases)

Size (m^2)	Number of sites*	Average Size (m^2)	Average W:L ratio
0–999	531	318 (8–996)	1.2 : 1 (0.1 : 1–4.3 : 1)
1000–1999	71	1,317 (1,000–1,930)	2.5 : 1 (0.2 : 1–4.8 : 1)
2000–10,000	23	3,157 (2,000–6,300)	2.3 : 1 (0.3 : 1–5.6 : 1)
>10,000	5	20,028 (10,140–44,000)	4.9 : 1 (2 : 1–10.2 : 1)

*69 sites with no size data available, figures in brackets give the maximum and minimum values.

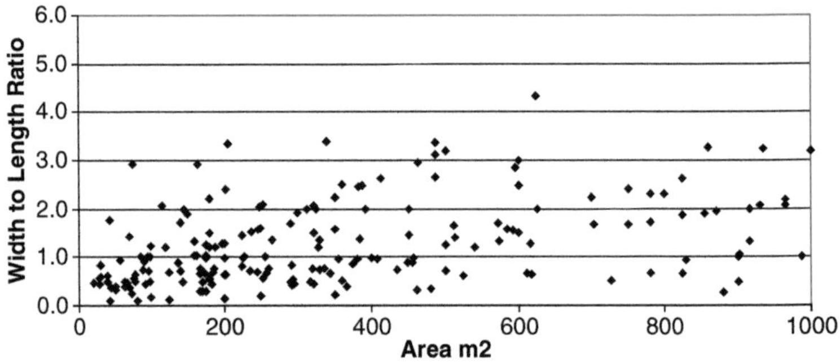

Fig. 9.3 Width to length ratios for HSSF tertiary treatment constructed wetlands in the UK up to 1000 m^2

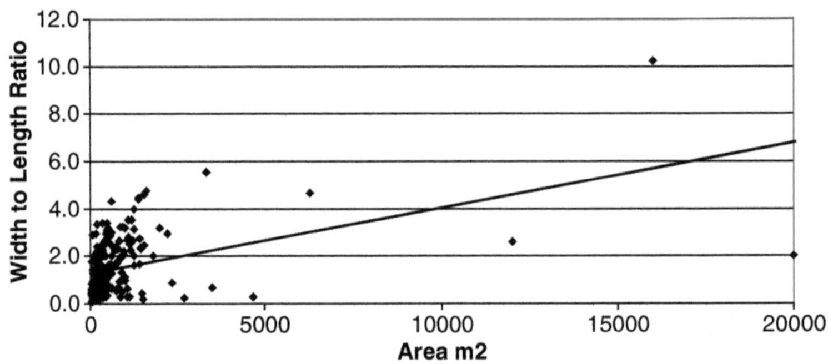

Fig. 9.4 Width to length ratios for all HSSF tertiary treatment constructed wetlands in the UK

indicating that there is no correlation between the size of the bed and the aspect ratio. There are some systems where the width to length (W:L) ratio is less than 1:1 (i.e. the length is longer than the width), but generally the W:L ratio is equal or greater than 1:1 for smaller systems. When plotting all the systems (Fig. 9.4), the data indicates that the W:L ratio increases with the size of the bed. However, this is possibly skewed by the few large (>10000 m^2) systems which have a W:L ratio of between 2:1 and 10:1.

Reed beds have a relatively large footprint and this is ultimately the deciding factor for UK treatment works where the available space is limited. Designing reed beds to specifically fit into the available space may compromise the treatment efficiency. Achieving optimum performance whilst keeping the footprint as low as possible is the key motivation behind the evolution of the HSSF concept in the UK.

UK systems are designed for stable period operation, with the input and output averaged (IWA, 2000). When designing primarily for the removal or organic matter, the surface area is usually calculated by the Kickuth Equation which is a combination of first order kinetics and plug-flow model (Sun et al., 2008). For tertiary HSSF systems, this equation has been used to derive a value of 1 m^2 pe^{-1}. Over time, a value of 0.5–0.7 m^2 pe^{-1} has been used following a re-evaluation of the rate constant in the design equation (Green & Upton, 1995; Cooper & Green, 1995; Cooper et al., 1996; Griffin & Pamplin, 1998). Griffin & Pamplin (1998) acknowledge that the beds are probably over sized, stating that this safety margin is valuable, especially in situations where the secondary effluent is poor. In 2008, Griffin, Wilson, and Cooper revised this statement stating that the reed beds are actually undersized and incapable of dealing with high flow rates associated with rainfall events. Lack of flow data has resulted in inadequate assessment of infiltration levels. A design assumption for infiltration of 30% is made for the daily domestic average flow rate, however there are many sites where the infiltration rate is greater than this leading to under-sizing of the reed bed.

Green & Upton (1995) describe the design of Severn Trent reed beds in more detail, but generally they follow the guidelines as suggested in the EC/EWPCA report from 1990. Many of the reed beds in the UK have been designed using this report as its basis. It is still relevant today, with many of the original design recommendations still being implemented in the basic design of reed bed treatment systems. The report suggests that the surface should be flat, and that the base slope should not exceed 1%. The average depth for these systems should be 0.6 m with a freeboard of 0.5 m above the surface of the gravel and the beds should be sealed using impermeable material such as a high or low density polyethylene liner, puddle clays (if the soil has a hydraulic conductivity of 10^{-8} m s^{-1}) or bentonite. A combination of geotextile and low density polyethylene (LDPE) liner is used today, largely due to cost. Inlet distribution recommendations suggest a simple pipe with tees which can be adjusted to produce even distributions, whilst the outlet pipe consists of a pipe with a 90° swivelling elbow to raise or lower the water level in the bed (although some utilise a weir plate system to control the water level in the system). Completed beds are still flooded for a period after planting to prevent weed ingress and rabbits grazing on new shoots.

However, there have been some recommendations made by the report that have not been followed. It is essential to get even distribution of flow at the inlet, the use of troughs with castellated and V-notch weirs was not recommended in the report as the weirs collect debris, and the troughs act as a sedimentation unit collecting

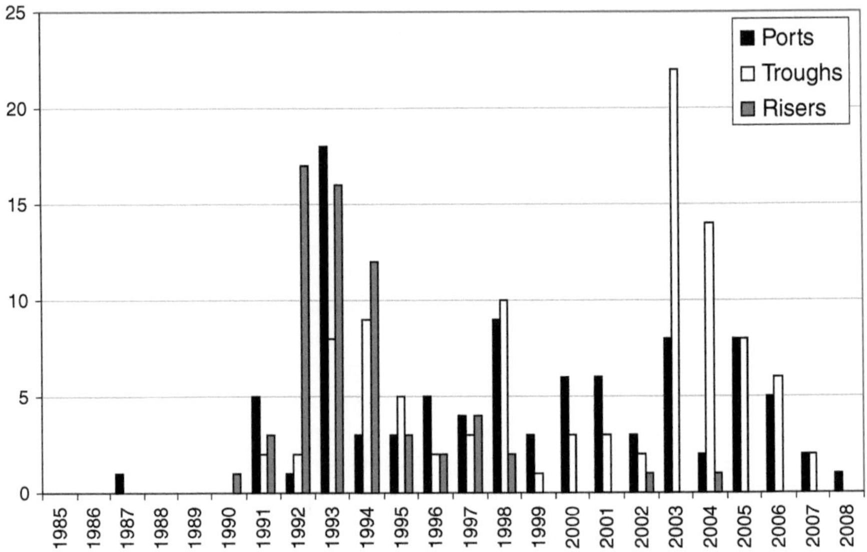

Fig. 9.5 Dispersal of troughs, risers and pipe (ports) inlet distribution structures installed on Severn Trent reed beds between 1985 to 2008 in the UK (data sourced from ARM Maintenance database)

sludge and grit despite this, Severn Trent have installed some trough systems with both castellated and V-notch weirs on their reed beds.

It is difficult to determine the predominant distribution structure in the UK as a whole as the inlet design features such as distribution structure is not listed on the CWA database. Data from the ARM maintenance database give an idea of the types of structures used on Severn Trent reed beds (Fig. 9.5). This graph indicates that risers were popular in the early years; however this has been surpassed by the use of troughs and pipe (port) structures. None of the inlet structures available are ideal, as they are all prone to blockage and collecting rags which result in uneven distribution of flow onto the bed (Griffin et al., 2008). Ease of maintenance for cleaning and removing blockages is an important factor in selecting an inlet structure. Backward-facing troughs were popular on Severn Trent sites, where the effluent is distributed away from the media, and flows under the trough into the bed. These systems have a propensity to clog, causing the effluent to flow away from the reed bed flooding paths. This can be easily remedied by turning the troughs 180° so that the distribution is directly onto the bed.

The CWA database and ARM's Maintenance database indicate that a significant proportion of reed beds are constructed with a gravel media. Initially the EC/EWPCA report (1990) recommended 3–6 mm and 5–10 mm gravel size. This has been increased to 10 mm single size, but in practice, this is more likely

10–14 mm. Griffin et al. (2008) recommends increasing the gravel size to 10–12 mm to increase the asset life of wetland systems.

A case study described by Griffin and Pamplin (1998) stated that the hydraulic conductivity of the first of Severn Trent's new generation of tertiary treatment reed beds at Leek Wooton, remained in the design range of 10^{-2} m s^{-1} with the gravel mix showing little sign of solids accumulation after 7 years of operation. This gave confidence that the asset life of reed bed treatment systems would reach 20 years without remedial work. This particular site was refurbished for the first time in 2006, not quite making the 20 year life expectancy being refurbished after 16 years. Griffin et al. (2008) acknowledges that although many of the Severn Trent reed beds are reaching the end of their design asset lives, there are many where this is not being achieved. The effect of this is to reduce the average design life of beds from 15 to 8 years, raising questions about the cost effectiveness of reed bed treatment solutions (Griffin et al., 2008). In an ARM internal study of bed ages at refurbishment for Severn Trent sites, the maximum bed age at refurbishment is 16.5 years, with the minimum at 2 years. The average bed age at refurbishment is 10 years, half that of the predicted 20 year expected life. However, there are approximately 11% of Severn Trent beds that after 15–18 years of operation (as of 2009) have not yet been refurbished.

Reed beds for sewage tre atment in the UK are generally planted with *Phragmites australis*. These would have been originally planted with rhizome segments, however plant growth developed very slowly, hampered by competition with weeds. Since 1987, reeds have been grown from seed and seedlings planted in a density of 4 plants per square meter (Cooper & Green, 1995).

Reed beds designed by ARM use LDPE or butyl liners. During the refurbishment of reed beds it has been noticed on some sites that the macrophyte rhizomes have punctured the liner along the sides of the systems, causing potential leaking of the system (Fig. 9.6). To prevent this, ARM now sandwich the liner in between two geotextile layers, to provide a double barrier against punctures from rhizomes.

Fig. 9.6 Roots and rhizomes of *Phragmites australis* puncturing LDPE liner at the base and side of the reed bed

9.5 Scaling up the Design

ARM have been designing reed beds for over 20 years and during this time, have been able to draw upon practical experience in the design stage, and adapting the design parameters of HSSF reed beds to apply effective tertiary treatment to larger works treating >10000 pe. ARM has had influence on the design and construction phases for the three largest reed beds in England treating tertiary sewage. The first built in 1996 for Southern Water at 20000 m^2, the second built in 2004 for Anglian Water at 12000 m^2 and the third built in 2008 for Thames Water at 16000 m^2. These three systems provide tertiary sewage treatment for between 24000 and 26370 pe. Table 9.2 presents the similarities between these three systems and how the design has been adapted over time from 1996 when the first system was installed to the last in 2008. Redgate Mill and Walton on the Naze are perhaps the most significant sites to demonstrate design evolution, as they are treating similar flows for similar population equivalents. The total surface area has reduced by 40% from Redgate Mill to Walton on the Naze, and in addition, the number of cells (compartments) has doubled from 4 to 8 to ensure better hydraulic efficiency. Due to the flow requirements the Berkhamsted system has to treat, the aspect ratio for this site gives an anomaly when compared to the normal range. This system has been designed to cope with 11500 $m^3\ d^{-1}$ due to high infiltration rates to the treatment works. In order to prevent scouring and channelling, the system has been compartmentalised into 10 cells, with an aspect ratio of over 10:1. The width of the reed bed (where distribution takes

Table 9.2 Data from 3 systems providing tertiary sewage treatment for >24000 pe in the UK

	Redgate Mill	Walton on the Naze	Berkhamsted
Commissioned	1996	2004	2008
Number of cells	4	8	10
Bed area (m^2)	5,000	1,500	1,600
Total area (m^2)	20,000	12,000	16,000
Average flow ($m^3\ d^{-1}$)	5,800	6,000	11,500
pe	25,000	26,370	24,000
Pretreatment	Percolating filter/humus tank	Percolating nitrifying filters/humus tank	Percolating filters, SAF
Inlet distribution	Risers	Pipes with tees	Pipes with tees
Macropyhtes	*Phragmites australis*	*Phragmites australis*	*Phragmites australis*
Design requirements	Solids and BOD	Solids and pathogens	Solids and BOD
Consent BOD	10 mg L^{-1}	15 mg L^{-1}	10 mg L^{-1}
Consent SS	20 mg L^{-1}	30 mg L^{-1}	15 mg L^{-1}
Granular medium	5–20 mm	10–14 mm	10–14 mm
Inlet gravel	60–100 mm	50–100 mm	50–100 mm
Length (m)	50	25	12.5
Width (m)	100	62.5	128
Width:length ratio	2:1	2.5:1	10.1
Average depth (m)	0.6	0.6	0.55
Slope (%)	1.0	1.0	1.0

place) has increased. Solids drop out in first few meters of the reed bed, resulting in clogging and compromising distribution, by spreading the distribution over a wider area, this should help minimise the extent of clogging and extend the longevity of this reed bed. This evolution in reed bed design has effectively altered the dimensions of reed beds without compromising treatment efficiency. ARM are continuing to adapt reed bed designs to cope with the challenges of reed bed clogging and its implication on longevity and are currently testing a number of hybrid designs.

The 1990 EC/EWPCA report states that the guidelines are not aimed to recommend the best way to design the best working system as at that time, knowledge did not allow for this. Most of the ideas for new designs for constructed wetlands in order to improve performance had at that point not been in operation at full scale for long enough to provide accurate performance for these new systems. Nearly 30 years later, the stance on this should have changed. A large number of systems have been built and been fully operational for several years, however Sun et al. (2008) suggests that there is still little understanding about the mechanisms and kinetics of pollutant removal in these systems, leading to inaccurate design. He suggests two options to improve the design criteria. Firstly to acquire further information about the complex processes occurring in constructed wetlands to produce a more accurate design equation and secondly to statistically analyse performances of existing systems to allow designers to become more aware of the reliability of the current design criteria.

There are many full scale systems in operation globally that are not being monitored for performance. There is opportunity to learn from failing systems in order to "design out" problem areas, with possibly the most important factor in designing today's constructed wetlands being to prolong asset life by minimising the clogging of systems. Most HSSF designs are inadequate for nutrient removal due to oxygen limitations in the system. The European Water Framework Directive has marked out Nutrient Vulnerable Zones where discharge consents for nitrate will be tighter. Perhaps the time has come for a new design revolution and to move away from HSSF systems, and consider more widespread use of hybrid, vertical flow or engineered systems such as forced bed aeration to achieve ever tightening consent limits.

References

Akratos, C. S., & Tsihrintzis, V. A. (2007). Effect of temperature, HRT, vegetation and porous media on removal efficiency of pilot-scale horizontal subsurface flow constructed wetlands. *Ecological Engineering, 29*, 173–191.

Brix, H. (1994). Use of constructed wetlands in water pollution control: Historical development, present status, and future perspectives. *Water Science and Technology, 30*(8), 209–223.

Brix, H. (1997). Do macrophytes play a role in constructed treatment wetlands? *Water Science and Technology, 30*(5), 11–18.

Brix, H. (1998). Denmark. In J. Vymazal, H. Brix, P. F. Cooper, M. B. Green & R. Haberl (Eds.), *Constructed wetlands for wastewater treatment in Europe* (pp. 123–152). Leiden, Netherlands: Backhuys Publishers.

Cooper, P. (1999) A review of design and performance of vertical flow and hybrid reed bed treatment systems. *Water Science and Technology, 40*(3), 1–9.

Cooper, P., & Green, B. (1995). Reed bed treatment systems for sewage treatment in the United Kingdom – The first 10 years experience. *Water Science and Technology*, *32*(3), 317–327

Cooper.D., Griffin, P., & Cooper, P., (2005). Factors effecting the longevity of sub-surface horizontal flow systems operating as tertiary treatment for sewage effluent. *Water Science and Technology*, *51*(9), 127–135.

Cooper, P. F., Job, G. D., Green, M. B., & Shutes, R. B. E. (1996). *Reed beds and constructed wetlands for wastewater treatment.* Swindon: WRc Publications.

CWA (2008) *Constructed wetland association database v11.01*, March 2008. Rugeley, Staffordshire, UK.

Davison, L., Headley, T., & Pratt, K. (2005). Aspects of design, structure, performance and operation of reed beds – eight years' experience in northeastern New South Wales, Australia. *Water Science and Technology*, *51*(10), 129–138.

Decamp, O., & Warren, A. (2000). Investigation of *Escherichia coli* removal in various designs of subsurface flow wetlands used for wastewater treatment. *Ecological Engineering, 14*, 293–299.

EC/EWPCA Emergent Hydrophyte Treatment Systems Expert Contact Group, Water Research Centre (1990) *European design and operations guidelines for reed bed treatment systems*, Swindon: WRc.

Ellis, J. B., Shutes, R. B. E., & Revitt, D. M. (2003). *Guidance manual for constructed wetlands*. R&D Technical Report P2-159/TR2. Environment Agency for England and Wales.

García , J., Aguirre, P., Barragán, J., Mujeriego, R., Matamoros, V., & Bayona, J. M. (2005). Effect of key design parameters on the efficiency of horizontal flow constructed wetlands. *Ecological Engineering, 25*, 405–418.

Geller, G. (1997) Horizontal subsurface flow systems in the German speaking countries: Summary of long-term scientific and practical experiences: Recommendations. *Water Science and Technology, 35*(5), 157–166.

Green, M. B., & Upton, J. (1995). Constructed reed beds: Appropriate Technology for Small Communities. *Water Science and Technology, 32*(3), 339–348.

Griffin, P., & Pamplin, C. (1998). The advantages of a constructed reed bed based strategy for small sewage treatment works. *Water Science and Technology, 38*(3), 143–150.

Griffin, P., Wilson, L., & Cooper, D. (2008). Changes in the use, operation and design of subsurface flow constructed wetlands in a major UK water utility. *Proceedings of the 11th International Conference on Wetland Systems for Water Pollution Control* (pp. 419–426). Indore, India: Vikram University.

IWA Specialist Group on Use of Macrophytes in Water Pollution Control (2000). *Constructed wetlands for pollution control: Processes, performance, design and operation.* London: IWA Publishing.

Kadlec, R. H., & Knight, R. L. (1996) *Treatment wetlands* (1st ed.). Boca Raton, FL: CRC.

Kadlec, R. H. (2003) Effects of pollutant speciation in treatment wetlands design. *Ecological Engineering, 20*, 1–16.

Kadlec, R. H., & Wallace, S. D. (2009) *Treatment wetlands* (2nd ed.). Boca Raton, FL: CRC.

Knowles, P. R., Griffin, P., & Davies, P. A. (2010). Complementary methods to investigate the development of clogging within a horizontal subsurface flow tertiary treatment wetland. *Water Research*, *44*(1), 320–330.

Nuttall, P. M., Boon, A. G., & Rowell. M. R. (1997). Review of the design and management of constructed wetlands. *CIRIA Report 180.* London: Construction Industry Research and Information Association.

Pries, J. H. (1994) *Wastewater and stormwater applications in Canada.* Issues Paper No. 1994-1. Ottawa, Canada: North American Wetlands Conservation Council.

Rousseau, D. P. L., Vanrolleghem, P. A., & De Pauw, N. (2004). Model-based design of horizontal subsurface flow constructed treatment wetlands: A review. *Water Research, 38*, 1484–1493.

Sun, G., Cooper, D. J., & Zhaoxiang, Y. (2008). A statistical analysis on the removal or organic matter in constructed wetlands in the UK. *Environmental Technology, 29*, 1139–1144.

Sun. G., & Zhang G. (2008). The design of treatment wetlands in the UK: successes, failures, and alternative approaches. *Wetland Science, 6*, 343–350.

Vymazal. J. (1998). Czech Republic. In J. Vymazal, H. Brix, P. F. Cooper, M. B. Green & R. Haberl (Eds.), *Constructed wetlands for wastewater treatment in Europe* (pp. 95–121). Leiden, Netherlands: Backhuys Publishers.

Vymazal, J. (2005). Horizontal sub surface flow and hybrid constructed wetland systems for wastewater treatment. *Ecological Engineering, 25*, 478–490.

Vymazal, J. (2009). The use of constructed wetlands with horizontal sub-surface flow for various types of wastewater. *Ecological Engineering, 35*, 1–17.

Wallace S., Parkin, G., & Cross, C. (2001). Cold climate wetlands: Design and performance. *Water Science and Technology, 44*(11–12), 259–265.

Wallace, S. D., & Knight, R. L. (2006). *Small scale constructed wetland treatment systems: Feasibility, design criteria and O&M requirements*. Final Report, Project 01-CTS-5. Alexandria, VA: Water Environment Research Foundation (WERF).

WPCF (1990). *Natural systems for wastewater treatment (Manual of practice FD-16)* (1st ed.). Alexandria, VA: Water Pollution Control Federation.

Chapter 10
Treatment of Landfill Leachate in Aerated Subsurface Flow Wetlands: Two Case Studies

Jaime Nivala and Scott Wallace

Abstract Treatment of landfill leachate is challenging due to high concentrations of oxygen-demanding compounds, such as biochemical oxygen demand (BOD) and ammonia nitrogen. The limited oxygen-transfer capability of conventional subsurface flow treatment wetlands has lead to the development of alternative design configurations that improve subsurface oxygen availability. This chapter compares the treatment performance of two aerated subsurface flow constructed wetlands treating landfill leachate against other, non-aerated systems. Results from these pilot studies indicate that aerated subsurface flow treatment wetlands are a viable technology selection for removal of ammonia-nitrogen of landfill leachate.

Keywords Aeration · Ammonia · Landfill leachate · *P-k-C** model · Subsurface flow wetland

10.1 Introduction

Little thought was originally given to the long-term stability and practical implications of permanently storing solid waste in the ground. Original landfills were unlined, and only some addressed the handling and treatment of the resulting wastewater generated at the site (herein referred to simply as *leachate*). Leachate is produced when rainfall and/or groundwater combines with the waste; when left unmanaged, it can create an imminent threat to nearby groundwater and surface waters. This is no small problem, considering that over 14000 landfills were closed in the United States between 1978 and 1988 (Mulamoottil, McBean, & Rovers, 1998). The number of active landfills in the United States is steadily decreasing; 8000 active sites were reported in 1988, but only 1858 landfills remained open in

J. Nivala (✉)
Environmental and Biotechnology Center (UBZ), Helmholtz Center for Environmental Research (UFZ), Leipzig, Germany
e-mail: jaime.nivala@gmail.com

J. Vymazal (ed.), *Water and Nutrient Management in Natural and Constructed Wetlands*, DOI 10.1007/978-90-481-9585-5_10,

2001 (U.S. EPA, 2003). Landfills that have been closed (which by rough estimate is over 20000 in the United States alone) still require leachate collection and management. Leachate is notorious for containing high concentrations of organic matter and ammonia nitrogen (Table 10.1).

Since many of these sites are located in rural areas, connecting the landfill to the regional sewer is often economically and/or practically infeasible. Pumping and hauling leachate to the closest treatment facility is another option, but with no incoming revenue for closed facilities, who will pay the indefinitely recurring cost of transport and treatment? This is one situation in which onsite wastewater management becomes a favorable alternative. Subsurface-flow (SSF) constructed wetlands have been identified as a promising technologies for the treatment of landfill leachate (Kadlec & Knight, 1996). Treatment wetlands have a small ecological footprint, require a low level of operations and maintenance compared to some conventional wastewater treatment technologies, and have an aesthetic value similar to that of natural wetlands. The application of wetland technology for treating landfill leachate is still developing. There has been a call by academics and professionals alike for a better understanding of the movement, transformation, and removal of contaminants in these treatment systems through extensive and long-term studies (Mulamoottil et al., 1998).

In conventional (passive) subsurface flow treatment wetlands, the oxygen required for nitification and organic matter removal generally exceeds the amount of available oxygen in the system. As a result, passive treatment wetlands systems can discharge high levels of organic matter, and nitrogen in the form of ammonia. The limited oxygen-transfer capability of conventional subsurface flow treatment wetlands has lead to the development of alternative design configurations that improve subsurface oxygen availability, through means of frequent water level fluctuation (Austin, 2006; Behrends, 1999; Zoeller & Byers, 1999), shallow bed depth (García, Morató, Bayona, & Aguirre, 2004), or direct mechanical aeration of the gravel bed (Flowers, 2002; Ouellet-Plamodon, Chazarenc, Comeau, & Brisson, 2006; Wallace, 2001).

Subsurface flow wetlands equipped with mechanical aeration (Fig. 10.1) have been widely used in North America to increase oxygen transfer rates and sustain aerobic conditions in the substratum. This technology variant has been utilized to treat many types of wastewater, including domestic sewage, dairy waste, groundwater contaminated with petroleum hydrocarbons, and landfill leachate (Lockhart, 1999; Nivala, 2005; Kadlec & Wallace, 2005; Muñoz, Drizo, & Hession, 2006).

Treatment performance in constructed wetlands is a combined result of the internal mechanisms that store, transform, and remove organic matter and pollutants. Due to the limited amount of information available, system performance cannot be predicted with a high degree of accuracy. By necessity, current design tools use "lumped" parameters (such as first-order rate coefficients) to represent the aggregate impact of the internal mechanisms in wetland treatment systems. Because the combined effect of internal treatment processes varies from one system to the next, pilot studies that use site-specific wastewater (in this case, landfill leachate) can be used to substantiate parameters for full-scale designs. This paper summarizes results from two aerated subsurface-flow wetland pilot systems treating landfill leachate.

Table 10.1 Leachate characteristics from Landfills in North America

Landfill	BOD (mg L^{-1})	COD (mg L^{-1})	TSS (mg L^{-1})	NH_4–N (mg L^{-1})	NO_3–N (mg L^{-1})	Iron (mg L^{-1})	Reference
Fulton County, Indiana	390	1540	7.840	284	3	178	Kadlec (1999)
Sarnia, Ontario	407	1036	–	254	< 0.3	17.6	Kadlec (1999)
City Sand, Michigan	312	3203	241	2074	0	–	Kadlec (1999)
Saginaw, Michigan	729	–	–	322	0	22	Kadlec (1999)
Maidstone, Ontario	33	–	–	92.5	2.3	–	Kinsley, Crolla, and Higgins (2002)
Jones County, Iowa	116	781	186	212	2	21	This study
Beecher, Illinois	40	–	27	302	0.3	8.1	This study

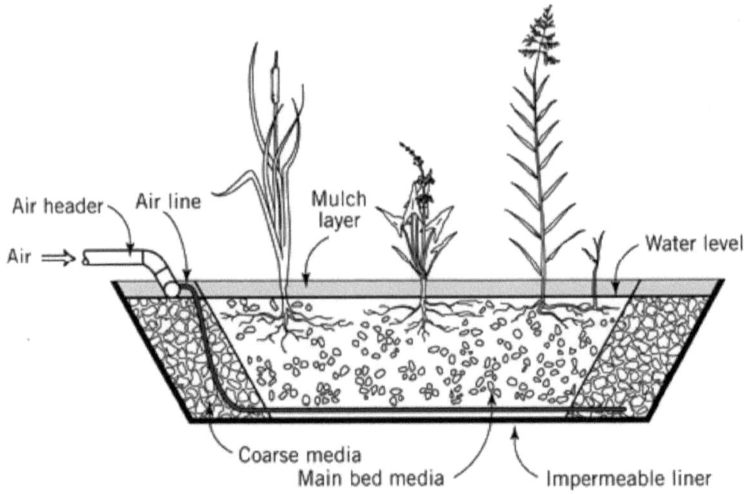

Fig. 10.1 Schematic of an aerated, horizontal subsurface-flow treatment wetland. Reprinted with permission from Kadlec & Wallace (2009)

10.2 Study Sites

10.2.1 Jones County Landfill

The Jones County wetland was installed at the local municipal landfill in 1999 to demonstrate the use of the technology for treatment of landfill leachate at small, rural landfills. The treatment system consists of one 15.5 m long by 6 m wide horizontal flow cell, lined with an impermeable liner. The design flow is 0.4 m^3d^{-1}. The cell is equipped with a patented aeration system (Wallace, 2001) and consists of a 30 cm layer of pea gravel ($d_{10} = 5$ mm) underneath a 15 cm layer of well-decomposed peat. The purpose of the peat layer is to protect the system against freezing in the wintertime. The system received high loads of iron between 1999 and 2002, which clogged the aeration system. The aeration system in the wetland was replaced in September 2002, and a pretreatment step was installed before the wetland cell, which included an aeration chamber and settling tank for iron removal. Additional site details are provided in Nivala, Hoos, Cross, Wallace, and Parkin (2007).

10.2.2 Beecher Landfill

The Beecher Landfill pilot-scale wetland system was constructed in 2008 as a part of an engineering feasibility study to validate a full-scale design. The treatment components were scaled down by a factor of 2000, which resulted in a design flow of 9.5 L d^{-1} for the pilot system. Based on experience with iron-related clogging at the Jones County pilot (Nivala et al., 2007), the Beecher system included pre-treatment for iron oxidation. Pretreatment was achieved using a 5-L aeration chamber followed by a 15-L sedimentation tank. The design was based on historical leachate data; design concentration for CBOD was 100 mg L^{-1} (treatment target: 15 mg L^{-1}) and ammonia-nitrogen design concentration was 300 mg L^{-1} (treatment target: 0.2 mg L^{-1}). The aeration system was sized to supply enough oxygen to satisfy the sum of the carbonaceous biochemical oxygen demand (CBOD) and nitrogenous oxygen demand (NOD). A two-cell, vertical-flow design was chosen to improve the hydraulic efficiency of the system. Each cell was approximately 55 L in volume and filled with river gravel. Additional components of the pilot system included a limestone cell and peat filter, but those results are not presented here. The pilot system was fed a mixture of water and leachate during start-up of the system. Following the acclimatization period, the pilot system received full-strength leachate. The leachate used in the pilot study was shipped directly from the Beecher Landfill every two weeks.

10.3 Methods

The first-order, tanks-in-series *P-k-C** model set forth by Kadlec & Wallace (2009) can be used to assess and compare performance amongst different wetland treatment systems. The *P-k-C** model is a compromise between accuracy and computational simplicity and can be represented by Eq. (10.1):

$$\frac{C_o - C^*}{C_i - C^*} = \frac{1}{(1 + k/Pq)^P} \tag{10.1}$$

where:

C_o = outlet concentration, mg L^{-1}
C_i = inlet concentration, mg L^{-1}
C^* = background concentration, mg L^{-1}
k = first-order areal rate coefficient, m yr^{-1}
P = apparent number of tanks-in-series
q = hydraulic loading rate, m yr^{-1}

The parameter P is a fitted parameter that accounts for both the hydraulic efficiency of the reactor (number of tanks-in-series, N) and weathering of a pollutant as it undergoes treatment in the wetland (Kadlec, 2003). Thus, P will be always less than or equal to N. The temperature dependence of the rate coefficient k is described by the modified Arrhenius equation (Eq. (10.2)):

$$k_T = k_{20}\,\theta^{(T-20)} \tag{10.2}$$

where:

k_T = first-order areal rate coefficient at temperature T, m yr^{-1}
k_{20} = first-order areal rate coefficient at temperature 20°C, m yr^{-1}
θ = temperature factor, dimensionless
T = operational water temperature, degrees Celsius

The first-order tanks-in-series *P*-*k*-*C** model offers a platform for comparing data from different treatment systems. Results for *k*-rates and θ-factors for both pilot systems were calculated using Solver™ in Microsoft excel. Background ammonia concentrations (C^*) for the Jones County and Beecher treatment systems were selected to be zero, and the *P*-value was chosen to be six, in order to allow direct comparison to the values reported in Kadlec & Wallace (2009).

10.4 Results and Discussion

10.4.1 Jones County Landfill

In 2002, after replacement of the clogged aeration system, the microbial community in the Jones County wetland took approximately 6 months to adapt to the ammonia loading. Figure 10.2 shows the ammonia and nitrate concentrations during the recommissioning period. Effluent ammonia concentrations decreased throughout the winter months despite water temperatures as cold as 2°C. It is interesting to note the corresponding increase in nitrate during this time, and the following decrease in nitrate as the denitrifying bacteria became established in 2003.

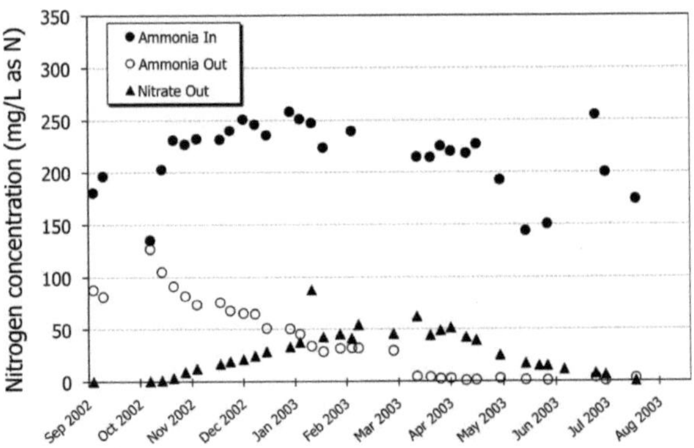

Fig. 10.2 Ammonia removal in the Jones County HSSF wetland

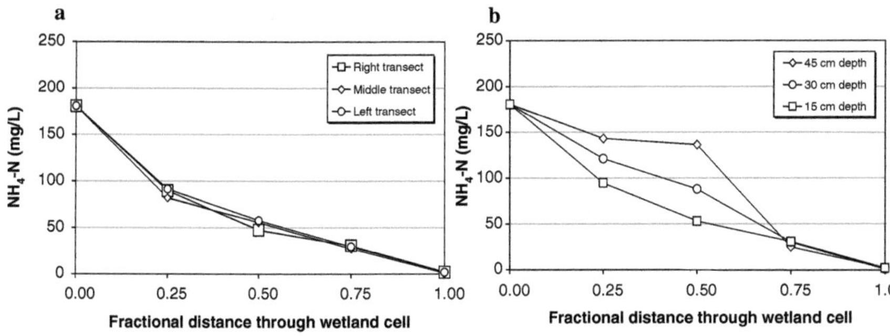

Fig. 10.3 Lateral (**a**) and vertical (**b**) ammonia profiles in the Jones County HSSF wetland

The Jones County system was also monitored intensively from August 2004 to May 2005. Internal sampling points located at 25, 50, and 75% through the wetland cell indicated that the wetland was well-mixed in the direction orthogonal to flow (Fig. 10.3a). However, a set of depth-specific sampling points along the centerline of the cell (15, 30, and 45 cm deep) indicated that there was a clear pattern of vertical stratification in the first half of the wetland cell (Fig. 10.3b). These observations were further validated by tracer study results that displayed similar spatial patterns (Nivala, 2005). Ammonia removal efficiency for the wetland system (on a concentration basis) was 98% during this monitoring campaign ($n = 25$).

10.4.2 Beecher Landfill

Analytical data was collected from February until April 2008 (Table 10.2). Samples were collected and analyzed weekly; the average ammonia-nitrogen concentration

Table 10.2 Selected results from the Beecher pilot study

	CBOD		NH_4-N		Iron		pH	
	C_{in} (mg L^{-1})	C_{out} (mg L^{-1})	C_{in} (mg L^{-1})	C_{out} (mg L^{-1})	C_{in} (mg L^{-1})	C_{out} (mg L^{-1})	SU	SU
Week 1	12.8	6.0	175	0.6	5.4	0.1	8.4	8.4
Week 2	9.5	3.4	224	0.1	3.7	0.1	8.6	8.4
Week 3	8.5	2.0	248	0.2	3.6	0.1	8.4	8.4
Week 4	8.4	3.3	201	0.3	3.8	0.2	8.4	8.4
Week 5	7.8	3.1	162	0.3	3.1	0.3	8.7	8.8
Week 6	13.6	2.4	210	0.4	4.1	0.1	8.8	8.9
Week 7	8.9	2.2	249	0.3	4.2	0.2	8.8	8.9
Week 8	7.0	2.0	262	0.5	4.5	0.1	8.8	9.0
Week 9	8.6	3.8	231.0	0.4	3.2	0.2	8.7	9.0
Week 10	9.3	2.9	213.0	0.3	3.3	0.4	8.6	8.9
Week 11	24.0	11.7	160.0	0.3	2.5	0.2	8.5	8.8
Week 12	13.3	2.0	120.0	0.3	2.8	0.1	8.8	9.1
Average	11.0	3.7	204.6	0.3	3.7	0.2	8.6	8.7

Table 10.3 Aerated full-scale treatment wetland P-k-C^* results

Site Name and Location	Flow Direction	Wastewater Type	k_{20} (m/yr)	θ	Reference
Jones County Landfill, Iowa	Horizontal	Landfill Leachate	10.7	0.96	This study
Prinsburg, Minnesota	Horizontal	Municipal Wastewater	9.2	1.08	Wallace and Nivala (2008)
St. Croix Chippewa, Wisconsin	Horizontal	Municipal Wastewater	13.3	1.04	Wallace and Nivala (2008)
			11.4	1.01	Kadlec and Wallace (2009) 50th Percentile

Table 10.4 Aerated pilot-scale treatment wetland *P-k-C** results

Site name and location	Flow direction	Wastewater type	k_{20} (m yr^{-1})	θ	Reference
Beecher Landfill, Illinois	Vertical	Landfill Leachate	137	1.03	This study
North Glengarry, Ontario	Vertical	Municipal Wastewater	818	1.02	Wallace et al. (2006)
Rosebel, Suriname	Vertical	Gold Mine	518	1.02	Wallace et al. (2006)

entering the wetland cells was 204.5 mg L^{-1}; effluent concentration averaged 0.3 mg L^{-1}. Average ammonia removal efficiency (on a concentration basis) was 99.8%. Selected parameters from the Beecher pilot system are shown in Table 10.2 for the duration of the study.

10.4.3 P-k-C* Model Results

The Jones County system, which was operated outdoors under a range of temperatures, displayed an overall k_{20} of 10.7 m yr^{-1}, which is near the 50th percentile for non-aerated HSSF wetlands. The θ-factor of less than 1.0, however, indicates that ammonia removal in the Jones County wetland system was less sensitive to temperature than what has been reported in the literature for non-aerated wetlands. Because the Jones County wetland was often removing ammonia to less than one milligram per liter in the effluent, the reported k-value in Table 10.3 is a low estimate of the system's actual ammonia removal rate coefficient.

Table 10.4 shows that the Beecher system, which was operated indoors and under a small range of temperatures, displayed an overall k_{20} of 137.0 m yr^{-1} (ten-fold higher compared to that of the Jones County system) and a θ-factor of 1.027. In laboratory studies, wetland aeration has been found to increase nitrification rates up to twenty-fold (Kinsley et al., 2002). The results from the laboratory-scale Beecher pilot are consistent with this finding.

10.5 Conclusions

Results from these pilot studies indicate that aerated subsurface flow treatment wetlands are a viable technology selection for effective removal of ammonia-nitrogen of landfill leachate. Furthermore, the Beecher Landfill pilot study indicates that saturated vertical flow wetlands equipped with mechanical aeration are extremely efficient in removal of ammonia-nitrogen. Future research is required to better characterize the rate coefficients in alternative treatment wetland design configurations.

References

Austin, D. C. (2006). Influence of cation exchange capacity (CEC) in a tidal flow, flood and drain wastewater treatment wetland. *Ecological Engineering, 28*, 35–43.

Behrends, L. L. (1999). *Patent: Reciprocating subsurface flow constructed wetlands for improving wastewater treatment. U.S. Patent No. 5,863,433*. Washington, DC: U.S. Patent and Trademark Office.

Flowers, D. A. (2002). *Patent: Process and system for enhanced nitrogen removal in a wetland wastewater treatment facility. U.S. Patent No. 6,447,682*. Washington, DC: U.S. Patent and Trademark Office.

García, J., Morató, J., Bayona, J. M., & Aguirre, P. (2004). Performance of horizontal subsurface flow constructed wetland with different depth. In A. Liénard & H. Burnett (Eds.),

Proceedings of the 9th international conference on wetland systems for water pollution control (pp. 269–276). Avignon, France: Association Scientifique et Technique pour l'Eau et l'Environnement (ASTEE), Cemagref, and IWA, 26–30 September 2004.

Kadlec, R. H. (2003). Effects of pollutant speciation in treatment wetlands design. *Ecological Engineering, 20*(1), 1–16.

Kadlec, R. H., & Knight, R. L. (1996). *Treatment wetlands, first edition.* Boca Raton, FL: CRC Press.

Kadlec, R. H., & Wallace, S. D. (2009). *Treatment wetlands, second edition.* Boca Raton, FL: CRC Press.

Kinsley, C. B., Crolla, A. M., & Higgins, J. (2002). Ammonia reduction in aerated subsurface flow constructed wetlands. In T. S. A. Mbwette (Ed.), *Proceedings of the 8th international conference on wetland systems for water pollution control* (pp. 961–971). Tanzania: Comprint International Limited: University of Dar Es Salaam, 16–19.

Lockhart, A. (1999). *A comparison of constructed wetlands used to treat domestic wastes: Conventional, drawdown, and aerated systems.* M.S. Thesis, Department of Civil and Environmental Engineering, University of Iowa, Iowa City, IA.

Mulamoottil, G., McBean, E., & Rovers, F. A. (1998). *Constructed wetlands for the treatment of landfill leachates.* Boca Raton, FL: Lewis.

Munoz, P., Drizo A., & Hession, W. C. (2006). Flow patterns of dairy wastewater constructed wetlands in a cold climate. *Water Research, 40*(17), 3209–3218.

Nivala, J. A. (2005). *Treatment of landfill leachate using an enhanced subsurface-flow constructed wetland.* M.S. Thesis, Department of Civil and Environmental Engineering, University of Iowa, Iowa City, IA.

Nivala, J. A., Hoos, M. B., Cross, C. S., Wallace, S. D., & Parkin, G. F. (2007). Treatment of landfill leachate using an aerated, horizontal subsurface-flow constructed wetland. *Science of the Total Environment, 380*, 19–27.

Ouellet-Plamondon, C., Chazarenc, F., Comeau, Y., & Brisson, J. (2006). Artificial aeration to increase pollutant removal efficiency of constructed wetlands in cold climate. *Ecological Engineering, 27*, 258–264.

U.S. EPA (2003). Municipal solid waste in the United States: 2001 facts and figures executive summary, EPA/530/S-03/011, U.S. EPA Office of Solid Waste and Emergency Response.

Wallace, S. D. (2001). *Patent: System for removing pollutants from water. U.S. Patnet No. 6,200,469 B1.* June 23, 1998. Washington, DC: U.S. Patent and Trademark Office.

Wallace, S. D., & Kadlec, R. H. (2005). BTEX degradation in a cold-climate wetland system. *Water Science and Technology, 51*(9), 165–171.

Zoeller, K. E., & Byers, M. E. (1999). *Patent: Wastewater treatment system. U.S. Patent No. 5,897,777.* October 3, 1997. Washington, DC: U.S. Patent and Trademark Office.

Chapter 11
Nutrient Accumulation by *Phragmites australis* and *Phalaris arundinacea* Growing in Two Constructed Wetlands for Wastewater Treatment

Jan Vymazal and Lenka Kröpfelová

Abstract *Phragmites australis* and *Phalaris arundinacea* are two most commonly used plants in constructed wetlands for wastewater treatment in the Czech Republic. This chapter deals with development of biomass, nitrogen and phosphorus concentration in that biomass and nutrient standing stocks in *Phragmites* and *Phalaris* growing in two constructed wetlands. *Phragmites* aboveground biomass varies between 781 and 2532 g m^{-2} in Břehov and between 1309 and 2177 g m^{-2} in Mořina. *Phalaris* aboveground biomass varies between 1262 and 2265 g m^{-2} in Břehov and between 1194 and 1780 g m^{-2} in Mořina. These values are within the common range found in both natural and constructed wetlands for wastewater treatment. For both *Phragmites* and *Phalaris* the aboveground biomass was higher than belowground. Nitrogen and phosphorus concentrations in *Phragmites* aboveground biomass were comparable with concentrations found in both natural stands and constructed wetlands. On the other hand, N and P concentrations in *Phalaris* were lower than those found in constructed wetlands. Nitrogen standing stocks in *Phragmites* and *Phalaris* were comparable with those found in natural stands and constructed wetlands and for both species substantially more nitrogen was sequestered aboveground. When comparing *Phragmites* and *Phalaris* N standing stocks, *Phragmites* stocks were higher but only belowground standing stock was significantly higher. The *Phragmites* phosphorus standing stocks in our study are also comparable with standing stocks found in natural stands but they are also at the lower end of the range.

Keywords Constructed wetlands · Nutrients · *Phalaris arundinacea* · *Phragmites australis* · Wastewater

---

J. Vymazal (✉)
Faculty of Environmental Sciences, Department of Landscape Ecology, Czech University of Life Sciences in Prague, Prague, Czech Republic; ENKI, o.p.s., Dukelská 145, 379 01 Třeboň, Czech Republic
e-mail: vymazal@yahoo.com

J. Vymazal (ed.), *Water and Nutrient Management in Natural and Constructed Wetlands*, DOI 10.1007/978-90-481-9585-5_11,

11.1 Introduction

By far, the most commonly used plant for constructed wetlands with horizontal subsurface flow is *Phragmites australis* (Cav.) Trin. ex Steudel (Common reed). *P. australis* is a perennial and flood-tolerant grass that grows up to 8 m tall with an extensive rhizome system which usually penetrates to depths of about 0.6–1.0 m (Haslam, 1971; Thesiger, 1964). Common reed is a cosmopolitan grass occurring as a dominant component of freshwater, brackish water and in some cases also marine littoral communities almost all over the world (Dykyjová & Hradecká, 1976; Hocking, Finlayson, & Chick, 1983; Soetaert et al., 2004). It often occurs in large monospecific stands (Pénzes, 1960; Hocking et al., 1983). Its distribution is widespread throughout Europe, Africa, Asia, Australia and North America between 10 and 70° latitude (Hawke & José, 1996). *P. australis* has wide ecological amplitude (Haslam, 1973; Rodewald-Rudescu, 1974). The habitat description range from oligotrophic lakes (Björk, 1967) to heavily polluted lakes (e.g., Dykyjová & Hradecká, 1973).

While *Phragmites australis* is used in constructed wetlands with a horizontal subsurface flow throughout Europe, Canada, Australia and most parts of Asia, in the United States and New Zealand, Common reed is considered to be an introduced and invasive species and its use is usually restricted or prohibited.

Phalaris arundinacea L. (Reed canarygrass) is a 1–3 m tall, long-lived perennial grass with a C3 photosynthetic pathway (Kephart & Buxton, 1993; Lewandowski, Scurlock, Lindvall, & Christou, 2003). It produces a dense crown and prominent networks of vigorous roots and rhizomes, penetrating to a depth of about 30–40 cm, allowing for aggressive vegetative spread (Coops, van der Brink, & van der Velde, 1996; Kätterer & Andrén, 1999). Reed canarygrass is native to the temperate zones of the Northern Hemisphere and is widely distributed throughout Eurasia (Lavergne & Molofsky, 2004). It was originally introduced from Europe to the United States shortly after 1850 and has since spread throughout North America (Merigliano & Lesica, 1998); it is considered as an invasive species especially in anthropogenically disturbed areas (Lavergne & Molofsky, 2004; Kerchler & Zedler, 2004). *Phalaris* grows rapidly and tends to form monocultures (Marten, 1985; Miller & Zedler, 2003).

In the Czech Republic, *Phragmites australis* (Cav.) Trin. ex Steudel and *Phalaris arundinacea* L. have been used widely in HF CWs. These plants are used as monostands but very often they are planted together (Vymazal, 2002). The objective of the chapter is to evaluate and compare the accumulation of nutrients (standing stocks) in both plants.

11.2 Materials and Methods

The study was carried out at two constructed wetlands with horizontal subsurface flow Mořina and Břehov. CW Mořina (700 PE) was built in 2000 and it treats sewage from a separate sewer system. *Phalaris* and *Phragmites* are planted in bands

Fig. 11.1 Constructed wetland Břehov. *Phragmites* band in the middle, in between *Phalaris* bands. Photo Jan Vymazal

Fig. 11.2 Constructed wetland Mořina. *Phalaris* band in the middle, in between *Phragmites* bands. Photo Jan Vymazal

perpendicular to the water flow (Fig. 11.1). CW Břehov (100 PE) was built in 2003 and was designed to treat wastewater from a combined sewer system. *Phalaris* and *Phragmites* are planted in bands parallel to the water flow (Fig. 11.2).

Samples of biomass were taken from an area of 0.25 m^2 in four replicates (two samples in the inflow zone and two in the outflow zone) during the peak standing crop, i.e., in mid-July for *Phalaris* and at the beginning of September for *Phragmites*. The stems were clipped at the ground level and separated into stems, leaves including leaf sheaths and inflorescences (flowers). Belowground biomass was completely dug out from the sampled area and divided into roots and rhizomes and thoroughly washed with water. Both above- and belowground biomass was then dried at 70°C until reaching a constant weight and then weighed.

Five whole shoots of both plants from each sample were used to determine nitrogen and phosphorus concentrations in the biomass. We used whole plants because it is known that nutrient concentrations in various stem and leaf parts differ (Hocking, 1989). Dried biomass was ground and homogenized samples were mineralized in a pressure microwave apparatus MARS-5 using $HNO_3 + H_2O_2 + HF + H_3BO_3$ under high temperature and pressure (Sucharová & Suchara, 2006). Total phosphorus was determined spectrophotometrically using the molybdate-blue method (ČSN EN 1189, 2005). Total nitrogen was determined according to the ČSN EN 12260 (2003) after oxidation to nitrogen oxides using chemo-luminiscence detection. A comparable amount of the belowground biomass was processed with the same procedures as the aboveground biomass sample.

11.3 Results and Discussion

11.3.1 Biomass

Phragmites and *Phalaris* biomass values in CWs Břehov and Mořina are presented in Table 11.1. *Phragmites* aboveground biomass varied between 781 and 2532 g m^{-2} in Břehov and between 1309 and 2177 g m^{-2} in Mořina. Maximum aboveground biomass of *P. australis* is highly variable depending on latitude, climate, salinity, water depth, eutrophication and interactions among these factors (Soetaert et al., 2004). For example, Čížková (1999) reported the maximum seasonal biomass from 12 natural stands in nine European countries to be between 308 and 4165 g DM m^{-2}, Vymazal, Dušek, and Květ (1999) reported in their review biomass values from 22 sites around the world to be between 182 and 12696 g DM m^{-2} and Vymazal and Kröpfelová (2005) in their review reported a range between 413 and 9890 g DM m^{-2} for 12 natural stands from Europe, Asia and Australia. However, the most common values for maximum aboveground biomass found in natural stands are between 1000 and 3000 g DM m^{-2} depending on the trophic status.

The values of maximum aboveground biomass recorded for *Phragmites* growing in HF constructed wetlands are within the range found in natural stands. Vymazal et al. (1999) reported values from eight systems in Europe, North America, Australia and Africa between 788 and 6334 g DM m^{-2} and Vymazal and Kröpfelová (2008) reported values up to 11,280 g DM m^{-2}. However, the values found in CWs Břehov and Mořina are within the common range found in both natural and constructed wetlands for wastewater treatment.

The R/S (root/shoot) ratios for *Phragmites* (Table 11.1) were very low and varied between 0.34 and 0.83 in Břehov and between 0.44 and 0.47 in Mořina. The R/S ratio in natural stands is usually high, indicating high underground biomass. For example, Čížková (1999) in her review reported an average R/S ratio of 7.5 with values ranging between 1.5 and 21.9. However, it seems apparent that an R/S ratio in constructed wetlands for wastewater treatment is usually <1. Also, Vymazal and Kröpfelová (2008) reported R/S ratio <1 for four out of five systems in Austria,

Table 11.1 Dry biomass (g m^{-2}) of *Phragmites australis* (PG) and *Phalaris arundinacea* (PH) and Root/Shoot (R/S) ratio in constructed wetlands Mořina and Břehov

Locality	Plant	Leaves	Stems	Flowers	Aboveground biomass	Roots	Rhizomes	Belowground biomass	Total biomass	R/S
Břehov	PG 2004	436	345	0	781	146	499	645	1426	0.83
	PG 2005	1100	946	0	2046	75	830	905	2951	0.44
	PG 2006	1207	1280	45	2532					
	PG 2007	944	749	50	1743	87	877	964	2707	0.45
	PG 2008	960	1221	121	2302	138	643	781	3083	0.34
Břehov	PH 2004	446	905	67	1418	187	295	482	1900	0.34
	PH 2005	814	1389	62	2265	402	830	1232	3497	0.54
	PH 2006	622	951	63	1636					
	PH 2007	686	903	48	1637	301	373	674	2311	0.41
	PH 2008	572	663	27	1262	322	540	862	2124	0.68
Mořina	PG 2002	953	1147	77	2177	235	786	1021	3198	0.47
	PG 2004	940	1090	45	2075	208	712	920	2995	0.44
	PG 2006	886	996	36	1918					
	PG 2007	787	690	25	1502	66	968	1035	2537	0.64
	PG 2008	631	630	48	1309	104	768	872	2181	0.67
Mořina	PH 2002	713	990	77	1780	191	458	649	2429	0.36
	PH 2004	760	862	42	1664	158	427	585	2249	0.35
	PH 2006	828	991	80	1899					
	PH 2007	608	579	7	1194	101	143	243	1437	0.20
	PH 2008	732	688	24	1444	125	163	288	1732	0.20

Australia, Germany and USA. This may be due to several reasons such as continuous easy availability of nutrients in the root zone or stress caused by a high level of pollutants.

Phalaris aboveground biomass varied between 1262 and 2265 g m^{-2} in Břehov and between 1194 and 1780 g m^{-2} in Mořina. Maximum aboveground biomass of *Phalaris arundinacea* in natural stands is variable depending mostly on the trophic status of the stand. Vymazal (2005) in his review reported values between 440 and 2304 g DM m^{-2} with the lowest value recorded in a meso-eutrophic lake in Scotland (Ho, 1979a) and the highest value recorded for wet meadows in the Czech Republic (Hlávková-Kumnacká, 1980). Sparse results from HF constructed wetlands indicate that *P. arundinacea* maximum annual standing crop is within the range found in natural stands (Vymazal & Kröpfelová, 2008). The results obtained in Břehov and Mořina fit well within the range reported by Vymazal (2005).

In natural stands, the R/S ratio for *Phalaris arundinacea* is usually between 1 and 2. In constructed wetlands, however, the R/S ratio is usually <1.0 (Behrends, Bailey, Bulls, Coonrod, & Sikora, 1994; Bernard & Lauve, 1995). Similar results were observed in our study (Table 11.1). In Břehov, the R/S ratio varied between 0.34 and 0.68 while in Mořina this ratio varied between 0.20 and 0.36. Studies have shown that *Phalaris* responds positively to nutrient enrichment. The controlled experiments of Green and Galatowitsch (2001) and Maurer and Zedler (2002) showed that high nutrient treatments increased biomass of *Phalaris* and increased allocation to aboveground growth. Also Kätterer and Andrén (1999) reported that fertilization decreased the amount of biomass allocated belowground as compared to aboveground biomass.

11.3.2 Nitrogen

11.3.2.1 Concentration in Biomass

Nitrogen concentrations in aboveground and belowground biomass of *Phragmites* and *Phalaris* are shown in Table 11.2; average concentrations are shown in Fig. 11.3. The aboveground *Phragmites* N concentrations varied between 12.7 and 23.6 g kg^{-1} but there was no significant difference between Mořina and Břehov. The concentrations are within the range of 10.5 and 27.0 g kg^{-1} reported from natural wetlands (Allen & Pearasall, 1963; Auclair, 1979; Kaul, Trisal, & Kaul, 1980; Dykyjová & Květ, 1982; Boar, Crook, & Morris, 1989; Ennabili, Ater & Radoux, 1998). The values are also comparable with those reported from constructed wetlands. In Australia, Greenway (1996) and Adcock and Ganf (1994) reported the values of 18.4 and 23.0 g kg^{-1}, respectively. Peverly, Sanford, Steenhuis, and Surface. (1993) reported the value of 17.7 g kg^{-1} in the United States and Haberl and Perfler (1990) reported the values between 5.0 and 23.0 g kg^{-1} in Austria. Also, Vymazal et al. (1999) reported the N concentrations in aboveground *Phragmites* biomass in three constructed wetlands in the Czech Republic between 20.2 and 26.7 g kg^{-1}. However, Gries and Garbe (1989) reported values up to 35 g kg^{-1} and Vymazal (1995) reported values up to 30.4 g kg^{-1}. Belowground N concentration in

Table 11.2 Average nitrogen concentrations (g kg^{-1} DM) in aboveground (ABG) and belowground (BG) biomass of *Phragmites australis* and *Phalaris arundinacea* at constructed wetlands Mořina and Břehov

	Phragmites				*Phalaris*			
	Mořina		Břehov		Mořina		Břehov	
	ABG	BG	ABG	BG	ABG	BG	ABG	BG
2002	12.7	19.8			14.7	14.5		
2004	17.0	18.4	23.6	19.1	17.6	16.0	13.3	15.8
2005			23.3	19.1			14.2	16.6
2006	18.5		18.0		14.3		15.3	
2007	19.3	13.1	20.6	18.5	22.6	25.2	21.2	12.5
2008							16.3	12.0

Phragmites varied between 13.1 and 19.8 g kg^{-1} with the average values of 17.1 and 18.9 g kg^{-1} in Mořina and Břehov, respectively (Table 11.2, Fig. 11.3).

The aboveground *Phalaris* N concentrations varied between 13.3 and 22.6 g kg^{-1} but there was no significant difference between Mořina and Břehov. These values are substantially lower than values reported from other constructed wetlands for wastewater treatment. Bernard and Lauve (1995) reported N concentrations between 22.0 and 46.0 g kg^{-1} in a system treating landfill leachate in the United States (New York). Behrends et al. (1994) reported values between 22.0 and 33.0 g kg^{-1} in experimental constructed wetlands in Alabama, and Hurry and Bellinger (1990) found nitrogen concentrations up to 45.0 g kg^{-1} in treatment wetland in England. In enrichment experiments, Dubois (1994) reported the range of 16.0–36.0 g kg^{-1}. The concentrations found in our study were also lower as compared to values reported from natural stands. Bernard and Lauve (1995) reported N concentration in the range of 11.0–32.0 g kg^{-1} in New York state, Kline and Boersma (1983) found the range between 10.0 and 36.0 g kg^{-1} in Canada, and Hlávková-Kumnacká (1980) reported the nitrogen concentration range between 8.8 and 39.0 g kg^{-1} in the Czech Republic. However, Vymazal et al. (1999) reported comparable values between 10.2 and 29.0 g kg^{-1} from constructed wetlands Zásada and Chmelná in the Czech Republic. Belowground N concentrations in *Phalaris* varied between 12.0 and 16.6 g kg^{-1} (Table 11.2) with no statistical difference between Mořina and Břehov (Fig. 11.3). However, in Břehov, average N concentration in *Phalaris* belowground biomass was significantly lower than that in *Phragmites*.

11.3.2.2 Standing Stocks

In Tables 11.3 and 11.4, nitrogen standing stocks in *Phalaris* and *Phragmites* biomass in CWs Mořina and Břehov are shown. In Mořina (Table 11.3), aboveground N standing stocks in *Phragmites* varied between 27.7 and 35.5 g N m^{-2} with an average value of 31.9 g N m^{-2}. In *Phalaris*, nitrogen stocks varied only little between 26.1 and 29.2 g N m^{-2} with an average value of 27.4 g N m^{-2}.

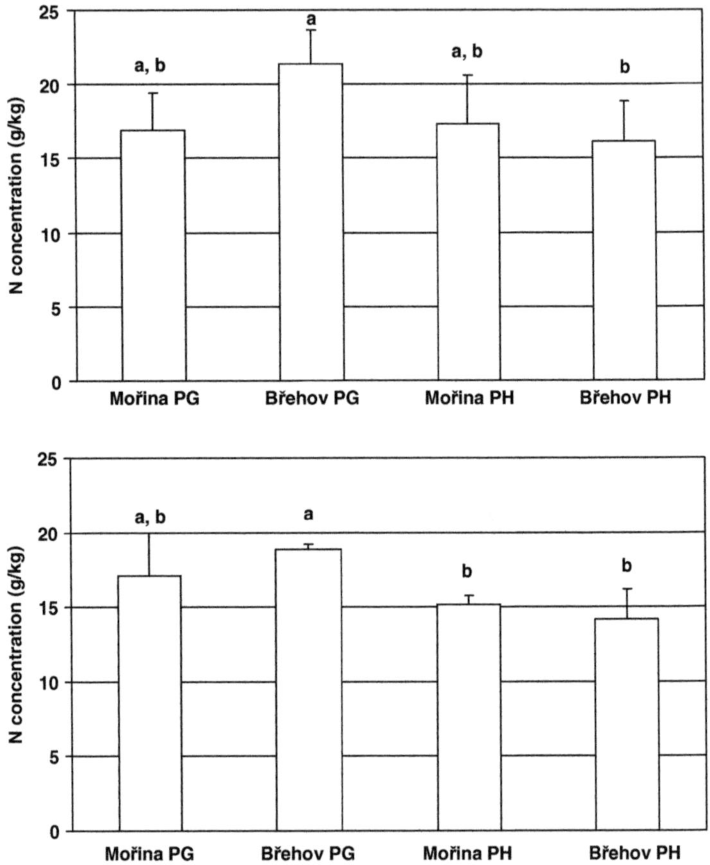

Fig. 11.3 Nitrogen concentrations in aboveground (*top*) and belowground (*bottom*) biomass of *Phragmites* (PG) and *Phalaris* (PH) in constructed wetlands Břehov and Mořina
Note: [a,b,c] Different letters indicate significant difference at $\alpha = 0.05$ between the means.

Belowground N standing stock was significantly lower in both *Phragmites* and *Phalaris* with respective average values of 16.9 and 7.5 g N m^{-2}. In both *Phragmites* and *Phalaris* substantially more nitrogen was sequestered aboveground (Table 11.3). The average aboveground/belowground N standing stock ratios were 1.86 and 4.38 for *Phragmites* and *Phalaris*, respectively. In natural stands, due to higher belowground biomass the aboveground/belowground standing stock ratio is usually <1. For example, Ennabili et al. (1998) reported the value of this ratio to be 0.81 for *Phragmites* growing in wetlands in Morocco. When comparing *Phragmites* and *Phalaris* N standing stocks, the *Phragmites* stocks are higher but only belowground standing stock was significantly higher (Table 11.3).

In Břehov (Table 11.4), the average *Phragmites* aboveground standing stock was higher (36.9 g N m^{-2}) than in Mořina while *Phalaris* standing stock was just slightly lower (26.3 g N m^{-2}) than in Mořina (Table 11.4). Also belowground standing

Table 11.3 Nitrogen standing stocks (g N m^{-2}) at CW Mořina. ABG = aboveground, BG = belowground

	Phragmites			*Phalaris*			*Phalaris/Phragmites*	
	ABG	BG	ABG/BG	ABG	BG	ABG/BG	ABG	BG
2002	27.7	20.2	1.37	26.1	9.42	2.77	0.94	0.47
2004	35.2	16.9	2.08	29.2	9.35	3.12	0.83	0.55
2006	35.5			27.2			0.77	
2007	29.0	13.6	2.14	26.9	3.71	7.25	0.93	0.27
Mean	31.9 [a]	16.9 [b]	1.86	27.4 [a]	7.5 [b,c]	4.38	0.87	0.43

[a,b,c] Different letters indicate significant difference at $\alpha = 0.05$ between the means.

Table 11.4 Nitrogen standing stocks (g N m^{-2}) at CW Břehov. ABG = aboveground, BG = belowground

	Phragmites			*Phalaris*			*Phalaris/Phragmites*	
	ABG	BG	ABG/BG	ABG	BG	ABG/BG	ABG	BG
2004	18.4	12.3	1.50	18.9	7.6	2.49	1.03	0.62
2005	47.7	17.3	2.76	32.3	20.5	1.58	0.68	1.18
2006	45.6			25.0			0.55	
2007	35.8	17.8	2.01	34.6	8.5	4.07	0.97	0.48
2008				20.6	10.4	1.98		
Mean	36.9 [a]	15.8 [b,c]	2.09	26.3 [a,c]	11.8 [b]	2.53	0.81	0.76

[a,b,c] Different letters indicate significant difference at $\alpha = 0.05$ between the means.

stocks in Břehov were comparable with those in Mořina. There was no statistically significant difference between Mořina and Břehov.

The results from CWs Mořina and Břehov (Tables 11.3 and 11.4) indicate that nitrogen standing stocks are within the range commonly reported for *Phragmites* and *Phalaris* from constructed wetlands for wastewater treatment (Table 11.5). Also, the results from our systems indicate that the standing stocks are very stable. The low *Phragmites* standing stock in Břehov in 2002 is influenced by the low aboveground biomass during the first growing season.

The *Phragmites* standing stocks found in our study are also comparable with standing stocks found in natural stands. Květ (1973), Dykyjová (1973a, 1973b, 1989) and Úlehlová, Husák, and Dvořák (1973) reported nitrogen aboveground standing stock for *Phragmites* in various localities in the Czech Republic in the range of 13.7–40.9 g N m^{-2}. Ho (1979b, 1981) reported the range of 34.3–63.4 g N m^{-2} for Scottish lochs, Kaul et al. (1980) reported the range of 15.5–38.2 g N m^{-2} from Kashmir, Boar et al. (1989) found the range of 16.6–20.7 g N m^{-2} in England and Ennabili et al. (1998) reported the value of 25.4 g N m^{-2} from Morocco. Hocking (1989) found aboveground N standing stock as high as 78.6 g N m^{-2} in enriched natural stands in Australia. On the other hand,

Table 11.5 Aboveground nitrogen standing stocks in *Phragmites australis* and *Phalaris arundinacea* reported in the literature from constructed wetlands for wastewater treatment

Location	Plant	Standing stock ($g\ N\ m^{-2}$)	Reference
Poland	*Phragmites*	58.7	Obarska-Pempkowiak (1999)
USA		51.2	Behrends et al. (1994)
Czech Republic		47.2	Vymazal et al. (1999)
Poland		46.8	Obarska-Pempkowiak and Gajewska (2003)
Australia		43.8	Headley (2004)
Austria		35.0	Haberl and Perfler (1990)
Poland		32.6	Obarska-Pempkowiak (1999)
Poland		29.2	Obarska-Pempkowiak and Gajewska (2003)
The Netherlands		27.0	De Jong (1976)
USA		26.9	Peverly et al. (1993)
Germany		23.1	Gries and Garbe (1989)
Australia		18.1	Adcock and Ganf (1994)
UK	*Phalaris*	46.7	Hurry and Bellinger (1990) multiple harvest
USA		37.6	Bernard and Lauve (1995)
USA		13.3	Behrends et al. (1994)
Czech Republic		12.2	Vymazal et al. (1999)

nitrogen standing stocks in aboveground *Phalaris* biomass in natural stands are usually lower than those found in constructed wetlands. Klopatek (1978) reported the value of 12.3 $g\ N\ m^{-2}$ from Wisconsin, Bernard and Lauve (1995) found the range of 2.0–15.5 $g\ N\ m^{-2}$ in New York State and Lukavská (1989) reported the range of 10.9–14.1 $g\ N\ m^{-2}$ in several sites in the Czech Republic.

11.3.3 Phosphorus

11.3.3.1 Concentration in Biomass

Phosphorus concentrations in aboveground and belowground biomass of *Phragmites* and *Phalaris* are shown in Table 11.6; average concentrations are shown in Fig. 11.4. The aboveground *Phragmites* P concentrations varied between 1.31 and 2.18 $g\ kg^{-1}$ but there was no significant difference between Mořina and Břehov. The concentrations are within the range of 0.9 and 2.8 $g\ kg^{-1}$ reported from natural wetlands (Viljoen, 1976; Auclair, 1979; Wiltshire, 1981; Kaul et al., 1980; Dykyjová & Květ, 1982; Ennabili et al., 1998). The values are also comparable with those reported from constructed wetlands. In Australia, Greenway (1996) and Adcock and Ganf (1994) reported the values of 2.0 $g\ kg^{-1}$ and 1.8 $g\ kg^{-1}$, respectively and Ennabili et al. (1998) reported the value of 1.2 $g\ kg^{-1}$ in Morocco. Vymazal et al. (1999) reported the P concentrations in aboveground *Phragmites* biomass in three constructed wetlands in the Czech Republic to be between 1.8 and 2.7 $g\ kg^{-1}$ and Gries and Garbe (1989) reported a wide range of 0.4–3.4 $g\ kg^{-1}$. Belowground

Table 11.6 Average phosphorus concentrations (g kg^{-1} DM) in aboveground (ABG) and belowground (BG) biomass of *Phragmites australis* and *Phalaris arundinacea* at constructed wetlands Mořina and Břehov

	Phragmites				*Phalaris*			
	Mořina		Břehov		Mořina		Břehov	
	ABG	BG	ABG	BG	ABG	BG	ABG	BG
2002	1.43	1.81			1.95	2.31		
2004	1.31	1.51	2.18	2.54	1.68	1.92	1.34	2.52
2005			2.15	2.68			1.63	3.33
2006	1.70		2.04		2.64		2.43	
2007	2.20	2.71	2.10	2.73	2.71	2.80	2.03	2.54
2008							2.36	2.23

P concentration in *Phragmites* varied between 1.51 and 2.73 g kg^{-1} with average values of 2.01 and 2.34 g kg^{-1} in Mořina and Břehov, respectively (Table 11.6, Fig. 11.4).

The aboveground *Phalaris* P concentrations varied between 1.34 and 2.64 g kg^{-1} with the average P concentration in Mořina (2.25 g kg^{-1}) being significantly higher than in Břehov (1.96 g kg^{-1}) (Fig. 11.4). These values are comparable with values reported from natural stands in the Czech Republic – Hlávková-Kumnacká (1980) and Lukavská (1989) reported the values between 1.0 and 3.5 g kg^{-1}. On the other hand, phosphorus concentrations in *Phalaris* biomass from constructed wetlands have often been reported higher. Vymazal (1995) reported P concentrations up to 4.8 g kg^{-1} in two systems in the Czech Republic and Hurry and Bellinger (1990) reported values between 1.4 and 10 g kg^{-1} in England.

Belowground P concentrations in *Phalaris* varied between 1.92 and 3.33 g kg^{-1} (Table 11.6) with no statistical difference between Mořina and Břehov (Fig. 11.3).

11.3.3.2 Standing Stock

In Tables 11.7 and 11.8, phosphorus standing stocks in *Phalaris* and *Phragmites* biomass in CWs Mořina and Břehov are shown. In Mořina (Table 11.7), aboveground P standing stocks in *Phragmites* varied between 1.31 and 2.20 g P m^{-2} with an average value of 1.66 g P m^{-2}. In *Phalaris*, phosphorus stocks varied between 2.79 and 5.01 g P m^{-2} with an average value of 3.63 g P m^{-2} which was significantly higher than in *Phragmites*. In Břehov (Table 11.8), the average *Phragmites* aboveground standing stock varied very little between 2.04 and 2.18 g P m^{-2} with an average value of 2.12 g P m^{-2}. Average P standing stock in *Phalaris* was significantly higher (3.17 g P m^{-2}) and varied between 1.89 and 3.98 g P m^{-2}.

Belowground P standing stock was higher than aboveground in *Phragmites* and lower in *Phalaris*. (Tables 11.7 and 11.8). The average aboveground/belowground P standing stock ratios were 0.77 and 0.86 for *Phragmites* in Mořina and Břehov while for *Phalaris* the respective values were 3.19 and 1.49.

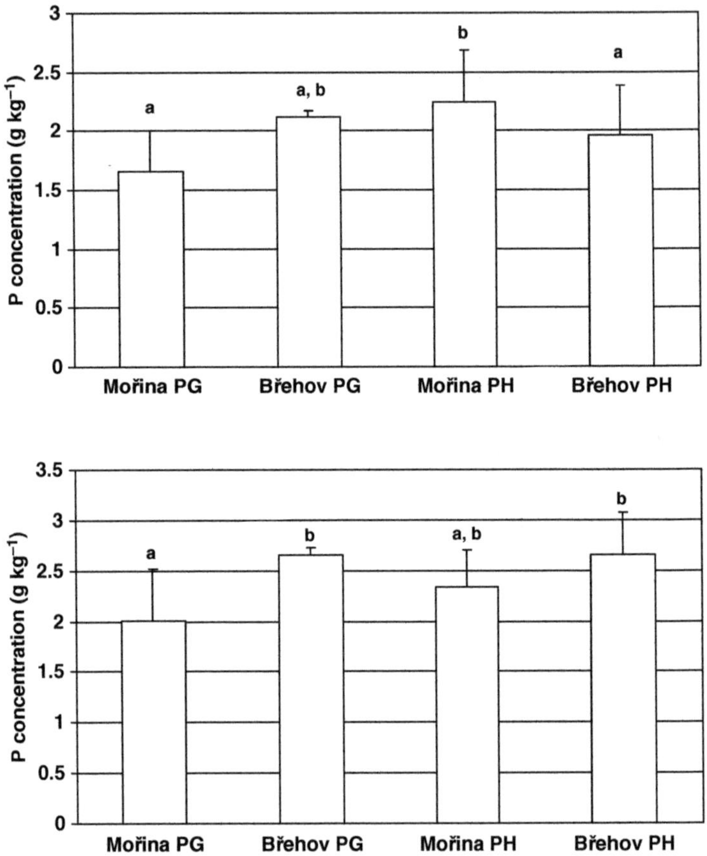

Fig. 11.4 Phosphorus concentrations in aboveground (*top*) and belowground (*bottom*) biomass of *Phragmites* (PG) and *Phalaris* (PH) in constructed wetlands Břehov and Mořina
Note: [a,b,c] Different letters indicate significant difference at $\alpha = 0.05$ between the means.

Table 11.7 Phosphorus standing stocks (g P m^{-2}) at CW Mořina. ABG = aboveground, BG = belowground

	Phragmites			*Phalaris*			*Phalaris/Phragmites*	
	ABG	BG	ABG/BG	ABG	BG	ABG/BG	ABG	BG
2002	1.43	2.01	0.71	3.48	1.50	2.32	2.43	0.75
2004	1.31	1.67	0.78	2.79	1.12	2.49	2.13	0.67
2006	1.70			5.01			2.95	
2007	2.20	2.70	0.81	3.24	0.68	4.76	1.47	0.25
Mean	1.66 [a]	2.13 [ab]	0.77	3.63 [b]	1.31 [a]	3.19	2.25	0.56

[a,b,c] Different letters indicate significant difference at $\alpha = 0.05$ between the means.

Table 11.8 Phosphorus standing stocks (g P m^{-2}) at CW Břehov. ABG = aboveground, BG = belowground

	Phragmites			*Phalaris*			*Phalaris/Phragmites*	
	ABG	BG	ABG/BG	ABG	BG	ABG/BG	ABG	BG
2004	2.18	2.18	1.00	1.89	1.21	1.56	0.87	0.56
2005	2.15	2.60	0.83	3.70	4.11	0.90	1.72	1.59
2006	2.04			3.98			1.95	
2007	2.10	2.78	0.76	3.32	1.71	1.94	1.58	0.62
2008				2.97	1.93	1.54		
Mean	2.12 [a]	2.52 [b]	0.86	3.17 [b]	2.24 [a,b]	1.49	1.53	0.92

[a,b,c] Different letters indicate significant difference at $\alpha = 0.05$ between the means.

Table 11.9 Aboveground phosphorus standing stocks in *Phragmites australis* and *Phalaris arundinacea* reported in the literature from constructed wetlands for wastewater treatment. In case of range, upper value is given

Location	Plant	Standing stock (g P m^{-2})	Reference
Poland	*Phragmites*	5.0	Obarska-Pempkowiak (1999)
Czech Republic		4.9	Vymazal et al. (1999)
USA		4.1	Behrends et al. (1994)
The Netherlands		3.5	De Jong (1976)
Australia		3.4	Headley (2004)
Poland		2.8	Obarska-Pempkowiak (1999)
Austria		2.4	Haberl and Perfler (1990)
Germany		1.6	Gries and Garbe (1989)
Australia		1.4	Adcock and Ganf (1994)
USA		0.7	Peverly et al. (1993)
UK	*Phalaris*	10.5	Hurry and Bellinger (1990) multiple harvest
USA		3.3	Bernard and Lauve (1995)
Czech Republic		1.8	Vymazal et al. (1999)
USA		1.7	Behrends et al. (1994)

The results from CWs Mořina and Břehov (Tables 11.7 and 11.8) indicate that phosphorus standing stocks are within the range commonly reported for *Phragmites* and *Phalaris* from constructed wetlands for wastewater treatment (Table 11.9). However, presented data for *Phragmites* are at the lower end of the range given in Table 11.9.

The *Phragmites* standing stocks in our study are also comparable with standing stocks found in natural stands, but they are also at the lower end of the range. Ho (1979b, 1981) reported the range of 3.44–8.94 g P m^{-2} for Scottish lochs and Hocking (1989) found aboveground P standing stock as high as 12.2 g P m^{-2} in enriched natural stand in Australia. Květ (1973), Dykyjová (1973a, 1973b, 1989) and Úlehlová et al. (1973) reported P aboveground standing stock for *Phragmites* in various localities in the Czech Republic in the range of 1.1–5.3 g P m^{-2}. Ennabili et al. (1998) reported the value of 2.76 g P m^{-2} from Morocco. Mason and Bryant

(1975) reported values up to 1.6 g P m^{-2} in England. Similar to nitrogen, phosphorus standing stocks in aboveground *Phalaris* biomass in natural stands are usually lower than those found in constructed wetlands. Klopatek (1978) reported the value of 1.92 g P m^{-2} from Wisconsin, and Lukavská (1989) reported the range of 1.24–2.01 g P m^{-2} in several sites in the Czech Republic.

11.4 Conclusions

Phragmites australis and *Phalaris arundinacea* are two most commonly used plants in constructed wetlands for wastewater treatment in the Czech Republic. *Phragmites* aboveground biomass varied between 781 and 2532 g m^{-2} in CW Břehov and between 1309 and 2177 g m^{-2} in CW Mořina. *Phalaris* aboveground biomass varied between 1262 and 2265 g m^{-2} in Břehov and between 1194 and 1780 g m^{-2} in Mořina. These values are within the common range found in both natural and constructed wetlands for wastewater treatment. For both *Phragmites* and *Phalaris* the aboveground biomass was higher than belowground. Nitrogen and phosphorus concentrations in *Phragmites* aboveground biomass were comparable with concentrations found in natural stands and constructed wetlands. On the other hand, N and P concentrations in *Phalaris* were lower than those found in constructed wetlands. Nitrogen standing stocks in *Phragmites* and *Phalaris* were comparable with those found in natural stands and constructed wetlands and for both species substantially more nitrogen was sequestered aboveground. When comparing *Phragmites* and *Phalaris* N standing stocks, *Phragmites* stocks were higher but only belowground standing stock was significantly higher. The results indicated that phosphorus standing stocks were within the range commonly reported for *Phragmites* and *Phalaris* from constructed wetlands for wastewater treatment and natural wetlands.

Acknowledgements The study was supported by grant No. 206/06/0058 "Monitoring of Heavy Metals and Selected Risk Elements during Wastewater Treatment in Constructed Wetlands" from the Czech Science Foundation and grant No. 2B06023 "Development of Mass and Energy Flows Evaluation in Selected Ecosystems" from the Ministry of Education, Youth and Sport of the Czech Republic.

References

Adcock, P., & Ganf, G. G. (1994). Growth characteristics of three macrophyte species growing in natural and constructed wetland systems. *Water Science & Technology, 29*, 95–102.

Allen, S. E., & Pearsall, W. H. (1963). Leaf analysis and shoot production in *Phragmites*. *Oikos, 14*, 176–189.

Auclair, A. N. D. (1979). Factors affecting tissue nutrient concentrations in a *Scirpus-Equisetum* wetland. *Ecology, 60*, 337–348.

Behrends, L. L., Bailey, E., Bulls, M. J., Coonrod, H. S., & Sikora, F. J. (1994). Seasonal trends in growth and biomass accumulation of selected nutrients and metals in six species of emergent aquatic macrophytes. In *Proceedings Of 4th International Conference Wetland Systems for Water Pollution Control* (pp. 274–289). Guangzhou: ICWS94 Secretariat.

Bernard, J. M., & Lauve, T. E. (1995). A comparison of growth and nutrient uptake in *Phalaris arundinacea* L. growing in a wetland and a constructed bed receiving landfill leachate. *Wetlands, 15*, 176–182.

Björk, S. (1967). Ecologic investigations of *Phragmites communis*. Studies in theoretic and applied limnology. *Folia Limnologica Scandinavica, 14*, 1–248.

Boar, R. R., Crook, C. E., & Morris, B. (1989). Regression of *Phragmites australis* reedswamp and recent changes in water chemistry in the Norfolk Broadland, England. *Aquatic Botany, 35*, 41–55.

Čížková, H. (1999). Growth dynamics and ecophysiology of *Phragmites* in relation to the climatic conditions in boreal-Mediterranean and oceanic-continental gradients. In H. Brix (Ed.), *Eureed II*, Final Project for contracts ENV4-CT95-0147 and IC20-CT-960020 (pp. 45–52). Denmark: University of Aarhus.

Coops, H., van der Brink, F. W. B., & van der Velde, G. (1996). Growth and morphological responses of four helophyte species in an experimental water-depth gradient, *Aquatic Botany, 54*, 11–24.

ČSN EN 1189. (2005). *Total phosphorus*. Czech Technical Norm.

ČSN EN 12260. (2003). *Total nitrogen*. Czech Technical Norm.

De Jong, J. (1976). The purification of wastewater with the aid of rush and reed ponds. In J. Tourbier & R. W. Pierson (Eds.), *Biological Control of Water Pollution* (pp. 133–139). Philadelphia, PA: Pennsylvania University Press.

Dubois, P. (1994). Uptake of macroelements by the helophyte *Phalaris arundinacea* L. *Aquatic Sciences, 56*, 70–79.

Dykyjová, D. (1973a). Accumulation of mineral nutrients in the biomass of reedswamp species. In S. Hejný (Ed.), *Ecosystem Study on Wetland Biome in Czechoslovakia* (pp. 151–161). Třeboň, Czech Republic (IBP/PT-PP Report No. 3): Institute of Botany, Czech Academy of Sciences.

Dykyjová, D. (1973b). Content of mineral macronutrients and emergent macrophytes during their seasonal growth and decomposition. In S. Hejný (Ed.), *Ecosystem Study on Wetland Biome in Czechoslovakia* (pp. 163–172). Třeboň, Czech Republic (IBP/PT-PP Report No. 3): Institute of Botany, Czech Academy of Sciences.

Dykyjová, D. (1989). Methods for determination of mineral nutrients pool and their cycling in the ecosystem. In D. Dykyjová (Ed.), *Methods for Ecosystem Studies* (pp. 414–435). Praha, Czech Republic: Academia (in Czech).

Dykyjová, D., & Hradecká, D. (1973). Productivity of reed-bed stands in relation to the ecotype, microclimate, and trophic conditions in the habitat. *Polish Archives of Hydrobiology*, 20, 111–119.

Dykyjová, D., & Hradecká, D. (1976). Production ecology of *Phragmites communis*. 1. Relation of two ecotypes to the microclimate and nutrient conditions of habitat, *Folia Geobotanica et Phytotaxonomica, 11*, 23–61.

Dykyjová. D., & Květ, J. (1982). Mineral nutrient economy in wetlands of the Třeboň Basin Biosphere Reserve, Czechoslovakia. In B. Gopal, R. E. Turner, R. G. Wetzel, & D. F. Whigham (Eds.), *Wetlands: Ecology and Management* (pp. 335–355). Jaipur, India: National Institute of Ecology and International Scientific Publications.

Ennabili, A., Ater, M., & Radoux, M. (1998). Biomass production and NPK retention in macrophytes from wetlands of the Tingitan Peninsula. *Aquatic Botany*, 62, 45–56.

Green, E. K., & Galatowitsch, S. M. (2001). Differences in wetland plant community establishment with additions of nitrate-N and invasive species (*Phalaris arundinacea* and *Typha* x *glauca*), *Canadian Journal of Botany, 79*, 170–178.

Greenway, M. (1996). Nutrient bioaccumulation in wetland plants receiving municipal effluent in constructed wetlands in tropical Australia. In *Proceedings of 5th International Conference on Wetland Systems for Water Pollution Control* (chapter II/I). Vienna, Austria: Universität für Bodenkultur.

Gries, C., & Garbe, D. (1989). Biomass, nitrogen, phosphorus and heavy metal content of *Phragmites australis* during the third growing season in a root zone waste water treatment. *Archiv für Hydrobiologie, 117*, 97–105.

Haberl, R., & Perfler, R. (1990). Seven years of research work and experience with wastewater treatment by a reed bed system. In P. F. Cooper & B. C. Findlater (Eds.), *Constructed Wetlands in Water Pollution Control* (pp. 205–214). Oxford: Pergamon Press.

Haslam, S. M. (1971). Shoot height and density in *Phragmites* stands. *Hydrobiologia (Bucharest), 12*, 113–119.

Haslam, S. M. (1973). Some aspects of the life history and autecology of *Phragmites communis* Trin.: A review. *Polish Archives of Hydrobiology, 20*, 79–100.

Hawke, C. J., & José, P. V. (1996). *Reedbeds. Management for Commercial and Wildlife Interests.* Bedfordshire: The Royal Society for the Protection of Birds, Sandy.

Headley, T. R. (2004). *Removal of nutrients and plant pathogens from nursery runoff using horizontal subsurface-flow constructed wetlands.* Dissertation, Southern Cross University, Lismore, NSW, Australia.

Hlávková-Kumnacká, H. (1980). *Production of Some Herbaceous Plant Species from Wetlands.* Dissertation, University of South Bohemia, České Budějovice, Czech republic (in Czech).

Ho, Y. B. (1979a). Growth, chlorophyll and mineral nutrient studies on *Phalaris arundinacea* L. in three Scottish lochs. *Hydrobiologia, 63*, 33–43.

Ho, Y. B. (1979b). Shoot development and production studies of *Phragmites australis* (Cav.) Trin. Ex Stuedel in Scottish lochs. *Hydrobiologia, 64*, 215–222.

Ho, Y. B. (1981). Mineral composition of *Phragmites australis* in Scottish lochs as related to eutrophication. I. Seasonal change in organs. *Hydrobiologia, 75*, 227–238.

Hocking, P. J., Finlayson, C. M., & Chick, A. J. (1983). The biology of Australian weeds. 12. *Phragmites australis* (Cav.) Trin. ex Steudel. *Journal of Australian Institute of Agricultural Science, 40*, 123–132.

Hocking, P. J. (1989). Seasonal dynamics of production, and nutrient accumulation and cycling by *Phragmites australis* (Cav.) Trin ex Stuedel in a nutrient-enriched swamp in inland Australia. I. Whole plants. *Australian Journal of Marine and Freshwater Res*earch, *40*, 421–444.

Hurry, R. J., & Bellinger, E. G. (1990). Potential yield and nutrient removal by harvesting of *Phalaris arundinacea* in a wetland treatment system. In P. F. Cooper & B. C. Findlater (Eds.), *Constructed Wetlands in Water Pollution Control* (pp. 543–546). Oxford: Pergamon Press.

Kätterer, T., & Andrén, O. (1999). Growth dynamics of reed canarygrass (*Phalaris arundinacea* L.) and its allocation of biomass and nitrogen belowground in a field receiving daily irrigation and fertilization. *Nutrient Cycling in Agroecosystems, 54*, 21–29.

Kaul, V., Trisal, C. L., & Kaul, S. (1980). Mineral removal potential of some macrophytes in two lakes of Kashmir. *Journal of Indian Botanical Sciences, 59*, 108–118.

Kephart, K. D., & Buxton, D. R. (1993). Forage quality responses of C3 and C4 perennial grasses to shade. *Crop Science, 33*, 831–837.

Kerchler, S. M., & Zedler, J. B. (2004). Multiple disturbances accelarate invasion of reed canary grass (*Phalaris arundinacea* L.) in a mesocosm study. *Oecologia, 138*, 455–464.

Klopatek, J. (1978). Nutrient dynamics of freshwater riverine marshes and the role of emergent macrophytes. In R. E. Good, D. F. Whigham & R. L. Simpson (Eds.), *Freshwater Wetlands: Ecological Processes and Management Potential* (pp. 195–216). New York: Academic.

Kline, P., & Boersma, K. (1983). The yield, nitrogen and nitrate content of reed canarygrass, meadow foxtail and timothy fertilized with nitrogen. *Canadian Journal of Plant Sciences, 63*, 943–950.

Květ, J. (1973). Mineral nutrients in shoots of reed (*Phragmites communis* Trin.). *Polish Archives of Hydrobiology, 20*, 137–147.

Lavergne, S., & Molofsky, J. (2004). Reed canary grass (*Phalaris arundinacea*) as a biological model in the study of plant invasions. *Critical Reviews in Plant Sciences, 23*, 415–429.

Lewandowski, I., Scurlock, J. M. O., Lindvall, E., & Christou, M. (2003). The development and current status of perennial rhizomatous grases as energy crops in the US and Europe. *Biomass and Bioenergy, 25*, 335–361.

Lukavská, J. (1989). *Influence of mowing on the wet grassland productivity.* Dissertation, University of South Bohemia, České Budějovice, Czech Republic (in Czech).

Marten, G. C. (1985). Reed canarygrass. In M. E. Heath et al. (Eds.), *Forages: The science of Grassland agriculture* (4th ed., pp. 207–216). Ames: Iowa State University Press.

Mason, C. F., & Bryant, R. J. (1975). Production, nutrient content and decomposition of *Phragmites australis* Trin. and *Typha latifolia* L. *Journal of Ecology, 63*, 71–95.

Maurer, D., & Zedler, J. B. (2002). Differential invasion of a wetland grass explained by tests of nutrients and light availability on establishment and clonal growth. *Oecologia, 131*, 279–288.

Merigliano, M. F., & Lesica, P. (1998). The native status of reed canary grass (*Phalaris arundinacea* L.) in the inland Northwest, USA. *Natural Areas Journal, 18*, 223–230.

Miller, R. C., & Zedler, J. B. (2003). Responses of native and invasive wetland plants to hydroperiod and water depth. *Plant Ecology, 167*, 57–69.

Obarska-Pempkowiak, H. (1999). Nutrient cycling and retention in constructed wetland systems in Darzlubie near Puck Bay, Southern Baltic Sea. In J. Vymazal (Ed.), *Nutrient Cycling and Retention in Natural and Constructed Wetlands* (pp. 41–48). Leiden: Backhuys Publishers.

Obarska-Pempkowiak, H., & Gajewska, M. (2003). Comparison of usefulness of three emergent macrophytes for surface water protection against pollution and eutrophication: case study, Bielkowo, Poland. In J. Vymazal (Ed.), *Wetlands – Nutrients, Metals and Mass Cycling* (pp. 129–142). Leiden: Backhuys Publishers.

Pénzes, A. (1960). Über die Morphologie, Dynamik und zönologische Rolle der sprosskolonien-bildenden Pflanzen. *Fragm. Florist. Geobot. (Kraków), 6*, 501–515.

Peverly, J., Sanford, W. E., Steenhuis, T. S., & Surface, J. M. (1993). *Constructed wetlands for municipal solid waste landfill leachate treatment.* Cornell University and U.S. Geological Survey Report No. 1011-ERER-MW-88.

Rodewald-Rudescu, L. (1974). *Das Schilfrohr Phragmites communis Trinius, Die Binnengewässer 27. Stuttgart.* Germany: E. Schweizerbart.

Soetaert, K., Hoffmann, M., Meire, P., Starink, M., van Oevelen, D., Van Regenmortel, S., et al. (2004). Modelling growth and carbon allocation in two reed beds (*Phragmites australis*) in the Scheldt estuary. *Aquatic Botany, 79*, 211–234.

Sucharová, J., & Suchara, I. (2006). Determination of 36 elements in plant reference materials with different Si contents by inductively coupled plasma mass spectrometry: Comparison of microwave digestions assisted by three types of digestion mixtures. *Analytica Chimica Acta*, 25, 163–176.

Thesiger, W. (1964). *The Marsh Arabs*. London: Longmans.

Úlehlová, B., Husák, Š., & Dvořák, J. (1973). Mineral cycles in reed stands of Nesyt fishpond in southern Moravia. *Polish Archives of Hydrobiology, 20*, 121–129.

Viljoen, L. (1976). Gebruike van *Phragmites australis* (Uses of *Phragmites australis*). *Hand. Wiedingsveren. South Africa, 11*, 19–22.

Vymazal, J. (1995). *Constructed Wetlands for Wastewater Treatment.* Czech Republic: ENVI Třeboň (in Czech).

Vymazal, J. (2002). The use of sub-surface constructed wetlands for wastewater treatment in the Czech Republic: 10 years experience. *Ecological Engineering, 18*, 633–646.

Vymazal, J. (2005). Removal of nitrogen via harvesting of emergent vegetation in constructed wetlands for wastewater treatment. In J. Vymazal (Ed.), *Natural and Constructed Wetlands: Nutrients, Metals and Management* (pp.209–211). Leiden: Backhuys Publishers.

Vymazal, J., & Kröpfelová, L. (2005). Growth of *Phragmites australis* and *Phalaris arundinacea* in constructed wetlands for wastewater treatment in the Czech Republic. *Ecological Engineering, 25*, 606–621.

Vymazal, J., & Kröpfelová, L. (2008). *Wastewater Treatment in Constructed Wetlands with Horizontal Sub-Surface Flow*. Dordrecht: Springer.

Vymazal, J., Dušek, J., & Květ, J. (1999). Nutrient uptake and storage by plants in constructed wetlands with horizontal sub-surface flow: a comparative study. In J. Vymazal (Ed.), *Nutrient Cycling and Retention in Natural and Constructed Wetlands* (pp. 85–100). Leiden: Backhuys Publishers.

Wiltshire, G. H. (1981). *Productivity of reed beds round a silt-laden dam.* Project NP14/106/3/1 Progress Report No. 3, CSIR, South Africa.

Chapter 12
Efficiency of Wastewater Treatment Constructed Wetlands during Non-Vegetation Season in the Czech Republic

Miloš Rozkošný

Abstract In the Czech Republic, water management authorities are required to maintain continuous treatment, throughout the year, of organic pollution and suspended solids at wastewater treatment plants (WWTP) with under 500 population equivalents (PE) and at WWTP 500–2000 PE also for the presence of ammonia nitrogen. Constructed wetlands built for wastewater treatment are not excepted. But their treatment efficiency during the winter period, when the macrophytic vegetation (mostly *Phragmites australis* and *Phalaris arundinacea*) is senescent, is often questioned. Under the climatic conditions of the Czech Republic (Central Europe, temperate climate), there is a non-vegetation period between half of October and the end of March. This non-vegetation period coincides with the winter period. In this chapter we assess the results of this treatment efficiency monitoring during the mentioned period of a year. The data were collected monthly at six constructed wetlands built for the load between 150–800 PE during the period 2000–2008. The main result is that the monitored constructed wetlands, which are representative samples of the Czech constructed wetlands, do not have statistically lower treatment efficiency during the non-vegetation period for the organic pollution and suspended solids. Only for the ammonia nitrogen, a significantly lower treatment efficiency during the non-vegetation period in comparison with the rest of year (vegetation period) was confirmed. The results have validated that constructed wetlands are applicable to wastewater treatment in villages under 500 PE and they can be used for the same purpose in the villages between 500 and 1000 PE, but the stable ammonia nitrogen treatment efficiency must be ensured by another wastewater treatment technology.

Keywords Constructed wetlands · Goodness-of-fit test · Mann-Whitney test · Non-vegetation period · Treatment efficiency · Wastewater treatment

M. Rozkošný (✉)
Výzkumný ústav vodohospodářský T.G.M., v.v.i., Mojmírovo nám.16, 612 00, Brno, Czech Republic
e-mail: milos_rozkosny@vuv.cz

J. Vymazal (ed.), *Water and Nutrient Management in Natural and Constructed Wetlands*, DOI 10.1007/978-90-481-9585-5_12,

12.1 Introduction

In the Czech Republic, based on Czech legislation (Anonymous, 2007), water management authorities require, throughout the full year, continuous treatment of organic pollution and suspended solids at wastewater treatment plants (WWTP) under 500 population equivalents (PE) and for WWTP 500–2000 PE also for ammonia nitrogen. Owners or operators of the treatment plant must guarantee proper treatment efficiency the entire year. Constructed wetlands (CW) built for wastewater treatment are not excepted. CW performance is affected by a range of factors such as operation mode (loading rate, continuous or batch-load) and environmental conditions (climate, season) (Chazarenc, Maltais-Landry, Troesch, Comeau, & Brisson, 2006). Temperature is one of the main characteristics affecting removal efficiency (Kuschk et al., 2003). Unlike the mechanical wastewater treatment plants, the treatment efficiency of constructed wetlands during the non-vegetation period, when the macrophytes (mostly *Phragmites australis* and *Phalaris arundinacea*) senesce, has been often evaluated in the Czech Republic (Rozkošný, 2008). We can take advantage of available results from cold climate areas, because the winter period comprises the main part of a non-vegetation period. Adaptation to a cold climate of wastewater treatment in horizontal subsurface flow constructed wetlands was summarized by Jenssen, Mæhlum, and Krogstad (1993), Mæhlum and Stålnacke (1999) and Mander and Jenssen (2002). Additional protection from freezing of reed beds may be obtained through specific design (bed depth) and with proper choice of materials for an insulation layer (snow, ice, straw, old dry macrophyte biomass, etc.) (Mæhlum & Jenssen, 2002; Kadlec & Wallace, 2009). The impact of such layers on the internal increase of temperature in the reed bed was reported by Kadlec et al. (2000) and Steinmann, Weinhart, and Melzel (2003). Detailed description of treatment process conditions in a cold climate was reported by Ouellet-Plamondon, Chazarenc, Comeau, and Brisson (2006). Under the climatic conditions of the Czech Republic (Central Europe, temperate climate), the non-vegetation period for most species is between half of October and the end of March (Rozkošný, Šálek, & Šálek, 2006). The aim of the chapter is to present results of an assessment of the efficiency of constructed wetlands treatment, including a comparison between vegetation and non-vegetation periods. Attention was paid to the constructed wetlands with capacity between 100 and 1000 PE.

12.2 Methods

The data were collected monthly during the period 2000–2008 at six constructed wetlands designed for a load between 150–800 PE (Table 12.1, Figs. 12.1 and 12.2). All CWs have a subsurface horizontal continuous flow and have been plated by *Phragmites australis*. This design is the most widely used CW type within the Czech Republic (Rozkošný, 2008; Vymazal, 2002). The reed bed's depth is 0.80 m in all cases. Usually, two distribution pipelines were built– the first one on the surface of the filtration medium (for vegetation period operation) and the second

Table 12.1 Basic information and design parameters of the monitored CWs. S = fine screens, ST = horizontal sand trap, SEPT = septic tank, IT = Imhoff tank, B = settling basin (pond), SWT = stormwater tank, PT = prismatic settling tank with side sludge digester chambers, Phragmites = *Phragmites australis* (Common reed)

	Dražovice	Myslibořice	Němčičky	Olší nad Oslavou	Rudíkov	Žernovník
Start of operation	2000	2004	1994 (end 2002)	1995	1996	1995
Design PE	780	480	640	267	675	150
Altitude	250	400	150	500	520	390
Type of sewerage	Combined	Combined	Combined	Combined	Combined	Combined
Measured average flow ($m^3\ d^{-1}$)	198.7	85.5	117.4	196.9	106.4	174.5
Pretreatment	S-ST-IT-SWT	S-ST-PT	S-ST-SEPT	S-ST-SEPT	S-ST-PT	B
Number of beds	3	2	2	3	2	3
Reed beds area (m^2)	3900	2400	1850	2260	1550	540
Area per PE (m^2)	5.0	5.0	2.9	8.5	2.3	3.6
Filtration material	gravel	gravel	gravel	gravel and sand	gravel	gravel
Plants	Phragmites	Phragmites	Phragmites	Phragmites	Phragmites	Phragmites
Final purification	Stabilisation pond	No	No	No	No	No

Fig. 12.1 Olší nad Oslavou CW during the winter. Photo author

Fig. 12.2 Inflow part of Dražovice CW final purification aerobic pond during the winter. Photo author

one about 0.20 m below the surface of the medium (for the non-vegetation period operation). Older datasets were available for two constructed wetlands located in the villages Němčičky and Rudíkov. Wastewater samples for determination of the ammonia nitrogen concentration were collected twice a month at Dražovice CW between 2005 and 2008. The following parameters were measured according to Czech Technical Norms: water and air temperatures, dissolved oxygen, pH, electrical conductivity, suspended solids (TSS), biochemical oxygen demand (BOD_5), chemical oxygen demand (COD), ammonia nitrogen ($N\text{-}NH_4^+$), total phosphorus (TP). The water temperature, air temperature and wastewater flow were measured continuously by data loggers at Dražovice over the same period. These parameters were measured also at the other localities, but not continuously, only once a month. More detailed description of monitoring programmes was mentioned by Rozkošný

(2008). Because there is a presumption that the treatment efficiency decreases in time, the obtained datasets were divided into 2-year periods for selected surveyed CWs and the treatment efficiency parameters were calculated for vegetation and non-vegetation parts of those periods.

Data obtained during the monitoring of the CWs (Table 12.1) from the effluent sampling points were evaluated with statistical tools. Data for selected parameters of water quality – BOD_5, COD, TSS, NH_4^+-N and TP – were divided into two parts: the first part corresponds with the vegetation period of a year and the second one corresponds with the non-vegetation part of a year. All the data sets were verified for normal distribution. Next, the statistical F-test (variance ratio test) was used to test the differences in variances of the data sets, and the t-test (goodness-of-fit test) was used for comparison of the average values calculated from the divided data sets. These analyses were done for the whole treatment plant system. Statistical analysis included also the Mann-Whitney non-parametrical test, comparing the long-term average values of BOD_5, COD, suspended solids, NH_4^+-N and TP loading of the reed beds only.

A survey of the reed bed vegetation and water balance, which was made at the Dražovice constructed wetland and at semi-operated experimental constructed wetlands located within the same regions as the monitored constructed wetlands (Rozkošný et al., 2006; Rozkošný, 2008), helped to adopt a decision about the beginning and end of the vegetation period. Evapotranspiration was estimated by measuring the total inflow and outflow and consequent water balance calculation. The evapotranspiration of *Phragmites australis* usually starts during March and ends in the second part of October at the monitored localities.

12.3 Results and Discussion

12.3.1 Water Temperature Measurement Within the Reed Beds

The assessment of the relationship between wastewater temperature and treatment efficiency was carried out for the dataset of the Dražovice CW. Water temperature within the wetland's reed beds was lower than 5°C in the period from half of December to half of March (Fig. 12.3) and exceeded 10°C usually between mid-May and half or end of October every particular year during the period 2006–2009. The temperature of water at the influent was about 2 or 3°C higher than inside the filtration medium close to the outflow shaft during the whole year (Fig. 12.3). Similar water temperature decrease along a reed bed was obtained by monitoring of the SSF wetland operating in a very cold climate in the USA (Kadlec et al., 2000). The course of water temperatures at the inflow and outflow shafts of the CWs at Myslibořice, Olší nad Oslavou and Žernovník villages are shown in Figs. 12.4, 12.5, and 12.6. The water temperatures decreased below 5°C between December and March within the reed beds during the monitoring period. The lowest winter temperatures were measured in reed beds of the CW Žernovník. These reed beds are

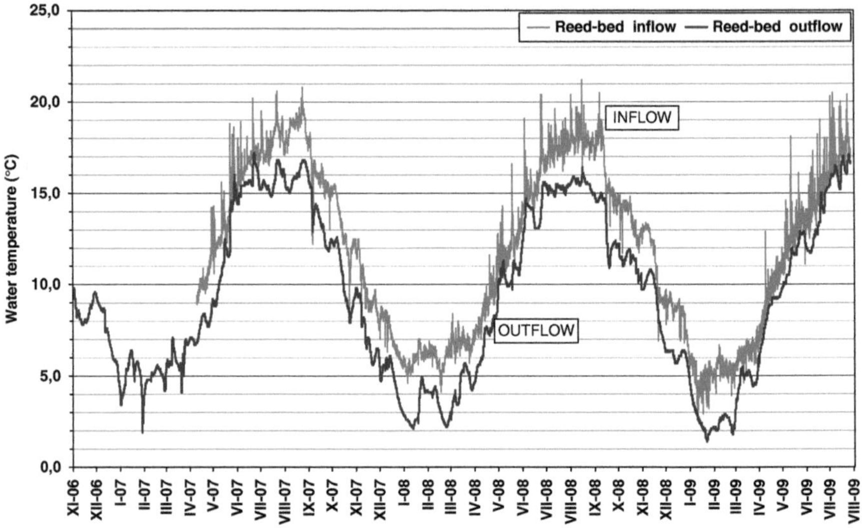

Fig. 12.3 Water temperature within the first reed bed of Dražovice CW

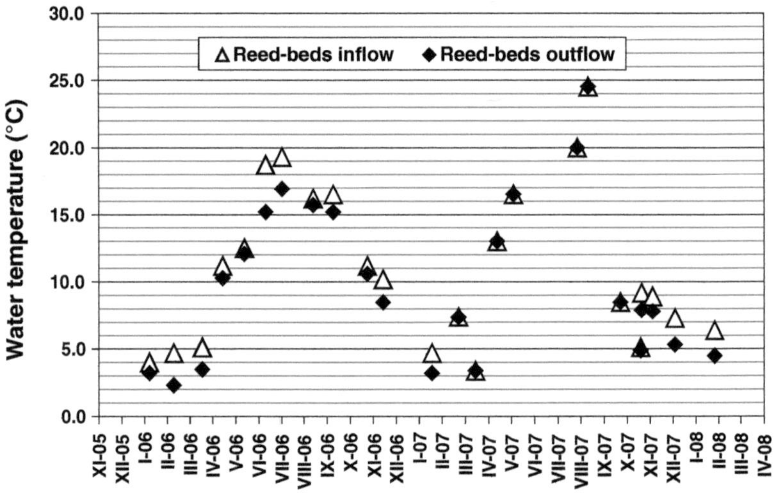

Fig. 12.4 Water temperature at the inflow and outflow shafts of Myslibořice CW

located in a cold basin which can cause a higher decrease in temperature. The annual cycle of wetland water temperatures and the relation to air temperatures, humidity and other climate characteristics are described by Kadlec and Wallace (2009).

Biological removal of N is most efficient at 20–25°C (Sutton, Murphy, & Dawson, 1975). Temperatures of water in a filtration beds influence both microbial activity and diffusion rates (Phipps & Crumpton, 1994). Water temperatures lower than 15°C have been shown to reduce the growth rate of nitrifying bacteria

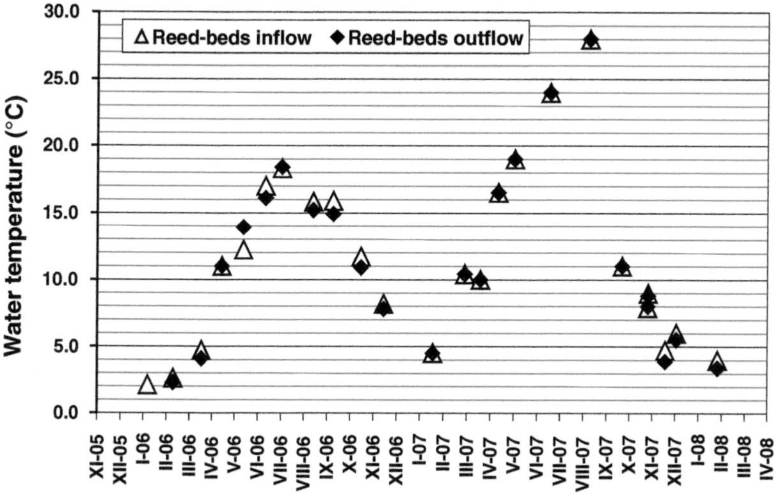

Fig. 12.5 Water temperature at the inflow and outflow shafts of Olší nad Oslavou CW

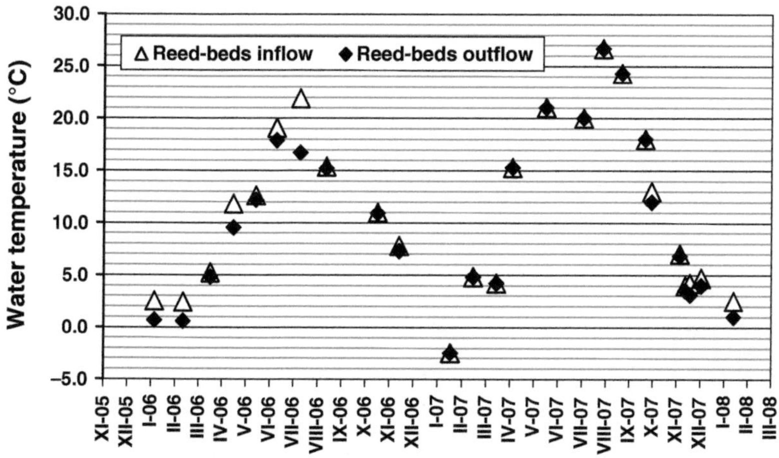

Fig. 12.6 Water temperature at the inflow and outflow shafts of Žernovník CW

(Spieles & Mitsch, 2000; Akratos & Tsihrintzis, 2007). However, nitrate removal can occur during the whole year even in cold climates and during a non-vegetation period (Brodrick, Cullen, & Maher, 1988; Phipps & Crumpton, 1994; Spieles & Mitsch, 2000). Generally, the nitrification process is stopped by a water temperature lower than 5°C (Šálek & Malý, 2001). For the denitrification processes, Brodrick et al. (1988) reported that significant denitrification occurred at temperatures as low as 5°C, but in the laboratory. Despite the low water temperature, biennial average ammonia nitrogen removal efficiency reached values between 12% and 32% (24 % on long-term average) with respective reed-beds outflow concentrations of

8.2 to 43.6 mg L^{-1} in the Dražovice CW during non-vegetation periods in the years from 2000 to 2008. Inflow concentrations of ammonia nitrogen varied from 3.6 to 65.9 mg L^{-1} during non-vegetation periods in the same years.

12.3.2 Hydraulic Load Assessment

Hydraulic loading for each of the monitored CWs was calculated from the flow measurement data. The recommended value of the hydraulic load for reed beds with a horizontal subsurface flow is between 2–8 cm d^{-1} (20–80 L m^{-2} d^{-1}) in the Czech Republic (Šálek & Tlapák, 2006). This range was exceeded in the Žernovník CW during its non-vegetation period; the average was eight times higher. Also, the peak values are higher than the recommended range in CW Dražovice (both periods) and again in the CW Žernovník CW (both periods) (Fig. 12.7). This was caused by improper functioning of the storm overflow structure during the snow melting period. This situation is often observed in combined sewage systems.

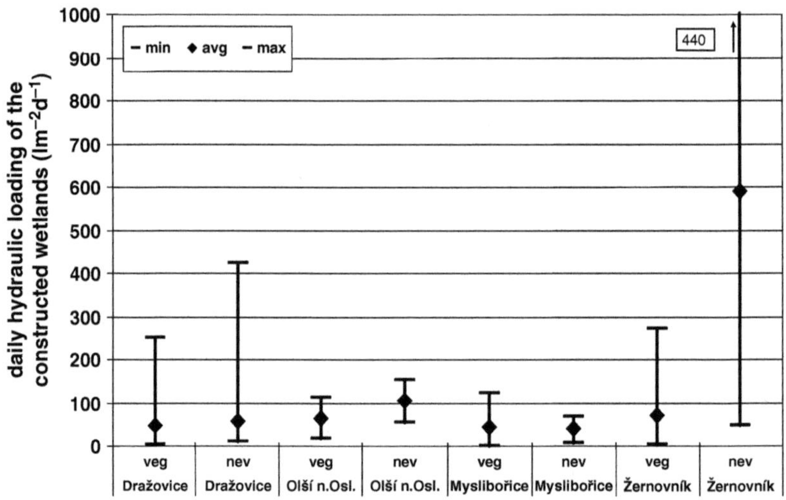

Fig. 12.7 The range of daily hydraulic loading (q) of the monitored constructed wetlands during the vegetation (veg) and non-vegetation (nev) periods

12.3.3 Treatment Efficiency Assessment

Comparison between input and output concentrations for selected parameters of water contamination is shown in Tables 12.3, 12.4, 12.5, and 12.6. The treatment efficiency changes in time were not calculated for the plants in the Myslibořice and Žernovník villages, because detailed monitoring of water quality had not been provided before our research started in 2006. The Tables present results of the whole treatment for all plant systems with inflow of raw sewage and outflow from the vegetated beds. Only in CW Dražovice is the final outflow represented by a discharge from a stabilisation pond with surface area 780 m^2 and water depth of about 1 m.

The assessment of total suspended solids removal from wastewater is not included in Tables 12.2, 12.3, 12.4, and 12.5. The treatment efficiency for this parameter ranges from 67% (in Myslibořice) up to 90% (Dražovice, Olší nad Oslavou, Rudíkov) and is especially dependent on the design and operation of sedimentation tanks. Seasonal decrease of treatment efficiency and increase of the suspended solids amount in outflow is often caused by the biological cycle of vegetation (particularly of algae) in stabilisation ponds and occurs typically at the end of May and at the end of the vegetation season in October. Such results were discovered during the monitoring of CW Dražovice. A statistically significant difference for the TSS was not proved by the research (Table 12.6). TSS removal higher than 95% was observed for horizontal subsurface CWs regardless of season, presence of plant or aeration (Chazarenc et al., 2006; Ouellet-Plamondon et al., 2006). Mean summer and winter TSS removal of almost 80% was recorded irrespective of individual CW's loading rate and plant presence (Tanner, Clayton, & Upsdell, 1995).

In most of the operational periods we have presented, the organic contamination expressed in terms of BOD_5 showed small differences during the year with moderately higher average treatment efficiency during the vegetation season, with the exception of CW Dražovice during the start-up period (Table 12.2). The start-up of a new wetland system is a critical time, because plants and microbes within the filtration medium must adjust to the hydrological conditions in wetland Kadlec et al. (2000). However, the situation in CW Dražovice could have been caused by the different hydraulic load during the two periods of the year. The peak values of the hydraulic load are almost twice as high in the non-vegetation period as in the vegetation one (Fig. 12.7), but the average daily hydraulic load values are similar. The results from CW Nemčičky revealed a loss of treatment potential due to clogging caused by the suspended solids and sludge released from the under-designed septic tank used for pretreatment of wastewater (Table 12.3). The whole constructed wetland was overloaded and the excessive BOD loading (concentration of BOD_5 over 1000 mg L^{-1}) occurred during autumn seasons. These factors led to the total clogging of reed beds and closing of the whole treatment plant after 8 years of operation. Calculated values of efficiency for the second parameter of organic contamination – COD – are very close during vegetation and non-vegetation periods, once again except for the Dražovice CW during the start-up period. No significant difference for BOD_5 and COD removal was found between vegetation and non-vegetation periods (Tables 12.2, 12.3, 12.4, and 12.5). Similar results were also presented by Akratos and Tsihrintzis (2007) and Steinmann et al. (2003) and their results indicate that the removal of organic matter is mostly a result of aerobic and anaerobic bacteria microbial activity, which function even in temperatures as low as 5°C (Steer, Fraser, Boddy, & Siebert, 2002; Vymazal, 2002). The decrease in removal

Table 12.2 Comparison of the treatment efficiency of CW Dražovice during vegetation and non-vegetation period – 2000/2001, 2002/2003 and 2006/2007. Eff. = treatment efficiency

	Vegetation period							Non-vegetation period (incl. winter)						
	Inlet (mg L^{-1})			Outlet (mg L^{-1})				Inlet (mg L^{-1})			Outlet (mg L^{-1})			
Parameter	min	avg	max	min	avg	max	Eff. (%)	min	avg	max	min	avg	max	Eff. (%)
Two years long period of the monitoring after the building of the WWTP – 2000/2001														
BOD_5	10	46.5	145	4.4	22.8	115	51	12	77	310	6	19	81	75
COD	27	121	248	20	92	364	24	44	183	748	23	76	222	59
NH_4^+-N	4.8	20.4	50.7	5.4	13.6	21.8	34	9.6	27.7	51.2	11.3	19.2	29.2	31
Two years long period of the monitoring – 2002/2003														
BOD_5	8.4	31.9	69	3.3	13.7	19	57	61	66	71	27	30	33	55
COD	52	85	139	31	55	78	35	107	118	128	61	64	66	46
NH_4^+-N	20.8	34.1	57.2	17.9	21.1	26.3	38	23.3	37.1	52.8	14.7	22.9	27.3	38
Two years long period of the monitoring – 2006/2007														
BOD_5	22	78.1	166	7	15.4	24.2	80	6.7	58.9	125	5.0	16.9	26.3	71
COD	52	133	255	41	53	69	60	31	116	177	29	42	49	64
NH_4^+-N	13.0	34.0	69.3	9.9	17.2	29.2	49	3.6	37.7	55.3	10.6	24.2	29.4	36

Table 12.3 Comparison of the treatment efficiency of CW Němčičky during vegetation and non-vegetation period – 1994/1995 and 1998/1999. Eff. = treatment efficiency

	Vegetation period							Non-vegetation period (incl. winter)						
	Inlet (mg L^{-1})			Outlet (mg L^{-1})				Inlet (mg L^{-1})			Outlet (mg L^{-1})			
Parameter	min	avg	max	min	avg	max	Eff. (%)	min	avg	max	min	avg	max	Eff. (%)
Two years long period of the monitoring after the building of the WWTP – 1994/1995														
BOD_5	91.5	467	1546	18.5	58.0	128	88	145	808	2230	39, 8	190	352	77
COD	175	595	1749	63	143	422	76	310	1087	3574	145	605	436	72
NH_4^+-N	22.6	50.1	80.9	11.2	19.4	28.0	61	33.3	73.7	112.1	10.7	36.1	60.4	51
Two years long period of the monitoring after the given period of operation – 1998/1999 (the WWTP closed in 2002)														
BOD_5	18	107	270	13	74	207	31	40	107	202	13	133	380	−24
COD	104	252	470	89	165	219	35	138	218	394	104	224	468	−3
NH_4^+-N	9.0	27.7	58.3	13.6	32.1	56.8	−16	14.5	42.5	87.6	27.0	42.7	64.5	−1

Table 12.4 Comparison of the treatment efficiency of CW Olší nad Oslavou during vegetation and non-vegetation period – 2002/2004 a 2006/2008. Eff. = treatment efficiency

	Vegetation period							Non-vegetation period (incl. winter)						
	Inlet (mg L^{-1})			Outlet (mg L^{-1})				Inlet (mg L^{-1})			Outlet (mg L^{-1})			
Parameter	min	avg	max	min	avg	max	Eff. (%)	min	avg	max	min	avg	max	Eff. (%)
Two years long period of the monitoring after the given period of operation – 2002/2004														
BOD_5	3.2	32.4	140	1.1	4.3	23.0	87	7.0	15.7	35.0	1.1	3.3	11.0	79
COD	12	80	269	8	34	113	58	21	58	104	9	33	59	43
NH_4^+-N	1.0	7.2	11.2	0.4	2.6	17.5	64	2.3	5.0	6.9	0.4	1.5	3.5	70
Two years long period of the monitoring after the given period of operation – 2006/2008														
BOD_5	5.9	19.2	40.0	0.2	1.4	2.6	93	3.7	15.9	38.0	0.7	1.2	1.8	93
COD	21	58	114	8	19	48	67	15	36	60	8	17	33	53
NH_4^+-N	2.8	10.6	45.8	0.1	3.1	14.9	71	0.7	3.9	8.5	0.1	2.4	11.5	38

Table 12.5 Comparison of the treatment efficiency of CW Rudíkov during vegetation and non-vegetation period – 1997/1998 and 2000/2001. Eff. = treatment efficiency

Parameter	Vegetation period – Inlet ($mg\ L^{-1}$) min	avg	max	Outlet ($mg\ L^{-1}$) min	avg	max	Eff. (%)	Non-vegetation period (incl. winter) – Inlet ($mg\ L^{-1}$) min	avg	max	Outlet ($mg\ L^{-1}$) min	avg	max	Eff. (%)
Two years long period of the monitoring after the building of the WWTP – 1997/1998														
BOD_5	14	170	760	6	14	27	92	52	220	540	7	21	47	90
COD	103	335	1172	40	86	146	74	134	356	622	52	86	144	76
NH_4^+-N	25	53	86	30	43	62	22	19	49	82	27	40	54	18
Two years long period of the monitoring after the given period of operation – 2000/2001														
BOD_5	22	109	311	1	11	34	90	24	98	343	3	13	55	87
COD	155	254	513	41	71	143	72	113	243	674	23	61	100	75
NH_4^+-N	16	55	155	21	38	55	31	9	41	72	7	37	50	10

Table 12.6 Summary of the statistical analysis results based on Mann-Whitney test

Parameter	TSS		BOD_5		COD		NH_4^+- N		TP	
Period	veg	non	veg	non	veg	non	veg	non	veg	non
	mg L^{-1}		mg L^{-1}		mg L^{-1}		mg L^{-1}		mg L^{-1}	
Reed beds Inflow										
Dražovice	52	61	46.7	62.2	113	126	29.70	29.43	5.36	5.06
Myslibořice	78	73	101.6	133.4	175	214	33.45	36.65	6.39	6.35
Olší nad Osl.	11	11	10.4	11.8	37	33	**11.23**	**5.28**	**1.81**	**0.97**
Žernovník	**68**	**37**	**71.2**	**43.0**	**158**	**86**	**41.45**	**20.10**	**7.05**	**2.98**
Reed beds Outflow										
Dražovice	11	12	20.4	27.8	66	74	**21.49**	**28.34**	4.39	5.11
Myslibořice	6	6	**14.3**	**42.1**	54	74	**32.83**	**43.87**	5.34	6.26
Olší nad Osl.	5	2	1.4	1.2	19	17	3.05	2.43	**2.05**	**0.74**
Žernovník	14	9	12.2	5.5	**55**	**29**	29.67	24.07	**5.93**	**3.64**

A significant difference in average values is highlighted (boldface)

in winter observed by Chazarenc et al. (2006) was mainly due to COD seasonal trends associated with a decrease of both internal loading and biodegradation rate (Barber, Leenheer, Noyes, & Stiles, 2001). A very significant difference between treatment efficiency during vegetation and non-vegetation periods was found in Němčičky CW during the period 1998–1999 (Table 12.3). The assessment of Olší nad Oslavou CW, where the sewage is highly diluted by ballast water, shows high treatment efficiency not only for organic pollution, but also for ammonia nitrogen (Table 12.4).

The comparison of ammonia nitrogen treatment efficiency between the Dražovice and Rudíkov CWs shows a positive impact of final purification provided by the stabilisation pond (Tables 12.2 and 12.4). Anaerobic conditions predominate in the filtration medium of reed beds of constructed wetlands with horizontal subsurface continuous flow, when the filtration medium is saturated by wastewater. This is the main factor affecting the ammonia nitrogen treatment efficiency, especially during the non-vegetation period, when the plants are dormant and do not take nitrogen from the wastewater. Brix (1994) suggested that nitrification can be limited seasonally. Efficiency of ammonia nitrogen removal is influenced by the temperature of wastewater and amount of dissolved oxygen, which is used for nitrification (Brix, 1994; Stein & Hook, 2005; Ouellet-Plamondon et al., 2006; Akratos & Tsihrintzis, 2007). The ammonia-N removal efficiency is higher during the vegetation period in the monitored CWs. This was confirmed by statistical analysis described in the text below. Similar results were reported by Kadlec et al. (2000) for a northern-climate CW treating septic tank effluent. The average outflow concentration of the ammonia-N was significantly higher in winter and spring. The average seasonal water temperature in filtration medium was calculated 2°C in winter and 5°C in spring in opposite to the 15°C in summer.

12.3.4 Results of Statistical Analysis

The statistical analyses (*F*-test and *t*-test) were performed for the wastewater treatment including mechanical pre-treatment with raw sewage as inflow. The STATGRAPHICS software was used for the analysis. The calculations did not show any significant difference for suspended solids and COD in any of the monitored CWs at $\alpha = 0.05$. There was a significant difference for BOD_5 in Myslibořice and Žernovník CWs, for $NH_4{}^+$-N in CWs Dražovice, Olší nad Oslavou and Žernovník, for TP in CW Olší nad Oslavou. A higher average inflow TP concentration in CW Olší nad Oslavou was measured during its vegetation period, probably due to higher dilution of wastewater by ballast waters and water from snow melting during the early spring period. The same situation in hydraulic loading caused significant differences in BOD_5 treatment efficiency in Myslibořice and Žernovník CWs.

The results of the Mann-Whitney non-parametrical test, used for the reed bed's treatment efficiency, showed a significant difference for all parameters in the inflow of Žernovník CW (Table 12.6). This result can be correlated with the high difference in seasonal hydraulic loading (Fig. 12.7) for both average and peak values. Treatment efficiency depends on actual hydraulic conditions and wastewater flow, which is typical for combined sewage systems of small villages. Contrary to this fact the statistical analysis did not show a significant difference for ammonia nitrogen outlet concentration in Žernovník and Olší nad Oslavou CWs (Table 12.6). Results of the analysis for the Dražovice and Myslibořice CWs validated the results of analysis mentioned above and the conclusion that there is a significant difference in the ammonia nitrogen removal capacity and treatment efficiency between the two periods of a year.

12.4 Conclusions

Monitoring of water temperature and evapotranspiration survey helped to determine the borders between vegetation and non-vegetation periods under the climatic conditions of the Czech Republic, specifically within the Jihomoravský Region and Region Vysočina. The non-vegetation period usually ends during March and vegetation period lasts usually until October for macrophytes in southern Moravia. Wastewater measurement provided datasets for hydraulic loading calculation of the monitored CWs. Consequential statistical analysis and calculation of treatment efficiency showed that the rate of hydraulic loading is a very important factor for the final assessment of treatment efficiency. Knight, Ruble, Kadlec, and Reed (1993) and Kowalik and Obarska-Pempkowiak (1994) derived equations of the correlation between outlet concentration, inlet concentration and hydraulic loading for organic pollution, suspended solids and ammonia nitrogen. An efficient hydraulic residence time (HRT) for organic matter removal was shown to be similar for different water temperatures (Akratos & Tsihrintzis, 2007), but an efficient HRT depends on the water temperature for nitrogen. Braskerud (2002) found some effect of the hydraulic load on nitrogen removal. Chazarenc et al. (2006) demonstrated that mass removal of COD is proportional to mass loading rates. The effect of hydraulic loading and retention time was examined by Huang, Reneau, and Hageborn (2000). An important decrease of the constructed wetland treatment efficiency in time for organic pollution, suspended solids and ammonia nitrogen was not proved by the research conducted in Dražovice, Olší nad Oslavou and Rudíkov CWs. However, a high decrease of treatment efficiency can be caused by clogging and inadequate design and operation of pretreatment stage of CW as it was presented for Němčičky CW. The final purification by use of a stabilisation pond has a positive effect on the ammonia nitrogen treatment efficiency, but the biological processes occurring in such a pond can contribute to the lower treatment efficiency for BOD_5 and suspended solids during certain parts of the vegetation period. Steinmann et al. (2003) suggested the use of a rock filter to eliminate suspended solids (algae). The statistical analysis comparison of the results from vegetative and dormant seasons did not show a significant difference between these periods for organic pollution, phosphorus and suspended solids. However, there was a significant difference in treatment for ammonia nitrogen with lower efficiency during the non-vegetation season. The results validated the contention that constructed wetlands are suitable treatment technology for point-pollution sources under 500 PE and they can be used for the category 500–1000 PE provided that steady ammonia nitrogen treatment efficiency is ensured by another wastewater treatment technology.

Acknowledgements This research was financially supported by the Ministry of the Environment of the Czech Republic (research project MZP0002071101).

References

Akratos, S. C., & Tsihrintzis, A. V. (2007). Effect of temperature, HRT, vegetation and porous media on removal efficiency of pilot-scale horizontal subsurface flow constructed wetlands. *Ecological Engineering, 29*, 173–191.

Anonymous. (2007). *Czech surface water quality standards – government regulation no. 229/2007 Coll.* Ministry of the Environment, the Czech Republic: Praha.

Barber, L. B., Leenheer, J. A., Noyes, T. I., & Stiles, E. A. (2001). Nature and transformation of dissolved organic matter in treatment wetlands. *Environmental Science and Technology, 35*, 4805–4816.

Braskerud, B. C. (2002). Factors affecting nitrogen retention in small constructed wetlands treating agricultural non-point source pollution. *Ecological Engineering, 18*, 351–370.

Brix, H. (1994). Use of constructed wetlands in water pollution control: historical development, present status and future perspectives. *Water Science and Technology*, 30, 209–223.

Brodrick, S. J., Cullen, P., & Maher, W. (1988). Denitrification in a natural wetland receiving secondary treated effluent. *Water Research, 22*, 431–439.

Chazarenc, F., Maltais-Landry, G., Troesch, S., Comeau, Y., & Brisson, J. (2006). Effect of loading rate on hydraulic and treatment performance of CWs treating anaerobic sludge storage tank supernatant. In V. Dias & J. Vymazal (Eds.), *Proceedings of 10th International Conference Wetland Systems for Water Pollution Control* (pp. 1697–1705). Lisbon, Portugal: MAOTDR.

Huang, J., Reneau, R., & Hageborn, C. (2000). Nitrogen removal in constructed wetlands employed to treat domestic wastewater. *Water Research, 34*, 2582–2588.

Jenssen, P. D., Mæhlum, T., & Krogstad, T. (1993). Potential use of constructed wetlands for wastewater treatment in northern environments. *Water Science and Technology, 28*, 149–157.

Kadlec, R. H., Knight, R. L., Vymazal, J., Brix, H., Cooper, P., & Haberl, R. (2000). *Constructed wetlands for pollution control. Processes, performance, design and operation.* Scientific and technical report No.8. London: IWA Publishing.

Kadlec, R. H., & Wallace, S. (2009). *Treatment wetlands* (2nd ed.). Boca Raton: CRC Press.

Knight, R. L., Ruble, R. W., Kadlec, R. H., & Reed, S. (1993). Wetlands for wastewater treatment: performance database. In G. A. Moshiri (Ed.), *Constructed wetlands for water quality improvement* (pp. 35–58). Boca Raton: CRC Press.

Kowalik, P., & Obarska-Pempkowiak, H. (1994). Zasady pracy malych hydrobotanicznych oczyszczalni ścieków [Principles of small hydrobotanical wastewater treatment plants.] Materialy informacyjne 28. Falenty: Wydawnictwo IMUZ (in Polish).

Kuschk, P., Wiebner, A., Kappelmeyer, U., Weibbrodt, E., Kastner, M., & Stottmeister, U. (2003). Annual cycle of nitrogen removal by a pilot-scale subsurface horizontal flow in a constructed wetland under moderate climate. *Water Research, 37*, 4236–4242.

Mæhlum, T., & Jenssen, P. D. (2002). Design and performance of integrated subsurface flow wetlands in a cold climate. In U. Mader & P. D Jenssen (Eds.), *Constructed Wetlands for Wastewater Treatment in Cold Climates* (pp. 69–86). Southampton: WIT Press.

Mæhlum, T., & Stålnacke, P. (1999). Removal efficiency of three cold-climate constructed wetlands treating domestic wastewater: Effects of temperature, seasons, loading rates and input concentrations. *Water Science and Technology, 40*, 273–281.

Mander, Ü., & Jenssen, P. D. (Eds.). (2002). *Constructed wetlands for wastewater treatment in cold climates. Advances in Ecological Sciences* (Vol. 11). Southampton: WIT Press.

Ouellet-Plamondon, C., Chazarenc, F., Comeau, Y., & Brisson, J. (2006). Artificial aeration to increase pollutant removal efficiency of constructed wetlands in cold climate. *Ecological Engineering, 27*, 258–264.

Phipps, R. G., & Crumpton, W. G. (1994). Factors affecting nitrogen loss in experimental wetlands with different hydrologic loads. *Ecological Engineering, 3*, 399–408.

Rozkošný, M. (2008). *Treatment efficiency assessment of the constructed wetlands and proposals of improvement of their function.* Dissertation, Department of Landscape Water Management, Faculty of Civil Engineering, Brno University of Technology, Czech Republic.

Rozkošný, M., Šálek, J., & Šálek, J. (2006). Water balance of the constructed wetlands – a study of the macrophyteś evapotranspiration. In V. Dias & J. Vymazal (Eds.), *Proceedings Of 10th International Conference Wetland Systems for Water Pollution Control* (pp. 123–130). Lisbon: MAOTDR.

Šálek, J., & Malý, J. (2001). Výzkum kyslíkového režimu a odstraňování amoniaku u vegetačních kořenových čistíren [Research of oxygen regime and ammonia nitrogen removal by the reed bed treatment plants]. In K. Valouchová (Ed.), *International seminar on water management* (pp. 187–192). Brno, zech Republic: FAST VUT Brno (in Czech).

Šálek, J., & Tlapák, V. (2006). *Natural treatment of polluted surface waters and wastewaters.* Praha, Czech Republic: ČKAIT (in Czech).

Spieles, D. J., & Mitsch, W. J. (2000). The effects of season and hydrologic and chemical loading on nitrate retention in constructed wetlands: a comparison of low- and high-nutrient riverine systems. *Ecological Engineering, 14*, 77–91.

Steer, D., Fraser, L., Boddy, J., & Siebert, B. (2002). Efficiency of small constructed wetlands for subsurface treatment of single-family domestic effluent. *Ecological Engineering, 18*, 429–440.

Stein O. R., & Hook P. B. (2005). Temperature, plants, and oxygen: How does season affect constructed wetland performance? *Journal of Environmental Science and Health, A40*, 1331–1342.

Steinmann, C. R., Weinhart, S., & Melzel, A. (2003). A combined system of lagoon and constructed wetland for an effective treatment. *Water Research, 37*, 2035–2042.

Sutton, P. M., Murphy, K. L., & Dawson, R. N. (1975). Low-temperature biological denitrification of wastewater. *Journal of Water Pollution Control Federation, 47*, 122–134.

Tanner, C. C., Clayton, J. S., & Upsdell, M. P. (1995). Effect of loading rate and planting on treatment of diary farm wastewaters in constructed wetlands – I. Removal of oxygen demand, suspended solids and faecal coliforms. *Water Research, 29*, 17–26.

Vymazal, J. (2002). The use of sub-surface constructed wetlands for wastewater treatment in the Czech Republic: 10 years experience. *Ecological Engineering, 18*, 633–646.

Chapter 13
Constructed Wetlands in the Czech Republic: 20 Years of Experience

Jan Vymazal

Abstract Constructed wetlands (CWs) have been used in the Czech Republic since 1989. At present, about 250 systems are in operation. All systems have been designed with horizontal subsurface flow and most CWs have been designed to treat municipal or domestic wastewater from both separate and combined sewerage. The majority of constructed wetlands were built either as on-site treatment systems or for small communities up to 500 PE (population equivalent). However, the numbers obtained from the survey carried out in 2009 indicate that over the years the size distribution has changed. During the first 10 years, constructed wetlands for small villages between 100 and 500 PE prevailed but after 1999 small on-site systems prevailed. The average specific area of filtration beds is 5.1 m^2 PE^{-1} but the specific area varies between 0.7 and 16.7 m^2 PE^{-1}. Crushed rock and gravel 4–8 mm or 8–16 mm are the most commonly used filtration material. *Phragmites australis* and *Phalaris arundinacea* are the most frequently used plants. Removal of BOD_5, COD and suspended solids is very high with respective median treatment efficiencies of 90, 81 and 90%. Removal of nutrients is much lower but there are no discharge limits for treatment plants up to 500 PE in the Czech Republic. Therefore, constructed wetlands in the Czech Republic are designed according to the requirements for the removal of organics and suspended solids. Two decades of constructed wetlands application in the Czech Republic have proved that CWs are a viable option for wastewater treatment in small settlements and individual houses.

Keywords Constructed wetlands · Czech Republic · Horizontal subsurface flow · On-site treatment · Wastewater treatment

---

J. Vymazal (✉)
Faculty of Environmental Sciences, Department of Landscape Ecology, Czech University of Life Sciences in Prague, Prague, Czech Republic; ENKI, o.p.s., Dukelská 145, 379 01 Třeboň, Czech Republic
e-mail: vymazal@yahoo.com

J. Vymazal (ed.), *Water and Nutrient Management in Natural and Constructed Wetlands*, DOI 10.1007/978-90-481-9585-5_13,

13.1 Introduction

Wetlands have been intensively studied in the Czech Republic for more than four decades In the late 1960s and early 1970s, numerous estimates of biomass, production, and nutrient content in helophyte communities proved their manifold connections with the availability of nutrients. The experiments were carried out in various natural wetlands including heavily polluted sites near wastewater treatment plant outfall (e.g., Dykyjová, 1971, 1980; Dykyjová & Květ, 1978, 1982; Hejný, 1973; Květ, 1973). However, the results of these experiments were not used for any kind of wastewater treatment design. Interest in the use of wetland macrophytes for wastewater treatment was revived in the Czech Republic only at the end of the 1980s. After a period of experiments in a small-scale constructed wetland (Vymazal, 1990) the first full-scale constructed wetland for the treatment of runoff waters from a dung-hill was built in 1989. Due to lack of rainwater and thus runoff in the summer of 1989, it was decided to use the system for the treatment of sewage from an adjacent village. Despite the fact that the treatment system was built with little knowledge about constructed wetlands (CWs) and was originally designed for a different type of wastewater, the treatment effect was very high (Vymazal, 1993, 1998). However, the appearance of the system, which was really far from neat, became a pretence for negative opinions on constructed wetlands given by various organizations (hygiene service, water inspection, the Ministry of the Environment etc.). Unfortunately, the results themselves were not taken into consideration and, as a result, in 1990 none and in 1991 only four constructed wetlands for wastewater treatment were built and put in full operation (Vymazal, 2006). However, after 1992 a steep increase in constructed wetland construction occurred in the Czech Republic (Vymazal, 1995, 1996a, 1996b, 2001, 2002) and since 1989 about 250 systems have been put in operation according to the survey performed in 2008 (Fig. 13.1).

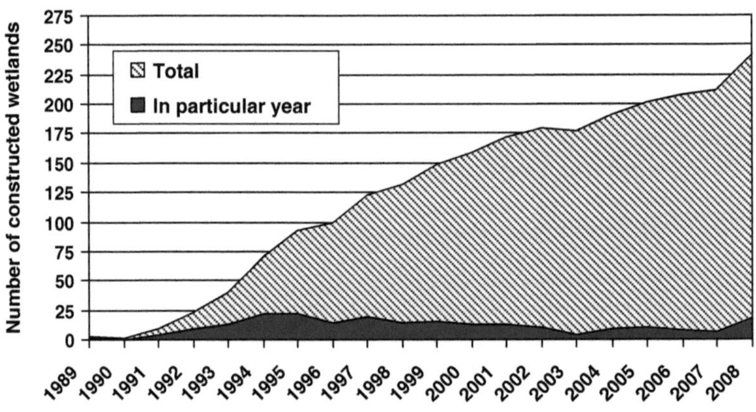

Fig. 13.1 Number of constructed wetlands put in operation in the Czech Republic. For 14 systems starting data were available in the survey

13.2 Size of Constructed Wetlands

The majority of constructed wetlands were built either as on-site treatment systems (<10 PE, 74 systems, Fig. 13.2) or for small communities (100–500 PE, 79 systems, Fig. 13.3). The size distribution of Czech constructed wetlands is shown in Fig. 13.4. However, numbers presented in Fig. 13.5 indicate that over the years the size distribution has changed. During the first 10 years constructed wetlands for

Fig. 13.2 Example of on-site constructed wetland for 4 PE. Kostelec nad Černými lesy, Czech Republic. Photo author

Fig. 13.3 Constructed wetland Cep, Czech Republic for 300 PE. Photo author

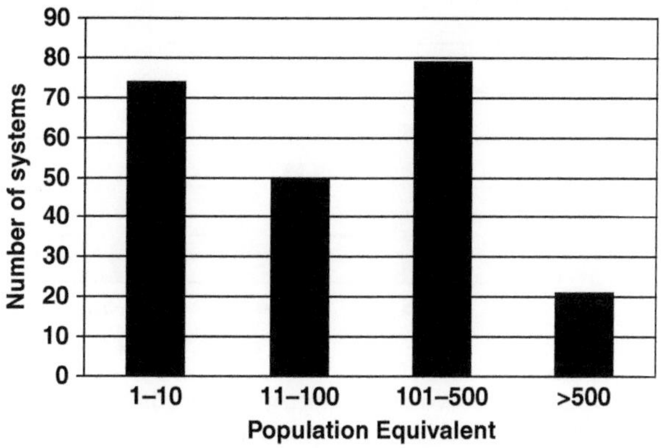

Fig. 13.4 Distribution of constructed wetlands in the Czech Republic according to the designed PE

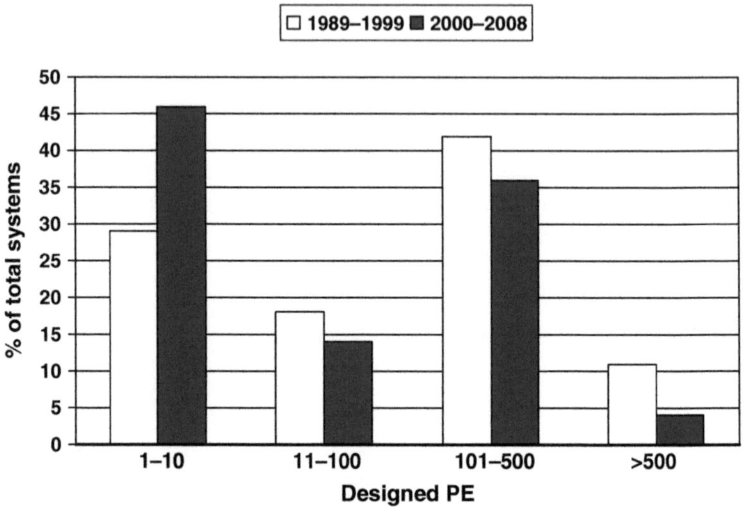

Fig. 13.5 Size distribution of constructed wetlands in the Czech Republic built during the periods 1989–1999 and 2000–2008

small villages between 100 and 500 PE prevailed but after 1999 small on-site systems prevailed. The average specific area of filtration beds is 5.1 m^2 PE^{-1} (median 5.0 m^2 PE^{-1}) but the specific area varies between 0.7 and 16.7 m^2 PE^{-1} (Fig. 13.6). These numbers, however, are based on the design number of PE. It is important to realize that at present, the number of PE does not equal the number of connected people (CP). The unit (PE = 60 g BOD_5 d^{-1}) is based on a large amount of data which was elaborated during the 1940s and the 1950s (Imhoff, 1960). Since then "household technology" has substantially changed (e.g., use of dishwashers or

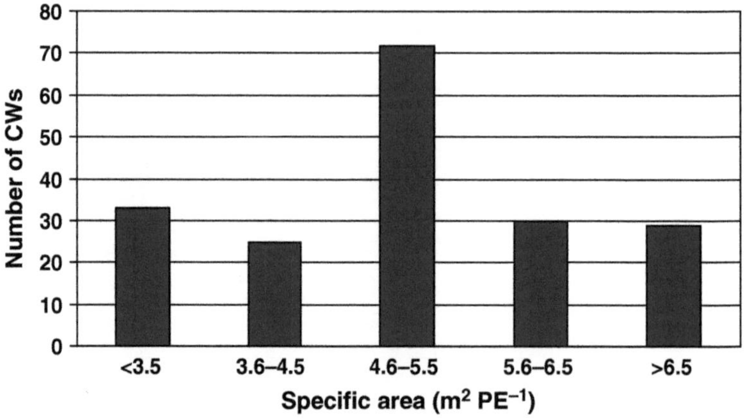

Fig. 13.6 Distribution of constructed wetlands in the Czech Republic according to the specific area. Data not available from all systems

washing machines, lower consumption of sugar, fats, etc.). All these factors resulted in a lower production of BOD_5 per person. Based on field measurements, the Czech Ministry of the Environment developed an equation for the relationship between PE and CP for settlements <2000 CP: PE $= 0.2764$ CP$^{1.1484}$. According to this equation, for example, 100 CP = 55 PE (Vymazal & Kröpfelová, 2007).

13.3 Other Design Parameters

All systems were designed with horizontal subsurface flow (HF CWs). Most constructed wetlands in the Czech Republic are designed to treat municipal or domestic wastewater from both separate and combined sewerage. Other types of wastewater include those from dairies, bakeries, abattoir facilities, and goat farms; one CW is a part of the system designed for the treatment of landfill leachate and one CW was designed to treat stormwater runoff only. With few exceptions each CW has been designed as a secondary treatment step.

Czech constructed wetlands use coarse media – gravel, crushed rock or gravel-sand. The early systems usually used fractions of 4–8 mm. At present, gravel or crushed rock with a fraction of 8–16 mm is often preferred because it has been shown that this fraction provides sufficient hydraulic conductivity and supports a healthy macrophyte growth and good treatment efficiency (Vymazal, 2006). The depth of filtration media commonly varies between 0.6 and 0.8 m.

The most important effects of macrophytes in relation to the treatment process in sub-surface horizontal flow CWs are the physical effects on plant tissue (i.e., erosion control, filtration effect, provision of surface area for attached microorganisms, insulation of the bed surface). The metabolism of the macrophytes (plant uptake, oxygen release and release of antibiotics) is of lesser importance in the HF CWs (Vymazal et al., 1998). More than a decade of experience has shown that, under the climatic conditions of the Czech Republic, the most important role of plants in CWs with subsurface horizontal flow is insulation of the bed during the winter period (Vymazal, 2006).

The most frequently used plant in the Czech CWs is *Phragmites australis* (Common reed) which is used either singly or in combination with other macrophyte species but predominantly with *Phalaris arundinacea* (Reed canarygrass). Other species used singly are *Phalaris arundinacea, Glyceria maxima* (Mannagrass), and *Typha latifolia* (Common cattail). The density of 4–8 seedlings per 1 m^2 is sufficient for dense vegetation cover (Vymazal, 2006). Weeds, if present, are restricted to vegetated bed margins (Voborník, 2009). Vegetation is not harvested at the end of the growing season because litter provides an excellent insulation for the surface of vegetated beds during periods of cold weather. The harvest usually takes place in early spring when heavy frosts are not likely. Small on-site systems very often use combinations of decorative plants such as *Iris pseudacorus* (Yellow flag), *Iris sibirica* (Blue flag), *Filipendula ulmaria* (Queen of the meadow), *Epilobium hirsutum* (Hairy willow-herb) or *Lythrum salicaria* (Purple loosestrife) (Vymazal, 2006).

13.4 Treatment Performance

13.4.1 Organics and Suspended Solids

Removal of organics and suspended solids is presented in Table 13.1. It is obvious that treatment performance is very good and median outflow concentrations of 9.6, 42 and 8.8 mg L^{-1} for BOD_5, COD and TSS, respectively are well within the discharge limits for small sources of pollution. Removal of organics and suspended solids is high for both separate and combined sewer outfalls with median removal efficiencies for BOD_5, COD and suspended solids amounting to 90, 81 and 90%, respectively. Also, detailed studies of a long-term performance of Czech CWs treating both diluted and "regular" strength sewage have shown that the removal of organics and suspended solids is very steady and do not deteriorate over the years of operation. However, there is also no apparent improvement with the length of operation (Vymazal, 2009a, 2009b). The numbers presented in Table 13.1 also show a wide variation in inflow concentrations of all parameters. This is a result of treatment of highly diluted wastewater by stormwater runoff and in many cases also drainage

Table 13.1 Treatment performance of constructed wetlands in the Czech Republic for organics and suspended solids during the period 1989–2008. Inflow (In) and outflow (Out) concentrations in mg L^{-1}, treatment efficiency (Eff) in %. n = number of annual means, CWs = number of constructed wetlands

	BOD_5			COD			TSS		
	In	Out	Eff.	In	Out	Eff.	In	Out	Eff.
Mean	171	14.4	85.3	372	52	75.6	189	12.0	82.7
Median	106	9.6	90.0	232	42	80.7	87.3	8.8	89.3
Min	3.0	1.0	6.7	9.2	2.6	−18.3	5.0	0.9	−113
Max	2540	114	99.7	8500	238	99.4	4230	262	99.8
n	423	423		396	396		410	410	
CWs	70			65			66		

Table 13.2 Loadings of constructed wetlands in the Czech Republic for organics and suspended solids during the period 1989–2008. Loadings in kg ha^{-1} d^{-1}. n = number of annual means, CW = number of constructed wetlands

	BOD_5			COD			TSS		
	In	Out	Removed	In	Out	Removed	In	Out	Removed
Mean	68.8	6.4	62.4	146	23	123	60.4	5.1	55.3
Median	39.9	3.7	35.7	95	18	74	31.7	3.2	29.3
Min	0.87	0.2	0.07	8.0	1.2	−4.1	2.3	0.4	−2.7
Max	1415	97	1318	3154	376	2278	435	150	427
n	273			258			257		
CWs	58			55			52		

waters. On the other hand, several systems are apparently under-dimensioned. High removal of organics and suspended solids from diluted wastewaters makes HF CWs an excellent treatment alternative, especially when organics and suspended solids are the major treatment target. The dilution of wastewater by rainwater and lower production of organics (see section 13.2) result in lower inflow organic loadings (Table 13.2). When real loadings are taken into consideration, the average specific area is 8.7 m^2 PE^{-1}.

13.4.2 Nutrients

Removal of nitrogen and phosphorus is low (Tables 13.3 and 13.4) but within the typical values reported from the literature for constructed wetlands with horizontal sub-surface flow (Vymazal, 2007; Vymazal & Kröpfelová, 2008). The major limitation for nitrogen removal in HF CWs is the lack of oxygen necessary for ammonia nitrification in filtration beds. Phosphorus removal is governed by sorption/precipitation with Fe, Ca and Al. The filtration materials used in the Czech Republic (crushed rock, gravel) do not provide this sorption capacity. It has been

Table 13.3 Treatment performance of constructed wetlands in the Czech Republic for nutrients during the period 1989–2008. For details see Table 13.1

	TP			TN			NH_4^+-N		
	In	Out	Eff.	In	Out	Eff.	In	Out	Eff.
Mean	6.8	3.7	36.4	49.8	25.4	44.9	31.1	18.2	33.2
Median	5.3	3.0	39.6	38.8	23.5	47.1	26.0	16.0	36.4
Min	0.4	0.01	−179	8.0	0.5	42.9	1.6	0.1	−357
Max	40	21.1	98.2	158	76.7	98.7	153	80	99.7
n	248	248		81	81		260	260	
CWs	52			22			53		

Table 13.4 Loadings of constructed wetlands in the Czech Republic for nutrients during the period 1989–2008. Loadings in g m^{-2} yr^{-1}, n = number of annual means, CW = number of constructed wetlands

	TP			TN			NH_4^+-N		
	In	Out	Removed	In	Out	Removed	In	Out	Removed
Mean	106	62	44.1	789	437	352	505	294	211
Median	79.2	49.1	25.8	672	393	304	380	215	129
Min	5.7	0.12	−70.8	144	35	−82	11.4	4.3	−64
Max	778	440	675	1937	1141	1289	4723	3262	2003
n	154	154		58	58		184	184	
CWs	45			17			53		

shown that sequestration of nutrients in aboveground plant biomass which can be harvested is usually low (<10% for N and <5% for P) as compared to inflow load for sewage (Vymazal & Kröpfelová, 2008). In the Czech Republic, there are no discharge limits for nutrients for wastewater treatment plants <500 PE; for treatment plants between 500 and 2000 PE discharge limits of 20 mg L^{-1} NH_4-N is set (periods with water temperature >12°C). However, HF CWs are not recommended in the Czech Republic for the sources of pollution where ammonia and/or phosphorus are the primary treatment target.

13.4.3 Trace Elements

Trace elements including heavy metals are effectively retained in HF CWs (Kröpfelová, Vymazal, Švehla, & Štíchová, 2009). The results indicated that HF constructed wetlands could be a very useful tool for the removal of trace elements such as aluminum, zinc, uranium, antimony, lead or copper, but at the same time it seems that some trace elements such as iron, manganese, selenium and cobalt are not retained efficiently or are washed out. It has also been shown that the amount of trace elements sequestered in the aboveground plant biomass is very low and the concentrations of trace elements including heavy metals in plants growing in HF CWs are comparable with those found in plants from natural stands (Vymazal, 2005; Vymazal, Švehla, Kröpfelová, & Chrastný, 2007, Vymazal, Kröpfelová, Švehla, Chrastný, & Štíchová, 2009).

13.5 Conclusions

Constructed wetlands (CWs) have been used in the Czech Republic since 1989. At present, about 250 systems are in operation. All systems have been designed with horizontal subsurface flow and most CWs have been designed to treat municipal or domestic wastewater from both separate and combined sewerage. The majority of constructed wetlands were built either as on-site treatment systems or for small communities up to 500 PE (population equivalent). However, the numbers obtained from the survey carried out in 2009 indicated that over the years the size distribution has changed. During the first 10 years, constructed wetlands for small villages between 100 and 500 PE prevailed but after 1999 small on-site systems prevailed. The average design specific area of filtration beds is 5.1 m^2 PE^{-1} but the specific area varies between 0.7 and 16.7 m^2 PE^{-1}. Crushed rock and gravel 4–8 mm or 8–16 mm are the most commonly used filtration material and *Phragmites australis* and *Phalaris arundinacea* are the most frequently used plants. Removal of BOD_5, COD and suspended solids is very high with respective median treatment efficiencies of 90, 81 and 90%. Removal of nutrients is much lower but there are no discharge limits for wastewater treatment plants up to 500 PE in the Czech Republic. Therefore, constructed wetlands in the Czech Republic are designed according to the requirements for the removal of organics and suspended solids. Two decades of constructed wetlands application in the Czech Republic have proven that CWs are a suitable option for wastewater treatment in small settlements and individual houses.

Acknowledgements The study was supported by grant No. 206/06/0058 "Monitoring of Heavy Metals and Selected Risk Elements during Wastewater Treatment in Constructed Wetlands" from the Czech Science Foundation and grant No. 2B06023 "Development of Mass and Energy Flows Evaluation in Selected Ecosystems" from the Ministry of Education, Youth and Sport of the Czech Republic.

References

Dykyjová, D. (1971). Production, vertical structure and light profiles in littoral stands of reed bed species. *Hidrobiologia* (Bucharest), 12, 361–376.

Dykyjová, D. (1980). Production ecology of *Acorus calamus*. *Folia Geobotanic and Phytotaxonomic.* (Praha), *15*, 29–57.

Dykyjová, D., & Květ, J. (Eds.). (1978). *Pond littoral ecosystems: Structure and functioning*. Springer: Berlin.

Dykyjová, D., & Květ, J. (1982). Mineral nutrient economy in wetlands of the Třeboň Basin Biosphere Reserve, Czechoslovakia. In B. Gopal, R. E. Turner, R. G. Wetzel, & D. F. Whigham (Eds.), *Wetlands ecology and management* (pp. 335–355). Jaipur, India: International Science Publishing, and Natinoal Institute of Ecology.

Hejný, S. (Ed.). (1973). *Ecosystem study on wetland biome in Czechoslovakia*. Czechoslovak IBP/PT-PP Report No. 3, Třeboň, Czech Republic.

Imhoff, K. (1960). *Taschenbuch der Stadtenwässerung*. München: R. Oldenburg Verlag.

Kröpfelová, L., Vymazal, J., Švehla, J., & Štíchová, J. (2009). Removal of trace elements in three horizontal sub-surface flow constructed wetlands in the Czech Republic. *Environmental Pollution*, 157, 1186–1194.

Květ, J. (1973). Transpiration of South Moravian *Phragmites communis* littoral of the Nesyt Fishpond. *Studies Czech Academy of Science, 15*, 143–146.

Vobornïk, D. (2009). *Evaluation of Vegetation in constructed wetlands for wastewater treatment*. Diploma, Czech University of Life Sciences, Prague, Czech Republic (in Czech).

Vymazal, J. (1990). Use of reed-bed system for the treatment of concentrated wastes from agriculture. In P. F. Cooper & B. C. Findlater (Eds.), *Constructed wetlands in water pollution control* (pp. 347–358). Oxford: Pergamon Press.

Vymazal, J. (1993). Constructed wetlands for wastewater treatment in Czechoslovakia: State of the art. In G.A. Moshiri (Ed.), *Constructed wetlands for water quality improvement* (pp. 255–260). Boca Raton: CRC Press/Lewis Publishers.

Vymazal, J. (1995). Constructed wetlands for wastewater treatment in the Czech Republic-state of the art. *Water Science and Technology*, 32(2), 357–364.

Vymazal, J. (1996a). Constructed wetlands for wastewater treatment in the Czech Republic: The first 5 years experience. *Water Science and Technology*, 34, 159–164.

Vymazal, J. (1996b). The use of subsurface-flow constructed wetlands for wastewater treatment in the Czech Republic. *Ecological Engineering*, 7, 1–14.

Vymazal, J. (1998). Czech Republic. In J. Vymazal, H. Brix, P. F. Cooper, M. B. Green, & R. Haberl (Eds.), *Constructed wetlands for wastewater treatment in Europe* (pp. 95–121). Leiden: Backhuys Publishers.

Vymazal, J. (2001). Constructed wetlands for wastewater treatment in the Czech Republic. *Water Science and Technology*, 44, 369–374.

Vymazal, J. (2002). The use of sub-surface constructed wetlands for wastewater treatment in the Czech Republic: 10 years experience. *Ecological Engineering*, *18*, 633–646.

Vymazal, J. (2005). Removal of heavy metals in a horizontal sub-surface flow constructed wetland. *Journal of Environmental Science and Health, A40* (6–7), 1355–1367.

Vymazal, J. (2006). The use of constructed wetlands for wastewater treatment in the Czech Republic. In A. R. Burk (Ed.), *Focus on Ecology Research* (pp. 175–196). New York: Nova Science Publishers.

Vymazal, J. (2007). Removal of nutrients in various types of constructed wetlands. *Science of the Total Environment, 380*, 78–65.

Vymazal, J. (2009a). Constructed wetlands with horizontal subsurface flow in the Czech Republic: Two long-term case studies. *Desalination and Water Treatment, 1*, 40–44.

Vymazal, J. (2009b). Horizontal sub-surface flow constructed wetlands Ondřejov and Spálené Poříčí in the Czech Republic – 15 years of operation. *Desalination, 246*, 226–237.

Vymazal, J., & Kröpfelová, L. (2007). Removal of organics and suspended solids in HF constructed wetlands in the Czech Republic: 15 years of experience. In Ü. Mander, M. Kõiv & C. Vohla (Eds.), *Proceedings 2nd Symposium Wetland Pollutant Dynamics and Control, WETPOL 2007* (pp. 339–341). Tartu, Estonia: Institute of Geography, Tartu University.

Vymazal, J. & Kröpfelová, L. (2008). Wastewater treatment in constructed Wetlands with horizontal sub-surface flow. Dordrecht: Springer.

Vymazal, J., Brix, H., Coopr, P. F., Haberl, R., Perfler, R. & Laber, J. (1998). Removal mechanisms and types of constructed wetlands for wastewater treatment. In J. Vymazal, H. Brix, P. F. Cooper, M. B. Green & R. Haberl (Eds.), *Constructed Wetlands for wastewater treatment in Europe* (pp. 17–66). Leiden: Backhuys Publishers,.

Vymazal, J., Švehla, J., Kröpfelová, L., & Chrastný, V. (2007). Trace metals in *Phragmites australis* and *Phalaris arundinacea* growing in constructed and natural wetlands. *Science of the Total Environment, 380*, 154–162.

Vymazal, J., Kröpfelová, L., Švehla, J., Chrastný, V., & Štíchová, J. (2009). Trace elements in *Phragmites australis* growing in constructed wetlands for treatment of municipal wastewater. *Ecological Engineering, 35*, 303–309.

Chapter 14
The Concept of a Sewage-Sludge Management System for an Individual Household

Hanna Obarska-Pempkowiak, Magdalena Gajewska, Ewa Wojciechowska, and Arkadiusz Ostojski

Abstract Individual farms in rural areas often face problems with domestic sewage collection and treatment. In many cases investment costs of sewerage systems and central wastewater treatment plants (WWTPs) are too high, due to long distances from one farm to another and terrain configuration. Treatment wetlands for individual farms can solve this problem. In the present study, an overview of individual treatment wetlands in Europe and Poland is reviewed. On this basis, the conception of domestic sewage and sewage sludge management for an individual farm was developed within the project "Innovative solutions for wastewater management in rural areas". The concept was aimed at preparation of a ready-to-implement solution of the problem of sewage and sludge management of a local community in a rural area by treatment wetlands serving individual households. The proposal of an innovative sanitary system is based on an idea of a closed cycle of matter in the environment. The nutrient substances: N, P and K compounds present in sewage and sludge should be recycled to the soil. Three configurations of hydrophyte beds will be tested in order to select the optimal solution. The implementation of the project has already started at the catchment area of river Borucinka at the municipality Stezyca, Pommerania Region, Poland.

Keywords Treatment wetlands · Individual household · Sewage treatment · Sewage sludge dewatering and stabilization

14.1 Introduction

The latest report of Environmental Protection Inspection regarding the condition of Polish waters defines protection of waters against eutrophication as one of the most significant environmental issues. Solution to eutrophication of waters in Poland

---

M. Gajewska (✉)
Faculty of Civil & Environmental Engineering, Gdansk University of Technology, Gdańsk, Poland
e-mail: mgaj@pg.gda.pl

J. Vymazal (ed.), *Water and Nutrient Management in Natural and Constructed Wetlands*, DOI 10.1007/978-90-481-9585-5_14,

demands strategies not only for big agglomerations but, first of all, for rural areas with dispersed farms and buildings. Due to the type of land usage in rural areas, surface waters are especially exposed to eutrophication. Also integrated management of water resources in such areas is a difficult task.

For the last 20 years, attempts to construct treatment wetlands (TW) in rural areas have been undertaken. It is estimated that the number of these facilities in Europe exceeds 10,000 and in Poland it is about 1000. The treatment wetlands are usually planted with reed or, sometimes, willow. The treatment wetlands are usually used at the second biological stage of sewage treatment, though very good performance of TW at the third stage of treatment has been proved. The sewage directed to the treatment wetlands is pre-treated in septic tanks, Imhoff tanks or sedimentation ponds. Although the treatment wetlands in Europe are usually designed to serve up to 500 inhabitants, most existing facilities receive sewage from less than 50 inhabitants or even from individual farms. Treatment wetlands for more than 1000 inhabitants are uncommon (Cooper, 1998; Vymazal, 1998).

The common problem that the owners of houses located in rural areas face is lack of a sewerage system. Due to local conditions and complexity of installation, it is very difficult to achieve satisfactory treatment results with small conventional WWTPs. Thus simple, reliable and cost-saving solutions in the field of sewage management in the rural areas are needed. Treatment wetlands can be an alternative solution to sewage management in small communities, schools, camping grounds and individual houses. These systems are not only simple, cheap and effective in operation, but they can provide landscape and educational values as well.

Within the research project "Innovative solutions for wastewater management in rural areas" (NORWET) the conception of sewage treatment and sewage sludge utilization in the treatment wetlands for individual households in rural areas was developed. The project was launched in the catchment area of river Borucinka at the municipality Stezyca, Pommerania Region. The idea was to prepare a ready-to-implement solution of sewage and sludge management problems of the local community in a rural area by treatment wetlands serving individual households. The local terrain conditions and dispersed development in the selected area do not allow for building sewerage system. The domestic sewage is collected at cesspools, that often leak into groundwaters.

On the basis of TWs application review in Poland and in Europe, three configurations of hydrophyte beds were proposed. The constructed wetlands will be implemented in the catchment area of river Borucinka at the municipality Stezyca, Pommerania Region.

14.2 Experiences from Treatment Wetlands Operation in Europe

According to Börner et al. (1998), treatment wetlands have been used for sewage treatment in Germany for over a 100 years. The experience from recent years has allowed for preparation of ATV – A 262 (1998) guidelines for design and operation of treatment wetlands. At present, there are several thousand treatment wetlands in operation in the rural areas of Germany (Börner et al., 1998). The analysis of operation of 107 treatment wetlands in Germany revealed that 24 vertical flow constructed wetlands (VF-CWs) facilities reached significantly better treatment effects than the other 83 horizontal flow constructed wetlands (HF-CWs) facilities (Börner et al., 1998; Kayser & Kunst, 2004). The average concentrations of COD and ammonia-N at the effluent of the VF-CW facilities were 68.2 mg and 9.5 mg L^{-1}, respectively, and were lower than the corresponding values for HF-CW facilities (102.5 mg $L^{-1}$and 36.0 mg L^{-1}, respectively). Whereas the total nitrogen concentration in the effluent from VF-CW beds was higher than that from HF-CW beds – 67.1 mg L^{-1} and 52.1 mg L^{-1}, respectively. This resulted from the fact that nitrate nitrogen was nine times higher in the effluent from VF-CW beds than from HF-CW beds (65.2 mg L^{-1} and 7.3 mg L^{-1}, respectively) (Börner et al., 1998, Table 14.1). Table 14.1 presents the concentrations of characteristic pollutants in sewage after VF-CW and HF-CW beds in Germany (Börner et al., 1998) and Austria (Haberl, Perfler, Laber, & Grabher, 1998).

According to Langergraber (2007), the effluent from one-stage VF-CW beds with specific area of 4 m^2 PE^{-1} (Person Equivalent) and the organic matter loading of 20 g m^2d^{-1}, can meet rigorous Austrian outflow standards (COD: 90 mg L^{-1}, BOD_5: 25 mg L^{-1}), regardless of season and air temperature. On the basis of results obtained in Austria and experience from other countries, the Austrian guidelines for designing and construction of treatment wetlands ÖNORM ÖN B 2505 (1995) were introduced (Haberl et al., 1998).

In France, the VF CWs have been used for treatment of raw sewage (without primary mechanical treatment) for over 20 years (Boutin et al., 1997; Molle, Lienard, Boutin, Merlin, & Iwema, 2004). According to Molle et al. (2004), two sequential VF-CW beds, periodically supplied with raw sewage, provide effective treatment. The unit area should be equal to 1.5 m^2 PE^{-1} for the first bed and only 1.0 m^2 PE^{-1} for the second bed. This configuration of VF-CW beds allows for reducing pollutant concentrations to the following level: COD: 60 mg L^{-1}, TSS: 15 mg L^{-1}, TKN: 8.0 mg L^{-1}. Molle et al. (2004) recommend that hydraulic loading of the beds working in batches is below 600 mm d^{-1}. With such operation conditions the beds provide good and stable pollutant removal efficiencies (Table 14.2).

Table 14.1 Comparison of average concentrations (mg L^{-1}) of characteristic pollutants in sewage treated on VF-CW and HF-CW beds in Germany (Börner et al., 1998) and Austria (Haberl et al., 1998)

	VF-CW bed		HF-CW bed	
Parameter	Germany	Austria	Germany	Austria
COD	68.2	37.0	102.5	49.0
$NH_4^+ - N$	9.5	7.5	36.0	15.4
$NO_3- - N$	65.2	35.0	7.3	8.0
N_{tot}	67.1	–	52.1	–
P_{tot}	3.2	–	5.0	–

Table 14.2 The average concentrations of pollutants and treatment efficiencies for sequential VF-CW beds (data from Molle et al., 2005)

COD		TSS		Kiejdahl nitrogen	
Effectivity (%)	Outflow (mg L^{-1})	Effectivity (%)	Outflow (mg L^{-1})	Effectivity (%)	Outflow (mg L^{-1})
VF-CW I bed					
82.0 ± 3.0	145.0 ± 24.0	89.0 ± 3.0	33.0 ±7.0	60.0 ± 6.0	35.0 ± 7.0
VF-CW II bed					
60.0 ± 8.0	55.0 ± 8.0	72.0 ± 7.0	11.0 ± 4.0	78.0 ± 7.0	6.0 ±2.0

The main advantage of this solution is that the first VF CW is fed directly with raw sewage. Since there is no primary sedimentation tank or Imhoff tank, no primary sludge is produced. In order to protect the system from clogging, the first-stage filter should be divided into three compartments and the second-stage filter into two compartments. The first-stage and the second-stage filters have different construction, since their functions are different. Sewage is supplied to the first VF-CW bed in one, relatively high, dose. The single compartment works for 3–4 days, and then the sewage is discharged to the second compartment. This method of sewage supply assures adequate increase of bacteria biomass, provides aerobic conditions and mineralization of a large quantity of organic substances deposited on the filter surface from the raw sewage. In the next stage, the wastewater is directed into one of the compartments of the second VF-CW bed, where further removal of organics and nitrification takes place. In France, over 200 treatment wetlands have been constructed according to this scheme (Molle et al., 2004). More than 60 of these facilities were built in 2003. In 2005, the analysis of operation of 81 treatment wetlands (53 with sequential VF-CW I and II beds) was performed. The treatment effectiveness of the analyzed facilities was very high: >91.0% for COD, 95.0% for TSS and 85.0% for TKN. Due to high treatment effectiveness, the concentrations of pollutants in discharged sewage were very low: COD: 66.0 mg L^{-1}, TSS: 15 mg L^{-1} and THN: 13.0 mg L^{-1}. The removal effectiveness of TKN in the first stage of treatment amounted to 50.0% (the applied loading was 25–30 g m^2 d^{-1}). The small amounts of organics and TSS that were discharged to the second stage of the treatment were easily removed. According to Molle et al. (2004) the optimal conditions for effective nitrogen removal are reached when the load discharged to the second stage of treatment is equal to 15 g m^{-2} d^{-1}.

A completely different approach is promoted in Norway (Jenssen, Maehlum, Krogstad, & Vråle, 2005; Heistad, Paruch, Vråle, Adam, & Jenssen, 2006), where very effective primary treatment of sewage is applied (Fig. 14.1). In the last compartment of the primary sedimentation tank (at the outlet), specially designed lammel constructions are installed, in order to enhance TSS removal. Then, the small portions of sewage are sprayed on the prefilter surface. The prefilter is a vertical flow bed filled with LECA material (Lightweight expanded clay

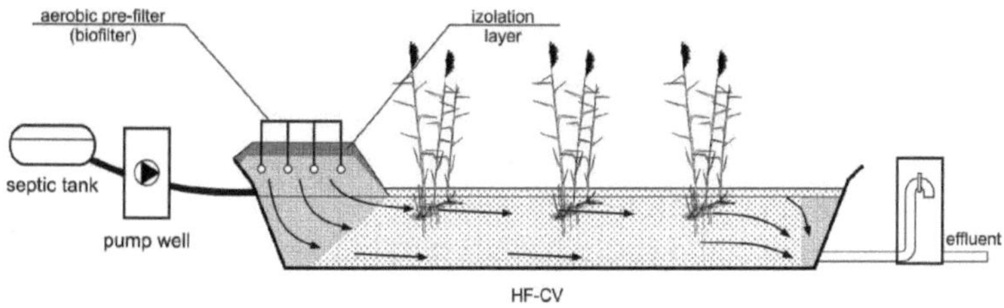

Fig. 14.1 Layout of CW system treating domestic wastewater in Norway (Gajewska, 2001)

aggregates), which has very good filtration properties. Another function of the prefilter is sewage aeration. Next, the sewage gravitationally outflows to the HF CW. Both beds are built as one compact unit. Because of the cold climate, HF-CW beds are deeper (0.9 m) than is usually recommended in other European countries (Cooper, 1998; Brix, 1996). Maehlum and Jenssen (2002) report the following removal efficiencies at the compact treatment wetlands: 50–90% for COD, 75–95.0% for BOD_7, over 50.0% for total nitrogen. The outflow concentrations were as follows: COD < 40 mg L^{-1}, BOD < 20 mg L^{-1}, $P_{tot} < 1$ mg L^{-1}, $N_{tot} <$ 30 mg L^{-1}.

14.3 Experience from Treatment Wetlands Operation in Poland

In Poland, from the late 1980s, several hundred treatment wetlands were built with horizontal subsurface flow of sewage (HF-CW) used in the second stage of domestic sewage treatment. These systems differ in size, from 5 PE (for individual households) to 2000 PE (for villages). Most of these facilities are not sufficiently monitored and therefore it is difficult to estimate their effectiveness. Many of the existing facilities were constructed without proper know-how and regardless of any guidelines. Only a few systems were designed according to the guidelines used in Great Britain or Denmark. The low effectiveness of nitrogen compounds removal achieved in Polish systems resulted in a negative opinion of treatment wetlands. The liberalization of outflow standards for facilities serving less than 2000 PE, which occurred in 2002 (the Regulation of Environment Ministry from 20th November 2002, Dz.U. Nr 212.1799), created more optimistic perspectives for applications of treatment wetlands in Poland.

Treatment wetlands serving individual households located near Lublin and Ostroleka were constructed under the UNEP WHO and Polish Ministry of Environmental Protection, Natural Resources and Forestry program "Sanitation of rural areas and proper agricultural practices". The systems near Ciechanow were designed and built by the Institute of Building, Mechanisation and Electrification of Agriculture in Warsaw, with some differences in the method of sewage distribution and in the filling materials. The specific area of treatment wetlands in the region of Ostroleka and Lublin (1, 2, 3, 4, 5, 6, 7) was 6.0 m^2 PE^{-1} and in the Ciechanow region (I, II, III and IV) – 4.5 m^2 PE^{-1}. The description and monitoring of the facilities were performed by Obarska-Pempkowiak and Gajewska (2005) and Obarska-Pempkowiak et al. (2003). Many of the analyzed facilities did not work properly. The main reason for their malfunction was improper operation of septic tanks, and lack of devices that would enable separation of grid and TSS from the sewage discharged to the treatment wetlands. As a consequence, the hydraulic capacity of constructed wetlands decreased, in some cases leading to a surface flow of sewage.

The individual household treatment wetlands monitoring results indicated that HF-CW facilities working at the second stage of sewage treatment provided effective removal of BOD_5, COD and TSS. The effectiveness of BOD removal varied from 25.6 to 99.1% (average 62.4%, Fig. 14.2) for organic loadings ranging between 11.2 and 115 kg $ha^{-1}d^{-1}$. However, the removal effectiveness of the total nitrogen was lower and varied from 22.4 to 84.2% (average 44.5%, Fig. 14.2), for loadings ranging between 8.5 and 34.0 kg ha^{-1} d^{-1}. In Table 14.3, removal of BOD_5 and nitrogen removal in the monitored treatment wetlands is presented.

Until 2004, the one-stage VF CWs were not used with the exception of pilot-scale research performed by, for example, Soroko (2001) or Kowalik, Mierzejewski, Randreson, and Williams (2004). The average removal of BOD_5 and total

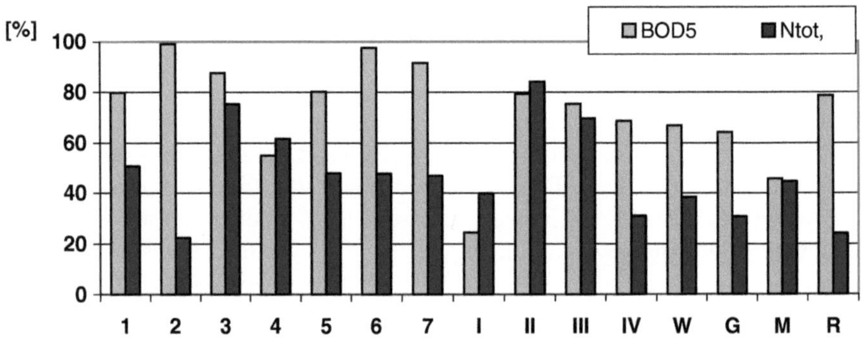

Fig. 14.2 The removal efficiencies of organics and total nitrogen at the individual household treatment wetlands. W = Wawrow, G = Gralewo, M = Malszyn, R =Rokitno

Table 14.3 Removal of BOD_5 and N_{tot} in monitored constructed wetlands. Values in g m^2d^{-1}

Individual household treatment wetlands					
Near Ciechanow	Near Lublin	Near Ostroleka	Near Ciechanow	Near Lublin	Near Ostroleka
BOD_5			Total-N		
I – 0.7	1 – 4.2	5 – 1.9	I – 2.2	1 – 0.6	5 – 0.8
II –3.4	2 – 8.2	6 – 2.7	II – 1.1	2 – 0.2	6 – 0.4
III –2.3	3 – 10.8	7 – 0.2	III – 0.7	3 – 0.1	7 – 0.9
IV –0.4	4 – 2.1		IV – 0.6	4 – 1.1	

nitrogen reported by Soroko (2001) amounted to 97.4 and 41.6%, respectively. Kowalik et al. (2004) reported a removal effectiveness of BOD_5 and N_{tot} of 89.1 and 76.1% at stage II of the treatment and 93.8 and 79.1% at stage III, respectively. The implementation of VF-CWs started in the Podlasie Region in 2004. The municipality Sokoly near Bialystok decided to launch proper sewage management. Since building and operation of a sanitary sewerage system were too expensive, the municipal authorities decided to build treatment wetlands for individual households. At present, in the Podlasie Region there are 600 treatment wetlands consisting of a one-stage bed with vertical flow of sewage in operation (Wasiak, 2008). The treatment wetlands are built by individual farmers on their own land, according to the conception and guidelines of the Institute of Applied Ecology in Skorzyn (Halicki, 2009). According to this conception, the treatment facility for a single family consists of a septic tank (retention time of 5 days), followed by a VF-CW bed, intermittently supplied with sewage by a pump. If denitrification of nitrogen compounds is required, a denitrification pond is built after a VF-CW bed. The edges of the pond are lined to a certain level (not the top), which allows for leaking of sewage into the ground. Sludge from the septic tank is dredged out twice a year. The treatment wetlands in Podlasie are monitored by several research institutions, however the monitoring results are not unanimous. According to the analyses reported by Wierzbicki, Eymontt, and Gutry (2008) the average outflow concentrations were: BOD_5 <13 mg L^{-1}, COD <93 mg L^{-1}, TSS < 42 mg L^{-1}.

Two treatment wetlands in the municipality of Sokoly monitored by Wierzbicki et al. (2008) effectively removed TSS and organics (COD and BOD_5). The effluent concentrations of these pollutants fulfilled the requirements given by the Environmental Ministry Regulation from 24th July 2006 (Table 14.4). However the removal effectiveness of nutrients, especially phosphorus, was unstable.

The analyses performed by authors of the article indicated that the monitored facilities are very effective in pollutants removal. The removal effectiveness of BOD_5 and COD varied from 86 to 98%, and from 79 to 94%, respectively (Table 14.5, Fig. 14.3).

Table 14.4 The average concentrations (mg L^{-1}) of pollutants in raw (*inflow*) and treated sewage (*outflow*) and the removal efficiencies (%) in the municipality Sokoly, Podlasie region (modified from Wierzbicki et al., 2008)

Parameter	Facility	TSS	COD	BOD_5	Ptot	Ntot	NH_4^+-N
inflow	farm I	239.3	707.3	443.3	30.8	94.6	75.2
	farm II	574.3	467.7	260.0	13.4	84.3	79.3
outflow	farm I	28.4 (88.1)*	78.3 (88.9)*	8.3 (98.1)*	17.9 (41.9)*	36.4 (61.5)*	4.9 (93.5)*
	farm II	90.4 (84.3)*	61.7 (86.8)*	13.3 (94.9)*	11.6 (13.4)*	38.7 (54.1)*	3.5 (95.6)*

*Removal efficiencies (%)

Table 14.5 The average concentrations of characteristic pollutants in raw and treated sewage from individual household treatment wetlands in the municipality Sokoly, Podlasie Region. Values in mg L^{-1} (except pH)

	J. Jamiolkowski		Stypulkowski		R. Jamiolkowki	
Parameter	Inflow	Outflow	Inflow	Outflow	Inflow	Outflow
pH	7.8	7.0	8.4	7.0	8.2	7.3
TSS	248.1	145.6	148.3	101.6	115.6	7.6
Organic SS	180.0	71.3	91.9	72.8	82.4	7.2
COD	931.4	191.5	517.0	62.5	722.6	37.1
BOD_5	496.1	25.9	180.5	26.7	274.6	4.6
TN	155.2	33.6	200.6	43.5	114.2	53.8
NH_4^+-N	100.0	4.3	145.5	7.5	87.5	1.3
Org-N	55.0	3.6	55.0	32.8	24.8	6.9
NO_3^- -N	0.2	25.7	0.1	3.2	1.9	45.6
TP	19.3	4.8	13.1	17.8	6.1	7.1

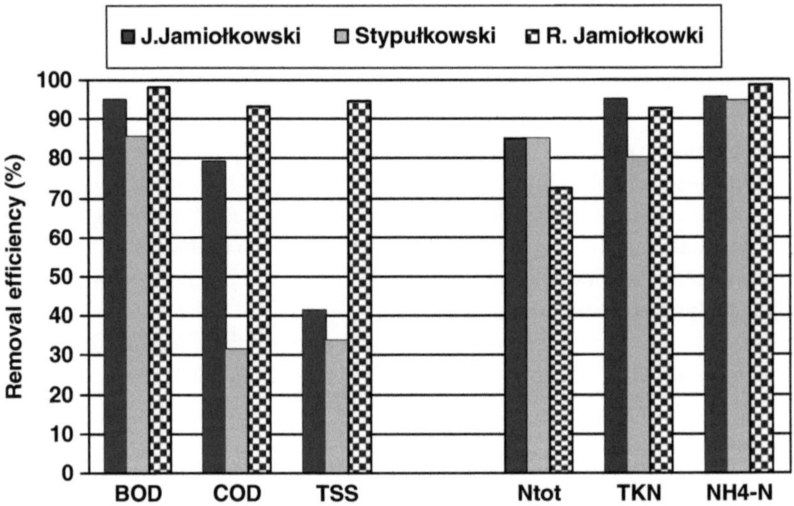

Fig. 14.3 The pollutant removal effectiveness in the treatment wetlands in the municipality Sokoly (Podlasie Region)

The results confirmed low effectiveness of total phosphorus removal, similarly as it was reported by Wierzbicki et al. (2008). Additionally, it was found that the effluent concentrations of TSS exceeded the admissible value of 50 mg L^{-1} (The Regulation of Environmental Ministry from 24th July 2006). The share of organic suspended solids in the total suspended solids at the effluent varied from 49 to 95%. Very good conditions for nitrification process existed in the treatment facilities. This is confirmed by very low concentrations of ammonia-N at the effluents of the three analyzed facilities. However, the removal of total nitrogen was substantially lower in comparison to ammonia nitrogen removal due to the high concentration of NO_3-N in the outflow, indicating that the denitrification pond fails to play its role.

Within the research project, "Innovative solutions for wastewater management in rural areas" (financed by the Polish Ministry of Science and Higher Education E033/P01/2008/02 and EOG Financial Mechanism and Norwegian Financial Mechanism Nr PL0271), researchers from Gdansk University of Technology, Department of Water and Wastewater Technology and Department of Sanitary Engineering are developing the conception of treatment wetlands for treatment of sewage from individual households. The aim of the research project is, among others, to develop the conception of innovative technology of sewage treatment and sewage sludge utilization in rural areas. Within the project, six treatment wetlands for individual households will be constructed. The most significant aspect of the research to be undertaken will be selection of the optimal configuration of hydrophyte beds, depending on local conditions. The three following configurations will be analyzed:

Configuration I: primary sedimentation tank with elongated hydraulic retention time (5–6 days), followed by VF-CW bed and a pond. The impact of retention time in the primary sedimentation tank on pollutant removal effectiveness in the VF-CW bed will be investigated.

Configuration II: screens/sieve followed by two sequential VF-CW beds – treatment wetlands working without a primary treatment of sewage will be tested.

Configuration III: (according to the guidelines of the Norwegian Partner of the Project) primary sedimentation tank, prefilter for more efficient primary treatment followed by the horizontal subsurface flow bed (HF-CW).

14.4 An Innovative Hydrophyte Sanitary System for Rural Areas

The proposal for an innovative sanitary system is based on an idea of a closed cycle of matter in the environment and the nutrient substances: N, P, K compounds present in sewage should be used as soil fertilizers. Compact VF-CW beds for individual households have recently been proposed as a solution for un-urbanized areas. These systems have been tested in Germany and Denmark. In Germany, about 100 households in the Wienhousen municipality near Hannover take part in the program of sewage and sludge management. Sewage from individual households is treated in individual treatment plants consisting of a septic tank and a VF CW. The sludge generated in the treatment process is periodically transported to the Central Utilization System (CUS) where it undergoes dewatering and stabilization. Sludge processing at the CUS is performed in reed beds, where intensified natural processes take place. The leachate generated in the sludge dewatering process is treated in a specially designed VF-CW and then it is discharged into surface water. The dewatered and stabilized sewage sludge becomes a valuable soil fertilizer for farm lands. In this way the matter cycle is closed. The sewage treatment technology as well as sludge processing is simple in operation and the method described has positive social, economic and environmental aspects. The strategy of sustainable development of the municipality is fulfilled.

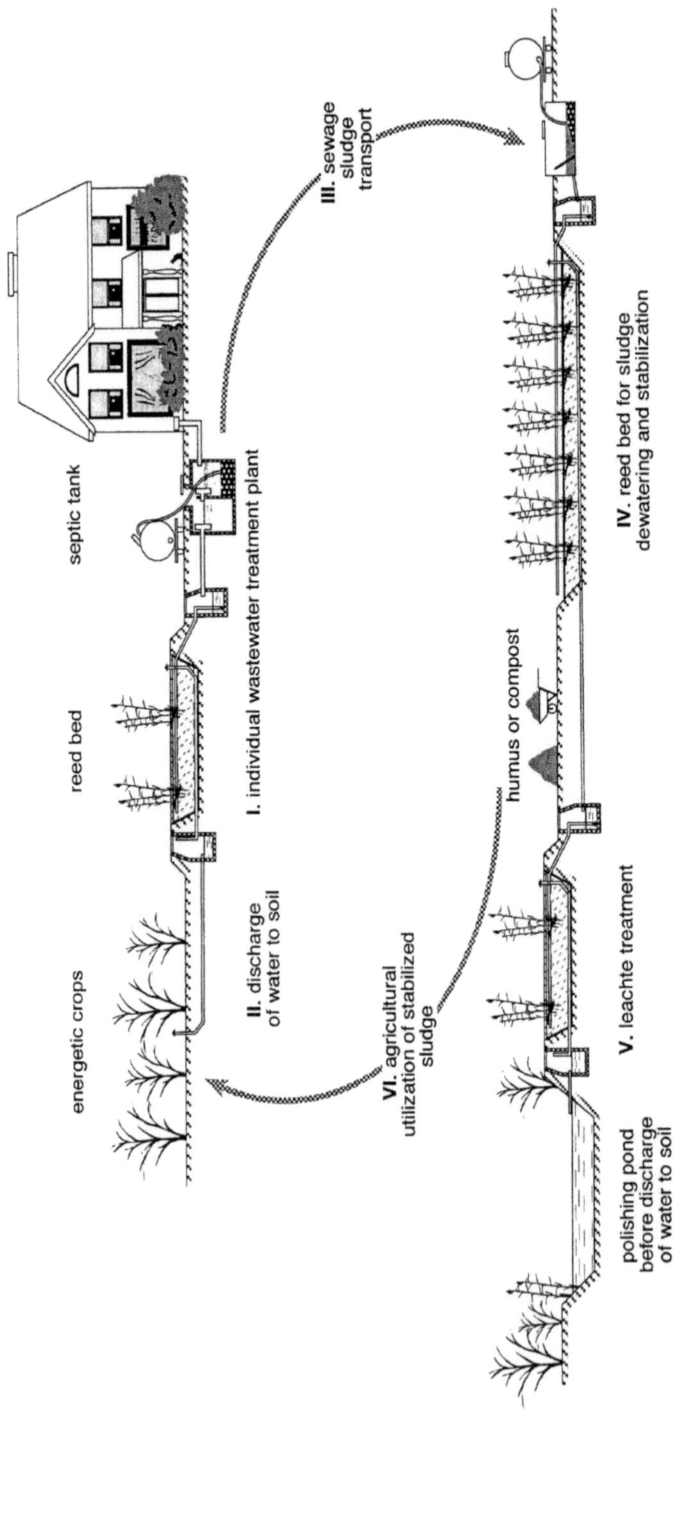

Fig. 14.4 The concept of sewage and sludge management for an individual household

According to the conception realized within the project "Innovative solutions for wastewater management in rural areas", another goal, apart from sewage treatment, is recycling of biomass or energy. Thus, the nutrient substances should be recycled to the soil in the form of compost or humus, or can be used as an alternative, renewable energy source. It is assumed that the sewage could also be used for watering of crops (for instance energetic crops or plants used for co-composting with sludge and organic fractions of domestic wastes) during the vegetation season. In Fig. 14.4 the concept of the proposed solution is presented.

14.5 Conclusions

1. Up till now, treatment wetlands with horizontal subsurface flow have been predominantly used for single households.
2. According to the conception of an innovative sewage-sludge utilization system for individual households, three configurations of hydrophyte beds will be tested in order to determine selection of the optimal solution.
3. Application of treatment wetlands in sewage-sludge utilization from individual households in rural areas allows for closing the water and nutrients cycle in the environment.

Acknowledgments Funding support from the EEA Financial Mechanism and Norwegian Financial Mechanism (PL 0271) and Polish Ministry of Science and Higher Education (E033/P01/2008/02) is gratefully acknowledged.

Reference

ATV Arbeitsblatt A262 (Juli 1998). Grundsätze für Bemessung Bau und Betried von Pflanzenbeeten für Kommunales Abwaser bei Ausbaugroßen bis 1000 Einwohnerwertte, ss. 2–10.

Börner, T., von Felde, K., Gschlössl, E., Gschlössl, T., Kunst, S., & Wissing, F. W. (1998). Germany. In J. Vymazal, H. Brix, P. F. Cooper, M. B. Green, & R. Haberl (Eds.), *Constructed wetlands for wastewater treatment in Europe* (pp. 169–190). Leiden: Backhauys Publishers.

Boutin, C., Lienard, A., & Esser, D. (1997). Development of a new generation of reed-beds filters in France: First results. *Water Science and Technology, 35*(5), 315–322.

Brix, H. (1996). Role of macrophytes in constructed wetlands. *Proceedings of 5th international conference on wetland system for water pollution control* (Vol. I, pp. 4–8). Vienna: Universitaet fuer Bandenkutur Wien and International Association on Water for water pollution control, Universitaet fuer Bandenkutur Wien and International Association on Water Quality.

Cooper, P. (1998). A review of the design and performance of vertical flow and hybrid reed bed treatment systems. In S. M. Tauk-Tornisielo, & E. Salati Filho (Eds.), *Proceedings of 6th International Conference On Wetland System for Water Pollution Control* (pp. 229–242). Sao Paulo: Universidade Estadual Paulista and IAWQ.

Gajewska, M. (2001). *Qualitative and quantitative transformations of nitrogen in hybrid condtructed wetlands*. Dissertation, Gdansk University of Technology, Poland.

Haberl, R., Perfler, R., Laber, J., & Grabher, D. (1998). Austria. In J. Vymazal, H. Brix, P. F. Cooper, M. B. Green, & R. Haberl (Eds.), *Constructed wetlands for wastewater treatment in Europe* (pp. 67–76). Leiden: Backhauys Publishers.

Halicki W. (2009). The schedule of natural wastewater treatment system implementation in podlasie region. Five years of experience. In D. Boruszko & W. Dąbrowski (Eds.), *Water, wastewater and solid waste. Low-cost wastewater and sludge treatment systems* (pp. 31–39). Białystok: Technical University of Bialystok (in Polish).

Heistad, A., Paruch, A. M., Vråle, L., Adam, K., & Jenssen, P. D. (2006). A high-performance compact filter system treating domestic wastewater. *Ecological Engineering, 28*, 374–379.

Jenssen, P. D., Maehlum, T., Krogstad, T., & Vråle, L. (2005). High performance constructed wetlands for cold climates. *Journal of Environmental Science and Health. Part A, 40*, 1343–1353.

Kayser, K., & Kunst, S. (2004). Processes in vertical – flow reed beds – nitrifiacation, oxygen transfer and soil clogging. In A. Lienard (Ed.), *9th International conference on wetland systems for water pollution control*, ASTEE (Association scientifique et technique pour l'eau et l;environnement) with scientific collaboration of Cemagref and Agence de l'Eau Rhone Mediterranee & Corse Ed (pp. 385–393). France: Alain Lienard Avignon.

Kowalik, P., Mierzejewski, M., Randreson, P., & Williams, H. (2004). Performance of subsurface vertical flow constructed wetlands receiving municipal wastewater. *Archives of Hydro-Engineering and Environmental Mechanics, 51*, 349–370.

Langergraber, G. (2007). Simulation of the treatment performance of outdoor subsurface flow constructed wetlands in temperate climates. *Science of the Total Environment, 380*, 210–219.

Maelum, T., Jenssen, P. D., (2002). Design and performance of integrated subsurface flow wetlands in cold climate. In Ü. Mander, & P.D Jenssen (Eds.), *Constructed wetlands for wastewater treatment in cold climates. Adfvances in Ecological Sciences* (Vol 11, pp. 69–86). WIT Press: Boston.

Molle, P., Lienard, A., Boutin, C., Merlin, G., & Iwema, A. (2004). How to treat raw sewage with constructed wetlands: An overview of the French systems. (*Proceedings*) *9th international conference on wetland system for water pollution control*. ASTEE (Association scientifique et technique pour l'eau et l;environnement) with scientific collaboration of Cemagref and Agence de l'Eau Rhone Mediterranee & Corse Ed (pp. 11–20). France: Alain Lienard Avignon.

Obarska-Pempkowiak, H., & Gajewska, M. (2003). The removal of nitrogen compounds in hybrid wetland systems. *Polish Journal of Environmental Studies, 12*(6), ss. 739–746.

Obarska-Pempkowiak, H., Gajewska, M. (2005). Recent developments in wastewater treatment in constructed wetlands in Poland. In A. Omelchenko, A. Alexander, W. Pivovarov, & J. Swindall (Eds.), *Modern tools and methods of water treatment for improving living standards* (vol. 48, pp. 279–295). Dordrecht: Springer (NATO Science Series: Series IV: Earth and Environmental Sciences, ISSN 1568–1238).

Soroko, M. (2001). Efficiency of organic matter and nitrogen removal in same treatment wetlands. *Woda – Środowisko – Obszary Wiejskie, 1, IMUZ, Zeszyt 1*, 173–186 (in polish).

Vymazal, J. (1998). Czech Republic. In J. Vymazal, H. Brix, P. F. Cooper, M. B. Green, & R. Haberl (Eds.), *Constructed wetlands for wastewater treatment in Europe* (pp. 95–121). Leiden: Backhauys Publishers.

Wasiak, A. (2008) Hydrophite WWTP step by step. In *Farmer handbook* (pp. 21–32) (in Polish) Oczyszczalnie roślinne, krok po kroku, Poradnik rolnika.

Wierzbicki, K., Eymontt, A., & Gutry, P. (2008). Monitoring report of ecological and sanitary effect assessment of single-family CWs in Sokoły, Rutka-Tartak, Jeleniewo, Punsk in Podlasie region (in Polish).

Chapter 15
Systematic Classification, Nomenclature and Reporting for Constructed Treatment Wetlands

Nat Fonder and Tom Headley

Abstract This paper proposes a structured foundation for classifying and naming different treatment wetland (TW) design alternatives, based on observable physical design traits. A classification hierarchy is organised like a polychotomous key, from general classification criteria to wetland type identification. The three characteristics are typical of all TW: the presence of macrophytic vegetation; the existence of water-logged or saturated substrate conditions for at least part of the time; and inflow of contaminated water with constituents to be removed. Treatment Wetlands are further classified based on hydrology and vegetation characteristics. Hydrological traits relate to water position, flow direction and degree of saturation. Based on the predominant position of water in the system, two main groups are identified: those with Surface Flow above a benthic substrate and those with Subsurface Flow through a porous media. The systems with surface flow are divided into three standards types, differentiated by vegetation type: Surface Flow (SF), Free-Floating Macrophyte (FFM), and Floating Emergent Macrophyte (FEM) TWs. Subsurface flow systems always contain sessile emergent macrophytes and are divided into four standard types, based on flow direction: Horizontal Flow (HF), Down Flow (DF), Up Flow (UF) and Fill and Drain (FaD) TWs. Standard types are described with their main applications. Nomenclature is also proposed for all associated variants. An overview of intensified variants, which have elevated energy, chemical or operational inputs in order to increase efficiency or overcome process limitations is provided. A list of the important design and operational information that should be provided when reporting about TW systems is also given.

Keywords Down flow · Fill and drain · Subsurface flow · Surface flow · Up flow

---

N. Fonder (✉)
Epuvaleau, npa, Hydrology and Rural Hydraulics Department, University of Liege, Gembloux Agro-Bio Tech, Gembloux, Belgium
e-mail: fonder@epuvaleau.eu

J. Vymazal (ed.), *Water and Nutrient Management in Natural and Constructed Wetlands*, DOI 10.1007/978-90-481-9585-5_15,

15.1 Introduction

Although some systematic approaches for classifying and naming treatment wetland types currently exist, the literature is riddled with a wide ranging and disconnected terminology. Numerous names are used, almost interchangeably to describe any given wetland variant, even if the physical design and operational characteristics are essentially the same. In some cases, the same name has been used for systems with very different design configurations. The applied terminology often varies by region, culture, discipline-base or the author's desire to give the impression that their design is new or innovative. This diversity in nomenclature within the treatment wetland field can lead to confusion, inefficiency and the general impression that our industry is lacking in organisation and professionalism, particularly to those new-comers or observers from other disciplines. Furthermore, there is a need for greater consistency in the fundamental design and operational information that is provided when reporting or publishing data from treatment wetlands.

The goals of this paper are: to review and clarify the current terminology used to define and describe constructed wetlands used for water treatment; to propose a standardised classification system that provides a framework for nomenclature in this field; and to summarise the important necessary details that should be provided when reporting about a treatment wetland system or study.

It is not the intention of this paper to provide a definitive nomenclature or to replace pre-existing terminology with a new system, but rather to make an initial step towards providing a structured foundation for classifying and naming different treatment wetland design alternatives. It is hoped that this can provide a platform for standardising the terminology that is used in our field. It is acknowledged that there will be existing and emerging applications that do not neatly fit the classification system given here. In many cases, authors will prefer a different name for their system; other, more common, names may be more appealing, user-friendly or politically attractive depending on the context or target audience. What we present here is intended to be a technical classification and terminology system that can be referred to in order to provide a clear and concise description of a given treatment wetland technology in a standardised way.

15.1.1 Methodology for Deriving Nomenclature

It is the belief of the authors that the most informative, technically correct and unambiguous system of nomenclature will be one based on easily definable physical traits of treatment wetland design. Thus, at the foundation of the proposed nomenclature is a classification system based on easily observable design traits. From this classification system, the various TW designs that require naming are identified. From the diverse range of TW systems identified, a sub-set of standard design types is defined based on fundamental differences in design and with consideration of the systems that are most commonly applied at full-scale today. These Standard Types form a central part of the nomenclature system and are assigned a route name and abbreviation. Design variants, which are considered to be modified versions of the Standard Types, are subsequently identified. The proposed nomenclature for these variants then takes the route name from the relevant Standard Type and couples it with a modifier which delineates the variant design. In some cases, a third level of common design variants can be defined either because they receive a very specific waste type (application-based variants) or have been intensified via elevated energy, chemical or operational inputs (intensified variants). Guiding principles in deriving the nomenclature have been to develop an efficient and clear terminology, eliminate redundant terms, and keep abbreviations as short as possible without sacrificing their meaning.

15.2 The Classification System

The general structure of the hierarchy is presented in Fig. 15.1. The hierarchy is organised like a polychotomous key, starting with general criteria and becoming progressively more specific at each step until a given wetland type can be identified. The key identifies individual TW unit types. A given TW system may be comprised of several TW units, with the units being either the same type or different types.

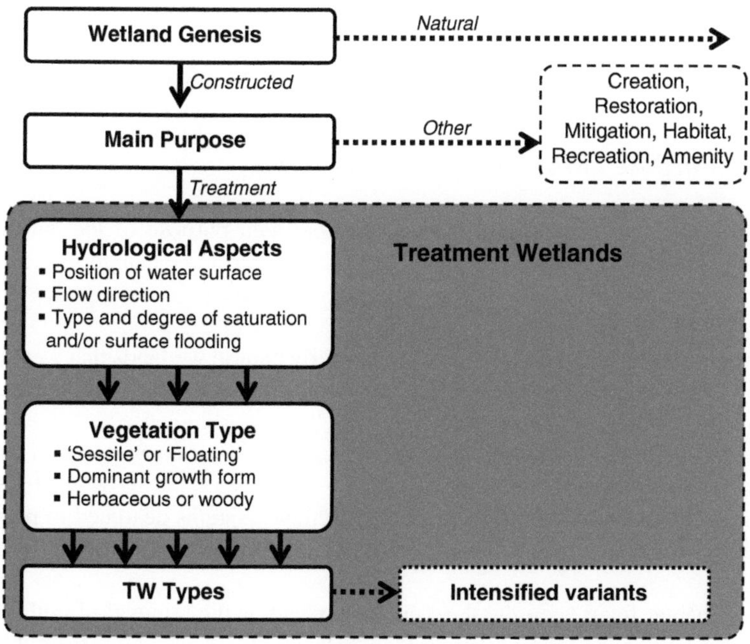

Fig. 15.1 The general structure of the classification hierarchy used to define treatment wetlands and their different types

A system for describing more complex TW systems is given in a subsequent section of this chapter (see Section 15.5.3).

The first step in the classification hierarchy is to define whether or not the system in question is a wetland. Wetlands can be defined as areas of land where the water table is at or near the surface for at least part of the year and are characterised by the presence of vegetation types and soil characteristics that have developed in response to the wet and saturated conditions (Kadlec & Wallace, 2008; Mitsch & Gosselink, 2007). In the context of this manuscript, closely related aquatic ecosystems that do not include higher plants (macrophytes), such as phytoplankton dominated ponds, are not included within the definition of wetlands.

15.2.1 Wetland Genesis

Wetlands can be first split into the two major types of Natural and Constructed wetlands. Natural Wetlands are only defined here as those wetland areas that exist in the landscape due to natural processes rather than having been created either directly or indirectly as a result of anthropogenic influences. This paper does not develop any classification or terminology for natural wetlands, as several systems already exist (see related references, e.g. Wetzel, 2001; Mitsch & Gosselink, 2007), but rather focuses on terminology related to constructed wetlands.

Constructed wetlands are artificially created ecosystems that would not otherwise exist without significant human intervention, such as earthworks or hydrologic manipulation. They are generally designed to mimic many of the conditions and/or processes that occur in natural wetlands (Vymazal & Kröpfelova, 2008).

15.2.2 Purpose of the Constructed Wetland

The next level of classification is based on the main purpose of the constructed wetland system. Constructed wetlands can be split into three categories according to their purpose:

- Restored Wetlands: Areas which were formerly natural wetlands that were lost or heavily degraded in the past and which, through human intervention, now support a near-natural wetland ecosystem.
- Created Wetlands: Non-wetland areas which have been converted to a wetland ecosystem by civil engineering works.
- Treatment Wetlands: Artificially created wetland systems designed to provide a specific water treatment function.

This chapter focuses on the third purpose, which is the removal of pollutants in Treatment Wetlands (TWs); a term which became widely used after the publishing of the text book "Treatment Wetlands: theory and implementation" by Kadlec and Knight (1996). Although the term "constructed wetland" is widely used to describe wetland systems built for water treatment purposes, we consider the term TW to be more specific and technically accurate. We suggest that the term "Treatment Wetland" is preferable when describing a wetland system constructed specifically for water quality improvement for the first time within a scientific publication.

We also acknowledge that natural or restored wetlands are in some cases used as receiving bodies for polluted waters and often provide important treatment functions. Such systems could also be considered "TWs". However, to avoid confusion we recommend that such systems be described as being used for water treatment or as "natural treatment wetlands".

15.2.3 Treatment Wetlands

Treatment wetlands are human-made systems designed to enhance and optimise certain physical and/or biogeochemical processes that occur in natural wetland ecosystems for the primary purpose of removing contaminants from polluted waters. As TWs can be constructed in a variety of hydrologic modes (Kadlec & Wallace, 2008), and numerous design variations have been developed for such a large array of pollutant removal mechanisms, there are a large number of design variants currently in use. Correspondingly, the commonly used terminology describing different TW systems is increasing and there is a need for a standardised set of terminology to clearly identify and define the most commonly used types in a technically accurate and consistent manner.

The first important question is: what constitutes a Treatment Wetland? Treatment Wetland designs range from those having standing water to those with no obvious surface water, and systems with vegetation types ranging from small free-floating species like duckweed, to forested swamps consisting of large trees like *Cypress* or *Melaleuca* species. Despite this diversity, three characteristics can be identified which are common to all TWs:

1. The presence of macrophytic vegetation that typically occur in natural wetlands;
2. The existence of water-logged or saturated substrate conditions for at least part of the time; and
3. The inflow of contaminated waters with constituents that are to be removed.

The first criterion excludes ponds and lagoons that consist primarily of microscopic algae (microphytes) without higher plants (macrophytes). However, it is acknowledged that ponds and TWs are closely related technologies often used in combination. One exception to this requirement for the presence of wetland plants is within the context of research, where "unplanted" versions are sometimes included to distinguish the effect of the plants in the system. Outside of this context, treatment units that do not include wetland vegetation, such as gravel or sand filters, should not be classified as a TW.

There may be some contention as to whether or not free-draining systems such as down flow wetlands, satisfy the second inherent requirement of a TW (the existence of water-logged conditions for at least part of the time). However, it is the opinion of the authors that even intermittently loaded down flow systems that are intended to operate in a primarily unsaturated mode, still experience ephemeral (periodic), localised water-logging of the substrate (e.g. during a loading event). Hence, such systems are considered to satisfy the requirement of partial water-logging.

15.2.3.1 Classification of Treatment Wetlands

The various TW designs that exist can be categorised based on two main physical attributes:

(a) Hydrology; and
(b) Vegetation characteristics.

These form the basis for the proposed hierarchical classification system which identifies TWs based on the six specific traits presented in Table 15.1. To classify a given TW unit using the hierarchy, the various physical traits are considered in

Table 15.1 Traits used to define the different classes of treatment wetland within the classification hierarchy

Physical attribute	Specific trait	Description	Defined classes for each trait	Sub-class
Hydrology	a. Water position	Position of water surface relative to soil or substrate	Surface flow[a]	–
			Subsurface flow[b]	–
	b. Flow direction	Predominant direction of flow through system	Horizontal	–
			Vertical[c]	Down Up Mixed
	c. Saturation of media[c]	Degree of saturation in media-based systems	Free-draining	–
			Intermittent	–
			Constant	–
	d. Surface flooding[c]	Type of surface flooding in media-based systems	None	–
			Ephemeral	–
			Permanent	–
Vegetation	a. Sessility[d]	Location of the roots: attached in the benthic sediments or floating	Sessile (benthic bound)	–
			Floating	–
	b. Growth Form	Dominant growth form of the vegetation in relation to the water	Emergent	Herbaceous Woody
			Submerged[d]	–
			Floating leaved[d]	–
			Free-floating[d]	

[a]majority of flow through a column of water overlying a benthic substrate
[b]majority of flow through a porous media
[c]only relevant to Subsurface Flow systems (by virtue of design, all Surface Flow TWs have horizontal flow, are constantly saturated and with a permanently inundated substrate)
[d]only relevant to Surface Flow systems (subsurface flow excludes submerged or floating plants)

a sequential order starting with the water position, followed by the flow direction, saturation degree, type of surface flooding and finally the vegetation sessility and growth form. Each step of the hierarchy gives rise to different classes of TW design.

The different traits considered in the classification hierarchy distinguish between key physical and operational aspects of the design which lead to fundamental differences between the various TW units in terms of their structure and functioning. Consequently, these traits have a major influence over the treatment conditions that develop within the TW unit and in determining its suitability for a particular application. It is therefore proposed that the technical nomenclature used to identify different TW units should reflect this classification based on physical attributes.

Water Position

Whether the TW has Surface Flow or Subsurface Flow has a major impact on the physicochemical conditions that develop within the treatment reactor and the types of vegetation that can be grown in the wetland. Surface flow TWs are defined as aquatic systems in which the majority of flow occurs through a water column overlying a benthic substrate. This is in contrast to subsurface flow TWs where the majority of flow is through a porous media. The term Surface Flow has been selected rather than the other commonly used term, Free Water Surface, for this category of TWs because it is a clear antonym for subsurface; the alternate type of flow position. The term "free water surface" is a somewhat more ambiguous combination of words, the meaning of which is not inherently clear, especially for people not familiar with the technology.

Flow Direction

By virtue of the design, essentially all surface flow TWs have flow in a predominantly horizontal direction (inlet and outlet horizontally opposed). In contrast, subsurface flow TW units have been designed with a range of different flow directions, with horizontal and vertical down-flow being the most common to date. Systems with up flow also exist along with an increasing number where the flow alternates between up and down flow (mixed). With regards to nomenclature, a distinction based on flow direction is therefore not necessary for surface flow systems (they all have "horizontal flow"). Similarly, the inclusion of a term to identify the flow direction (e.g. horizontal flow or down flow) should inherently imply that a system has subsurface flow, therefore rendering the term "subsurface" redundant in the nomenclature.

Media Saturation

This is a trait of particular relevance to units with subsurface vertical flow and refers to the saturation or moisture status of the media, which varies depending on the specific design. Systems with an outlet structure designed to hold water in the bed and maintain the media in a continuously saturated state are classed as having "constant" media saturation. By virtue of the design, all up flow systems have a constantly saturated media. Although down flow TWs can also be designed to maintain a constant saturation level, they most commonly operate in a free-draining mode, with a permanently open outlet at the bottom of the bed. Thus, such systems only experience rare and localised saturation of the media, such as during a loading event or within zones with large accumulations of water-retentive organic matter. In between constantly saturated and free-draining systems lie those TWs defined as having intermittent saturation, which refers to the situation where the media saturation varies periodically (e.g. TWs with alternating filling and draining cycles) or seasonally (e.g. TWs designed to evapotranspire all of the wastewater on an annual basis).

Surface flow TWs are invariably always designed to operate with a constantly saturated soil substrate, except if they are allowed to dry out for maintenance purposes. This trait is therefore not relevant to the classification of surface flow TWs.

Surface Flooding

Surface flooding refers to the degree to which the surface of the bed is flooded or inundated. It is an operational hydrologic trait only relevant to the classification of systems with subsurface vertical flow. By definition, surface flow TWs have a permanently flooded surface. In contrast, subsurface flow TWs with horizontal flow are virtually always designed with the intention of avoiding surface flooding.

Surface flooding in vertical flow units can be ephemeral (flooded for only a short period, often as a result of flood-loading), permanent (surface of the media is continuously flooded during normal operation) or non-flooded (entirely subsurface). Subsurface flow systems classed as having a permanently flooded media surface (e.g. some up flow TWs) are not considered to fall into the category of surface flow TWs because the majority of the flow still occurs through a porous media, as defined previously by the attribute: water position.

Vegetation Sessility

This is the first distinction made between TW units in the classification scheme based on vegetative traits. Sessile is a term used in the field of limnology to refer to vegetation that is anchored to the benthic environment, as opposed to floating. This trait is only relevant to TWs with surface flow because subsurface flow through a porous media precludes floating plants.

Vegetative Growth Form

The growth form of the dominant plants in the wetland is a major trait for distinguishing between the different surface flow TW designs. The vegetation growth forms used in this classification are based on those defined by Brix and Schierup (1989) and are divided into emergent, submerged, floating leaved and free-floating plants. Submerged and floating leaved plants come under the grouping of sessile vegetation. Emergent macrophytes typically grow in a sessile form. However, they can also grow on a buoyant mat that floats on the water surface which gives rise to a different type of TW design.

Emergent Vegetation Variants

The majority of emergent wetland plants are herbaceous (e.g. reeds, rushes and sedges). That is, they lack lignified tissues in their above-ground parts and consequently die down to ground level at the end of growing season. Due to their common use in TWs, herbaceous macrophytes are therefore the default type of emergent vegetation within the classification system. However, some TWs are dominated by woody species (trees and shrubs) which tolerate water-logged conditions (e.g. species of *Cypress* and *Melaleuca*). Because of the differences in growth cycles, shading, physical structure and element turn-over that can develop, TWs dominated by woody emergent vegetation form their own design variants within the classification system.

15.2.3.2 The Treatment Wetland Classification Tree

Applying the proposed classification system to the diverse range of TW designs in current use leads to the classification tree presented in Fig. 15.2. The standard types and their design variants are key-out at the base of the tree along with the proposed nomenclature and abbreviations. For the sake of clarity, redundant traits that do not play a role in identifying the different TW designs have been omitted from the tree diagram.

The classification tree highlights that the subsurface flow wetland group is a diverse class containing a broad range of designs differentiated essentially by hydrologic characteristics. In contrast, TWs with surface flow are classified based on vegetation characteristics.

Considering the current status of TW applications around the world and the fundamental design differences between the TW units identified using the classification system, we have distinguished between seven "Standard Types" of TW. These are denoted by a number above the nomenclature boxes in Fig. 15.2. The three standard types with surface flow are:

1. Surface Flow (SF) TW, dominated by emergent herbaceous macrophytes.
2. Free-Floating Macrophyte (FFM) TW containing free-floating vascular aquatic plants growing on the water surface.
3. Floating Emergent Macrophyte (FEM) TW with emergent macrophytes growing on a buoyant structure.

The four standard types with subsurface flow are:

4. Horizontal Flow (HF) TW.
5. Fill and Drain (FaD) TW in which the flow direction is mixed, normally alternating between up and down flow.

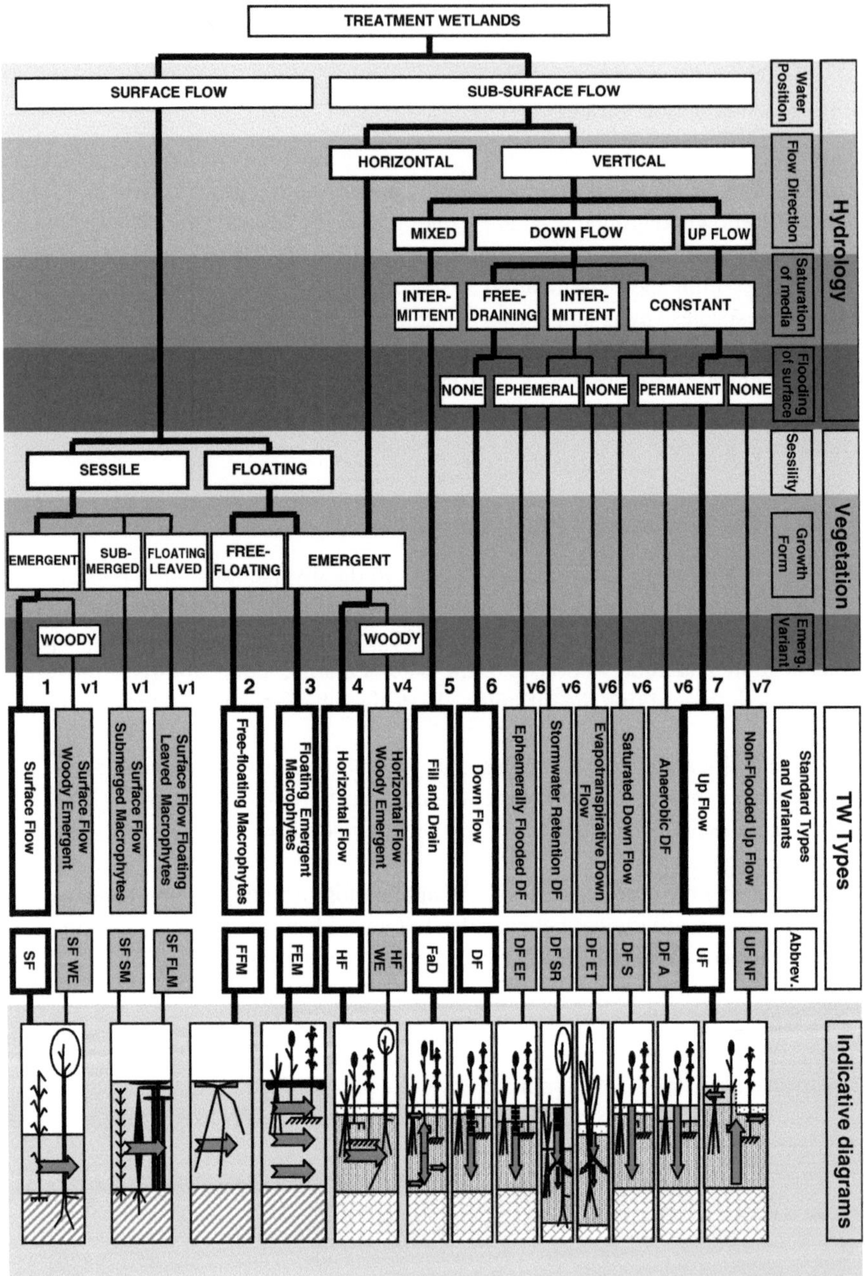

Fig. 15.2 Treatment Wetland classification tree

6. Down Flow (DF) TW, which is free-draining and without surface flooding.
7. Up Flow (UP) TW with a flooded surface.

There are numerous design variants identified in Fig. 15.2, which are considered modified versions of these more commonly applied standard types (e.g. based on atypical vegetation types, saturation degree or flow direction). Above the nomenclature box for each variant in the classification tree is a number denoting its respective standard type followed by a letter distinguishing it as a variant.

An additional sub-set of variants can be identified based on specific applications within the Ephemerally Flooded Down Flow (DF-EF) group of TWs (Fig. 15.3).

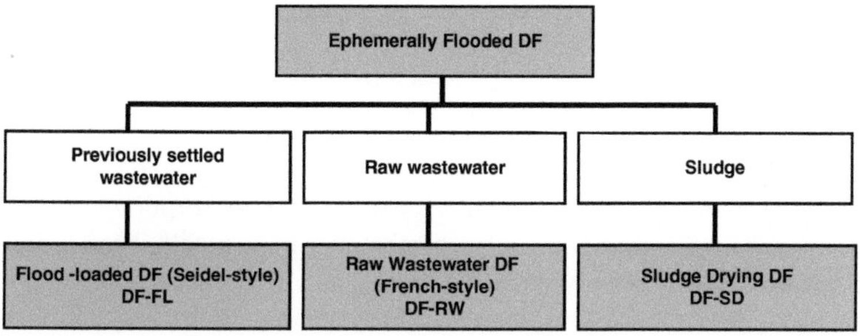

Fig. 15.3 The sub-classification of the Ephemerally Flooded Down Flow TW variants based on specific influent type

15.2.3.3 Intensified Treatment Wetland Sub-Variants

One special category of amended TW variants is what we have termed "Intensified" wetlands, which includes the increasing number of unit designs which are being developed to increase the efficiency and/or overcome process limitations (such as oxygen availability) of conventional zero- or low-energy input, passive wetland designs. After a given TW unit has been classified based on hydrologic and vegetation traits presented in Table 15.1, it can be further defined using the cross-cutting attribute of "Intensification" as outlined in Table 15.2. In principle, intensification represents a range of design or operational modifications that can be applied to virtually any of the TW types that have been classified in Fig. 15.2.

Intensification is generally achieved through increased inputs of electricity (e.g. aeration or recirculation pumping), physicochemical amendments (e.g. dosing with coagulants or flocculants, or use of highly sorptive media), or added operational effort or complexity (e.g. regular plant harvesting or cyclical resting). The use of a pump to move water uphill due to site constraints, or to load water into the wetland in a conventional manner (e.g. pulse loading of a IDF TW) is not considered to be a form on intensification. System designs incorporating specialised or

Table 15.2 Intensification classes that can be applied to enhance the performance of standard treatment wetland units

Type of intensified input	Intensification class	Example
Energetic	Aeration	Aerated subsurface flow TWs
	Pumping[a]	Fill and Drain TWs with reciprocation
Physicochemical	Sorptive media	Expanded clay, zeolites, bauxsol, chitinous material
	Chemical dosing	Alum, ferric chloride, oxidising agents
Operational	Frequent plant harvesting	Duckweed systems
	Cyclical resting	Routine alternation between multiple TW units in parallel
	Recirculation of flow [a]	DF TW with recirculation

[a]Energetic intensification via pumping is accompanied by modification of the physical design of the unit, whereas Operational intensification via flow recirculation (also involving pumping) does not necessitate a change in the physical design of the TW unit

synthesised media, such as those with exceptionally high sorption capacities, are considered to be intensified because such substrates have usually been heavily processed and received relatively intensive energy and/or chemical inputs during the manufacturing process and lead to an increased treatment efficiency of the TW for certain target contaminants. Intensification is an attribute that cuts across the final level of the classification hierarchy, in most cases representing a modification to one of the standard or variant types of TW identified in Fig. 15.2 and giving rise to an additional sub-set of TW variants.

The types of intensification described thus far can be considered Unit-based forms of intensification because they relate specifically to the design or operation of individual TW units. Another form of intensification can be identified as System-based Intensification, which relates to the way in which different TW units can be coupled together in various ways to form a treatment train with the aim of optimising the treatment efficiency of the overall system. Such systems are commonly referred to as "hybrid" or "integrated" systems. Further discussion about the description of such systems is provided in a subsequent section (see Section 15.5.3).

15.3 Definition and Nomenclature for Standard Treatment Wetland Units and their Variants

This section provides a description of the Standard TW Types and their variants that have been identified through the classification process, along with the recommended technical nomenclature (Fig. 15.2). Their current major applications are summarised, as well as currently used intensified variants.

15.3.1 Systems with Surface Flow

Systems where the majority of flow occurs through a water column above the surface of the benthic substrate are classified as Surface Flow TWs. These TWs are similar to many natural wetlands in that they contain areas of open water, floating vegetation, and emergent plants, either by design or as an unavoidable consequence of the design configuration.

15.3.1.1 Surface Flow Treatment Wetlands with Sessile Vegetation

This group of TWs includes all of those which have surface flow and contain macrophytes which have their roots bedded and anchored in the benthic substrate (sessile). Within this grouping are TWs that can be further classified based on the growth form of the dominant vegetation, with the major distinction being between those with emergent, submerged or floating leaved macrophytes.

Standard Type: Surface Flow TW

The most common type of TW with surface flow has water flowing above the surface of a permanently saturated soil substrate in a horizontal direction, with the predominant vegetation type being sessile herbaceous emergent macrophytes as depicted in Fig. 15.4 (unit 1 in Fig. 15.2). This represents one of the most widespread and commonly used TW designs worldwide. Hence, we have designated it as the central Standard Type for this group of systems and propose the name Surface Flow (SF) TW for this design. Note that redundant terms have been excluded from the nomenclature. Because all surface flow systems have flow in a predominantly horizontal direction, the term "horizontal" is not necessary. The presence of sessile emergent macrophytes is assumed to be the default condition.

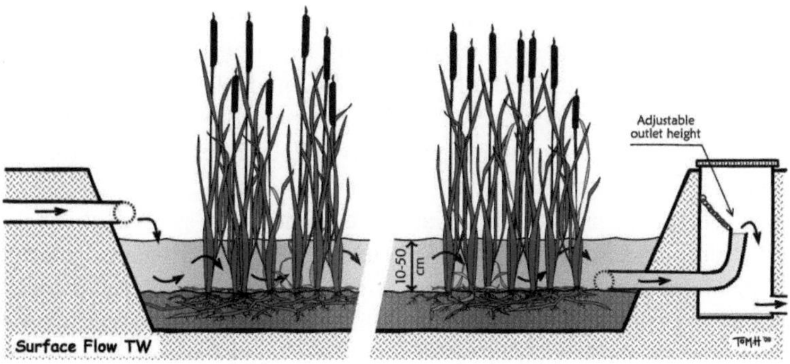

Fig. 15.4 Surface Flow (SF) TW

Common Applications

The most common application for SF TWs is to polish effluents from secondary treatment processes, such as natural systems (lagoons) or conventional systems (trickling filters and activated sludge). They are suitable in all climates, including very cold regions. Surface Flow TWs are a popular choice for the treatment of urban, agricultural, and industrial stormwaters because of their ability to deal with large variations in inflow rates and moderate, temporary changes in water levels. They are a frequent choice for treatment of mine waters, groundwater remediation and leachate. Surface Flow TWs are also typically the preferred design variant for large-scale applications (greater than one hectare) because at such scales the sand or gravel substrate used in TWs with subsurface flow becomes prohibitive due to cost and hydraulic limitations.

Significant ancillary benefits in the form of recreation and wildlife habitat can be provided by SF TWs. Operating costs of SF with emergent macrophytes are typically quite low and are usually capital cost-competitive with alternative technologies (Kadlec & Wallace, 2008).

Variants

All *variants* of the standard SF TW differ based on their vegetation growth form (units designated v1 on Fig. 15.2).

Variant 1a has woody, rather than herbaceous, emergent vegetation and is defined as a Woody Emergent Surface Flow (WE SF) TW. Note that the standard type "SF" forms the route of the abbreviation. This variant is typically limited to very large applications, often for tertiary polishing of effluents. They typically resemble natural wooded wetlands such as swamps, and are often associated with zones dominated by other vegetation types. Examples exist in the southern states of the USA such as Florida and Louisiana (dominated by *Cypress* species) and on the east coast of Australia (dominated by *Melaleuca* species).

Variant 1b is the submerged macrophyte surface flow (SF SM) TW which contains predominantly submerged macrophytic vegetation. These macrophytes have their photosynthetic tissue entirely submerged and typically only grow well in oxygenated water with good clarity. Thus, such systems cannot be used in wastewater with a high content of readily-biodegradable organic matter or high turbidity (Brix, 1993). They are often used for treating stormwater where the submerged leaves and associated biofilms provide an effective filter for removing fine suspended sediments.

Variant 1c has sessile floating leaved vegetation and is termed the Floating Leaved Macrophyte Surface Flow (SF FLM) TW. Floating-leaved macrophytes include plant species which are rooted in the benthic substrate with their leaves floating on the water surface, such as water lilies and some *Potamogeton* species. To date, the full scale application of SF FLM TWs is rare (Vymazal, 2009). They can have ornamental values when flowering water lily species are used. Floating leaved macrophytes are often used in combination with other emergent and submerged macrophytes (Kadlec & Wallace, 2008) either by accident or to create a more diverse ecosystem. Some floating leaved species have the ability to remove nutrients from the water column which can be beneficial in some cases (Greenway, 2003).

Intensification

Surface Flow TWs are intensified in rare cases where effluent is recirculated to the front of the system or where aeration is included at the inlet or in deepened zones which exclude macrophytes. Intensification of SF TWs has been attempted in the form of harvesting farms where the macrophytes are grown with the goal of biomass production (e.g. food for animals, energy biomass or production of ornamental flowers).

15.3.1.2 Surface Flow Treatment Wetlands with Floating Vegetation

A second sub-set of TWs with surface flow is identified as those where the vegetation is floating rather than sessile. These systems are classified as having surface flow, primarily in a horizontal direction, and remain permanently saturated. The buoyant vegetation means that these systems have more flexibility with regards to water depth as they are not limited by the flooding tolerance of the plants (as with sessile emergent macrophytes) or light penetration through the water column (as with submerged macrophytes).

These systems are further sub-classified into two main types based on whether they contain free-floating plants or emergent macrophytes growing on a floating mat.

Standard Types: Free Floating Macrophyte and Floating Emergent Macrophyte TWs

The first Standard Type has free-floating aquatic vegetation (unit 2 on Fig. 15.2) and is named the Free-Floating Macrophyte (FFM) TW (Fig. 15.5).

The second Standard Type of TW with surface flow and floating vegetation contain emergent macrophytes (normally sessile) growing on a mat or raft floating on the surface of a pond (unit 3 on Fig. 15.2) and is depicted by Fig 15.6. It is referred to here as the Floating Emergent Macrophyte (FEM) TW. Water is treated as

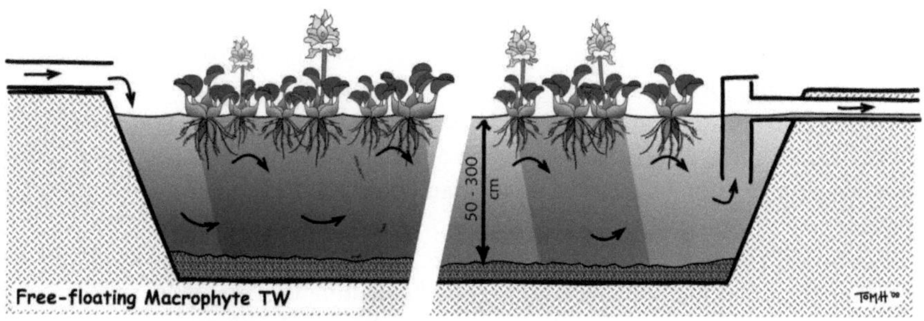

Fig. 15.5 Free-Floating Macrophyte (FFM) TW

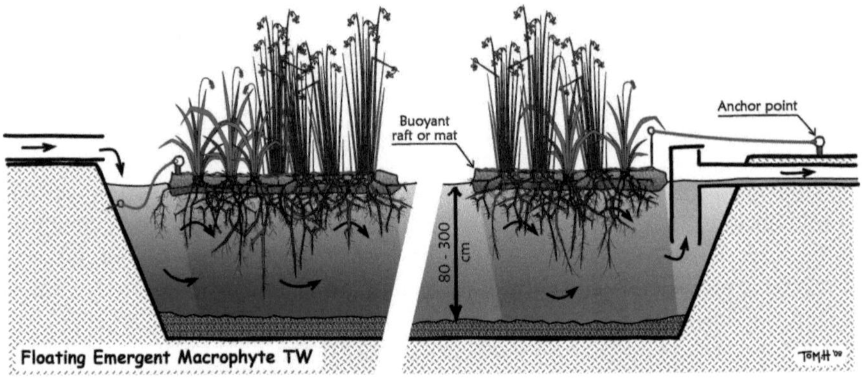

Fig. 15.6 Floating Emergent Macrophyte (FEM) TW

it moves through the floating mat and root-biofilm network that hangs in the water column beneath the mat. They are a relatively new and emerging TW type. They have a predominantly horizontal flow direction, with the inlet and outlet horizontally separated, combined with the capability to tolerate significant vertical water level variations due to the buoyant structure which holds the emergent macrophytes which would not otherwise survive such water level fluctuations.

The majority of flow in both of these TW units occurs above the surface of the soil substrate. They could therefore be classified as variants of the standard SF TW type. However, due to the unique interaction between the water and the floating plants and the specificity of their application, we have decided to classify them as a standard TW types. Hence the term "Surface Flow" has been dropped from the nomenclature and abbreviations for these two designs. In any case, the presence of surface water is an inherent requirement of floating vegetation.

Common Applications

Free-floating Macrophyte TWs are most commonly used for the treatment of municipal or industrial wastewaters. Some large-scale systems exist in the USA which have been designed specifically to facilitate regular harvesting of the rapidly growing biomass. There are also numerous examples of their use in tropical countries where suitable species occur naturally and productivity is particularly high. The uptake of this type of TW unit is hampered in many countries by the fact that many of the commonly used free-floating macrophyte species typically present significant weed problems outside of their natural range. Their free-floating nature can make them difficult to contain and easily transportable by waterfowl or flooding.

Floating-Emergent Macrophyte TWs are particularly suitable for drainage and runoff applications that are characterised by fluctuating flow regimes, such as treatment of urban stormwater, combined sewer overflows and agricultural drainage and runoff. This is due to their ability to tolerate variable and relatively deep water levels.

They can represent a relatively easy and inexpensive retro-fit option for upgrading existing aquatic systems, such as urban waterways and detention ponds. Because the plants are grown on a buoyant mat, they can be implemented in situations that would otherwise preclude the use of wetland plants due to excessive water depths or unsuitable substrate for planting, such as concrete lined urban canals. Another potential application is the upgrading of existing waste stabilisation ponds, where the root-biofilm network provides additional surface area for attached-growth processes, filtering of particulates. The shade provided by the floating mat can also be of benefit in preventing the excessive growth of phytoplanktonic algae. They have also been used for the remediation of eutrophic lakes and reservoirs.

Variants

Currently, no known variants of the FFM TW or FEM TW designs have been classified.

Intensification

A common form of intensification of FFM TWs is undertaken by the inclusion of regular vegetation harvesting in order to keep the floating plants in an active growth phase and optimise biomass production and uptake and removal of nutrients or other elements. This often requires specially developed harvesting machinery and adapted basin configurations.

Floating Emergent Macrophyte TWs have been intensified through the inclusion of aeration, inclusion of media with high sorption capacity within the floating mat or pumping of water from below to above the floating mat in order to enhance contact between water and the mat-root system. Systems are being developed in China for improving the water quality in polluted canals using commonly eaten plants such as Water Spinach and Watercress with the dual aims of enhancing the harvesting of nutrients while providing an incentive for local people to implement and maintain the FEM TW systems.

15.3.2 Systems with Subsurface Flow

The second major grouping of TWs is those with subsurface flow. This terminology refers to TWs where the majority of flow occurs through a porous media within which most of the treatment processes take place. Within this scheme of classification and nomenclature, a key difference between these systems and those with surface flow is that subsurface flow systems are media-based systems. In some cases there may be ephemeral or permanent flooding of the surface of the media, but this is primarily for the purpose of influent distribution or effluent collection; the majority of treatment still occurs within the porous media. Subsurface flow systems are sub-classified based on flow direction into those with a horizontal flow path and those where the flow is in a vertical direction.

15.3.2.1 Subsurface Flow TWs with Horizontal Flow

This grouping of TWs with subsurface flow represents those systems where the inlet and outlet are horizontally opposed. They are typically comprised of lined gravel, sand or soil based beds planted with sessile emergent vegetation. The wastewater flows through the rhizoshpere of the plants. Such systems generally have smaller surface areas (<0.5 ha) and higher hydraulic loading rates than SF TW (Xanthoulis et al., 2008).

Standard Type: Horizontal Flow TW

The standard and most commonly applied type TW with subsurface flow in a horizontal direction contains herbaceous emergent macrophytes as is termed the Horizontal Flow (HF) TW (unit 4 in Fig. 15.2). Because we have not used the term "horizontal" in the naming of surface flow TWs, its use here implies that the water position must be subsurface (Fig. 15.7). Consequently, the term "subsurface" becomes redundant within the nomenclature. The presence of herbaceous emergent macrophytes is assumed to be the default condition for the HF TW standard type, and is therefore omitted from the nomenclature. The most common species of herbaceous emergent macrophyte used in Europe is the Common Reed (*Phragmites australis*), while a range of other species are often used elsewhere, particularly those from the genera *Schoenoplectus*, *Cyperus*, *Typha*, *Baumea* and *Juncus*.

Common Applications

Horizontal Flow TWs are typically used to treat primary or secondary treated sewage prior to either soil dispersal or surface water discharge. In Europe, the systems tend to provide secondary treatment for village-sized communities of up to about 2000 population equivalents. In North America, they tend to provide tertiary treatment for larger populations (Xanthoulis et al. 2008). However, there are many

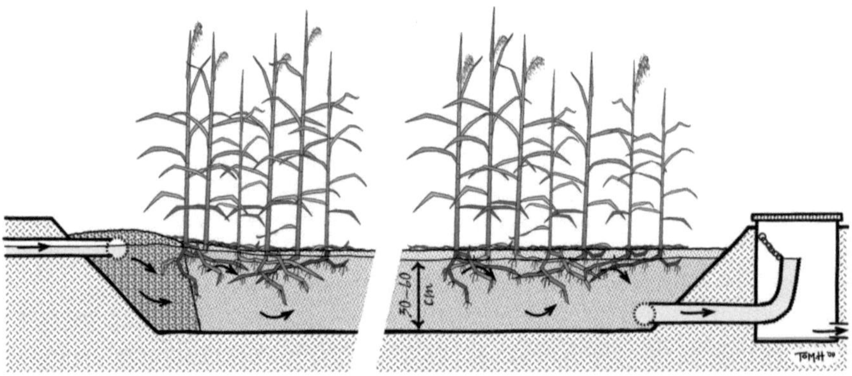

Fig. 15.7 Horizontal Flow (HF) TW, featuring herbaceous emergent macrophytes

other applications for specialty wastewaters from industry, and acid mine drainage. In general, HF TWs have been utilized for smaller flow rates than SF TWs, mainly because of cost and hydraulic limitations associated with flow through the porous media. These systems are capable of operation under colder conditions than SF TW, because of the ability to insulate the top surface and the thermal buffering provided by the substrate.

Variants

There is one identified variant of the HF TW which is different because of the vegetation type used (variant v4 on Fig. 15.2). This variant uses woody rather than herbaceous emergent vegetation, and is termed the Horizontal Flow Woody Emergent (HF WE) TW. Examples of HF WE TWs include systems in Australia used for on-site wastewater treatment and planted with *Melaleuca* species and systems planted with willows in Europe.

Intensification

The current intensification processes to boost the efficiency of HF TWs in done by means of aeration lines on the bottom of the bed (Intensified aerated HF TW), or specific substrates to enhance reactions like chemical precipitation, pH adjustment, or adsorption.

15.3.2.2 Subsurface Flow TWs with Vertical Flow

Vertical Subsurface Flow TWs generally consist of a bed of porous media (sand or gravel) through which the water moves in a vertical direction. In general, this group of TWs incorporates a diverse range of physical designs with respect to hydrologic characteristics. At the level of Flow Direction in the classification hierarchy (Table 15.1, Fig. 15.2), TWs with vertical flow are sub-divided into systems with downward flow, upward flow, and a combination of up and down flow ("mixed flow") which leads to the three standard types defined within this group of systems. Because of the diverse nature of TWs with a vertical flow direction, the term "vertical" is considered too vague to be meaningful with respect to the nomenclature and has therefore been dropped. All TWs with vertical flow fall within the subsurface category. Hence, the term "subsurface" is omitted from the relevant nomenclature. Similarly, the vegetation type is always emergent and does not play a role in the classification or naming of these systems.

Standard Types: Down Flow, Up Flow, and Fill and Drain TWs

The most common type of vertically flowing TW is the *Down Flow* (DF) TW (unit 6 in Fig. 15.2) which is free-draining (open outlet at the base of the bed) and remains unsaturated for most of the time. A network of pipes with multiple emitters is used to distribute the flow across the surface of the bed in a way that avoids surface flooding. The vegetation typically consists of herbaceous emergent macrophytes (Fig. 15.8).

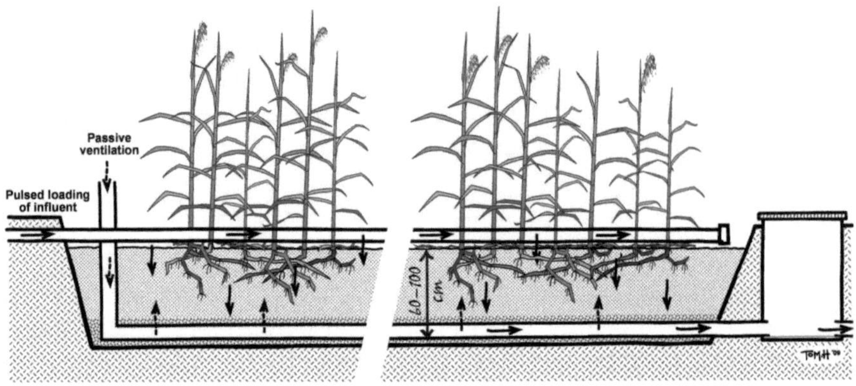

Fig. 15.8 Down Flow (DF) TW; free-draining and with herbaceous emergent macrophytes

Inlets are located vertically above the outlets. The bottom-most layer of media usually consists of coarse media with a network of perforated drainage pipes (Cooper, Job, Green, & Shutes, 1996), which are sometimes ventilated to the atmosphere to promote passive aeration of the substrate. Often the bed is planted with common reed (*Phragmites australis*), but a range of other emergent macrophytes are also used, as for HF TWs. Influent distribution pipes may be located above the system, or, in cold climates, buried within the granular media bed or under a layer of insulating mulch.

The second standard type of system with vertical subsurface flow is the *Up Flow* (UF) TW which features a constantly saturated media which is permanently flooded over the surface. The term "saturated" is redundant because the upward flow configuration dictates that the wetland must have a saturated bed. Wastewater is introduced at the bottom of the media bed via a series of distribution pipes and moves slowly upwards to the substrate surface. For practical reasons of conveying the effluent to the outlet, these systems have a flooded surface. However, we have classified these systems as having subsurface flow because the majority of important treatment processes are intended to occur within the saturated bed of media. These systems are sometimes referred to as Anaerobic Beds.

The third standard type of system within this category is the *Fill and Drain* (FaD) TW which is classified as having a mixed flow direction. Flow typically alternates between upward and downward flow, although it can sometimes be closer to diagonal flow depending on the relative location of the inlet and outlet. The media in these systems has an intermittent saturation level as it alternates between being saturated and unsaturated as a result of the filling and draining sequences. Normally the upper surface of the media is not flooded. These systems have been investigated for many years at mesocosm and pilot scale and their application at full scale is increasing due to the relatively high rates of oxygen transfer. Fill and Drain TWs can also provide the conditions necessary for complete nitrogen removal within the one reactor, with ammonia adsorption on the media during the filling stage, nitrification under aerobic conditions while the bed is drained, and denitrification with anaerobic condition and carbon source provided by the second filling sequence (Austin, 2006). Several Fill and Drain beds are typically incorporated in series.

Common names for this design include Tidal Flow and Fill and Draw wetlands. However, the term "tidal" can give the incorrect impressions that these systems have only two fill and drain cycles per day, may be somehow connected to the lunar cycle, or may be associated with marine or brackish water.

Common Applications

Down Flow TWs are very similar in design to intermittent sand filters (Liénard, Guellaf, & Boutin, 2001), which are widely used throughout the USA, Australia and New Zealand for decentralised wastewater treatment, except that DF TWs are planted with wetland vegetation. Down Flow TWs are used in many European countries particularly for achieving secondary treatment of primary settled sewage. Due to their higher oxygen transfer rates, DF TWs are becoming more common where discharge regulations require removal of ammonium. They are also a common design choice for effluents with a high carbonaceous or nitrogenous oxygen demand, such as landfill leachates and agricultural wastewaters.

Up Flow TWs are commonly applied where anaerobic treatment conditions are required, such as in some mining or industrial applications. They may be less prone to clogging than HF TWs because the influent solids and organic load is distributed over a much wider surface area.

Fill and Drain TWs are being increasingly applied for wastewaters with a high oxygen demand or where removal of total nitrogen is required. They often tend to have a smaller foot-print than other TW alternatives which makes them potentially suitable for arid regions where water loss via evapotranspiration can be a limitation.

Variants

Down Flow TW Variants

Several variants of the DF TW exist based on the degree of media saturation and the occurrence of surface flooding. The first variant includes a group of designs classed as *Ephemerally Flooded DF* (DF EF) TWs (Fig. 15.3). Like the standard DF TW, they are free-draining, but they are operated with ephemeral flooding of the media surface as a means of achieving distribution of the influent of the surface of the bed. This group contains three main sub-variants based on the type of inflow they receive. The first variant is similar to the standard Down Flow TW, except that distribution of the influent over the surface is achieved via flood-loading from point discharges, rather than the use of a network of distribution pipes with multiple emitters which maintain the influent essentially below the media surface. Hence, we have termed this design variant the Flood-loaded Down Flow TW (DF FL) TW. The distribution of the influent relies on the use of a hydraulically restrictive surface layer of media (usually sand). During loading events, the hydraulic conductivity of this surface layer is exceeded and the surface floods. The Flood-loaded DF design is typically used for treatment of wastewater that received at least primary settlement.

It was initially developed by Käthe Seidel in Germany during the 1970s and is still employed today in parts of Europe, particularly the UK (Kadlec & Wallace, 2008).

The second ephemerally flooded down flow sub-variant differs from the flood-loaded DF TW only by the fact it receives unsettled raw wastewater; hence the name Raw Wastewater Down Flow TW (DF RW) TW. This is a design variant developed and applied in France for the decentralised treatment of sewage. It is effectively another development of the original Seidel-style DF-FL TW approach. In order to manage the organic sludge layer that accumulates on the surface of the DF-RW beds and maintain permeability, systems employing this design typically have multiple DF-RW beds in parallel with a rotationally rested operation.

The final sub-variant of the DF-EF TW is also defined based on the waste type it receives; namely sludge. The Sludge-drying Down Flow (DF SD) TW is periodically flood-loaded with sludge for the purpose of dewatering and mineralisation. The vegetation, soil, sun, and gravity separate solids and liquids from the sludge. The solid fraction of the sludge stays on the surface of the bed while some of the water percolates down through a sand and gravel drainage layer. Substantial dewatering of the sludge occurs via the high evapotranspiration rates of the wetland plants. After each load, a dewatering period is allowed before a new layer of sludge is flood-loaded on top of the dewatered sludge. The plants progressively grow upwards as the stabilised sludge is gradually accumulated in the system. The roots provide drainage channels through the accumulated sludge and also contribute organic matter and enhance the physical structure of the dewatered sludge, thereby facilitating composting and stabilisation processes. This process continues until the bed is filled with dewatered, stabilised sludge (to a depth in the order of one metre) and has to be emptied (after about 10 years). The drainage water receives some treatment as it percolates through the layers of stabilised sludge, sand and gravel and is normally returned to the wastewater treatment plant for further treatment (Nielsen, 2003). They are commonly named Sludge Drying Reed Beds, which can cause confusion with HF TWs also commonly called Reed Beds in the UK. Both common terms are referring to two very different design configurations, and should therefore be avoided when describing TW systems.

Another variant of the DF TW identified in Fig. 15.2 is the *DF Stormwater Retention* (DF SR) TW, commonly referred to as a Bio-retention, Bio-filtration or Rain Garden systems within the stormwater management industry. These systems are typically bunded in order to accumulate runoff during rainfall events which then slowly percolates downwards through the porous substrate of sand or gravel. Hence they are classed as having an intermittent saturation level and ephemeral surface flooding. Woody emergent vegetation is often used in these systems because it tends to be more tolerant of the dry conditions that can develop within the filter between rainfall events.

The *Evapotranspirative DF* (DF ET) TW variant refers specifically to a design developed in Denmark which uses fast growing woody vegetation with a high evapotranspiration (ET) rate, such as willows, with the aim of evapotranspiring all of the wastewater on an annual basis. Primary treated wastewater is typically introduced near the upper surface of the substrate, giving rise to what could be considered a downward flow path. However, the actual flow direction is somewhat ambiguous because the systems do not have a through-flow of water and the efflux is to the atmosphere. They are classed as having an intermittent saturation level because the water level within the media varies seasonally depending on the balance between inflow, rainfall and evapotranspiration. Down Flow ET TWs are commonly referred to as Zero-discharge Willow Systems. They are typically deeper (approximately 2 m) than other TWs with subsurface flow, in order to provide storage capacity to accumulate water during periods of wet weather and low evapotranspiration.

The forth main variant of DF TW is the *Saturated Down Flow* (DF S) TW. The main difference to the Standard DF TW lies in the fact that the water level in the bed is maintained slightly below the upper surface of the media resulting in a "constantly saturated" bed (rather than free-draining). Consequently, these systems are very similar to HF TWs with regards to the biogeochemical conditions that develop within the wetland. However, because the influent is distributed across the entire upper surface of the wetland, the influent loading rate of organics and solids in the cross-sectional plain perpendicular to the flow direction are greatly reduced when compared to HF TWs. This means that the substrate clogging potential is substantially reduced compared to a HF TW receiving the same influent loading rate.

The final variant within the DF TW group is the *Anaerobic DF* (DF A) TW, which is defined as having constantly saturated media and a permanently flooded surface. Although they have a permanently flooded media surface, they are still classed within the subsurface flow group of TWs because the majority of flow and important treatment processes occur within the porous media. These systems are similar in application to Up Flow TWs and are often used to promote anaerobic treatment processes for mining and industrial applications.

Up Flow TW Variants

There is one main design variant for the Up Flow TW group. This is the Non-flooded Up Flow (UP NF) TW design which is identical to the standard UP TW except that the outlet collection pipes are configured in a way to avoid flooding of the upper surface of the media. Maintaining entirely subsurface flow in the up flow configuration can be challenging in terms of design and operation (Younger, Banwart, & Hedin, 2002). Hence, most examples of this variant have been only at mesocosm or pilot scale.

Fill and Drain TW Variants

The application at full-scale of this design type is still in its infancy. A range of different design variations based on flow direction have been trialled at mesocosm or pilot-scale, but no consistent variants are currently obvious.

Intensification

Down Flow TWs are often intensified operationally by the inclusion of recirculation of a portion of the effluent back to the pre-treatment stage in order to enhance treatment stability or achieve denitrification. Such systems are called Recirculating DF TWs.

Intensification of the Saturated Down Flow (DF S), Anaerobic DF (DF A) and Up Flow (UF) TWs has been used for treatment of mine waters by incorporation of selected media, such as a mixture of compost and limestone, as the substrate in order to promote anaerobic conditions and the production of alkalinity. In this case, the intensification refers to the use of specific substrates to enhance the treatment efficiency. Such a system is commonly termed an anaerobic wetland or alkalinity producing system (Kadlec & Wallace, 2008). These systems are often constructed without wetland vegetation to avoid plant mediated oxygen transfer to the substrate, in which case they would not be classified as a TW.

Another common form of intensification of the Saturated Down Flow (DF S) TW is to actively pump small bubbles of air into the saturated substrate via a network of aeration lines installed at the base of the bed in order to overcome oxygen transfer limitations. These systems can be termed Aerated DF-S TWs, which can be abbreviated to A-DF-S TW. Such A-DF-S TWs are often used for treatment of waters with a particularly high oxygen demand, either due to high concentrations of organic compounds (e.g. airport de-icing runoff) or total kjeldahl nitrogen (e.g. landfill leachates) (Wallace, Higgins, Crolla, Bachand, & Verkuijl, 2006). Such systems are sometimes further intensified through the use of pumping to recirculate a portion of the effluent back through the aerated bed. In keeping with the nomenclature protocol, these systems would be termed Recirculating Aerated Saturated Down Flow TWs (R-A-DF-S TWs).

An intensified version of the Fill and Drain (FaD) TW is what is commonly termed a Reciprocating Wetland, or if aligned with the proposed nomenclature: a Reciprocating FaD (R-FaD) TW. Intensification is achieved by repeatedly transferred the water in a Fill and Drain fashion between two partnered beds using pumps. A reciprocation cycle typically involves the pumping of the majority of water from one bed to an adjacent drained bed, followed by a rest period. After the rest period, the majority of water is pumped from the full bed back to the original bed which has been resting in a drained state. The number of reciprocation cycles per day is largely dependent on the oxygen demand of the wastewater and is commonly in the order of one complete cycle every 1 or 2 h. Hence, these systems are intensified by the elevated input of energy for pumping of the water.

15.4 Other Important Information that Should be Reported when Publishing

In addition to using technically accurate, informative and consistent nomenclature, there is a range of design and operational information that should be provided whenever reporting about a TW system or study. Such fundamental details, presented in a uniform way, will greatly improve the usefulness and readability of published TW data and studies and enhance the exchange of knowledge and information within our discipline. This will enable readers to more readily interpret and relate published data to their own experiences and allow data from different systems to be compared and contrasted in a more meaningful way. The following is provided as a guide to the key information that should be reported as a minimum wherever possible.

15.4.1 *Type of waste or source water*

Treatment wetlands are used to treat a wide range of polluted waters or wastes. Hence, the characteristics of the source water or waste should be described. The main categories of source waters and wastes are described below. In brief, the category and origin of the water or waste to be treated by the wetland should be clearly indicated, as well as a summary of general water quality and flow characteristics if not otherwise presented in the results section of the article.

15.4.1.1 Runoff and drainage waters

Rainfall or irrigation-derived surface (runoff) and drainage waters, include urban stormwater, agricultural runoff, mine water and leachates. The area and type of catchment should be specified (e.g. agricultural, urban, airport, or highway). These types of waters often tend to be characterised by variable hydrology and/or pollutant composition. This variability should be adequately described and summarised. For leachates, the material of percolation (e.g. landfill, compost, mining spoil); and for mining applications, details about the type of mining and processing operations contributing to the drainage water should be provided.

15.4.1.2 Human generated wastewaters

Human generated wastewaters are derived from human activities and can have their origins in agriculture, human settlements or industries. The agricultural system producing the effluent (e.g. fish, poultry, pig, or dairy) or the type of industry (e.g. chemical, petrochemical, tannery, pulp and paper, food processing, textile, abattoir, laundry) should be specified. For sewage effluents, the source of the wastewater (e.g. on-site, domestic, municipal) should be given, along with the level of previous treatment as described in the next section.

15.4.1.3 Non-effluent By-products

This category encompasses the non-effluent types of waste or by-products from other treatment processes, such as biosolids and sewage sludges. These wastes typically have relatively high solids content and would not normally be classified as effluent. The process from which the non-effluent by-product is derived from should be cited (e.g. activated sludge, septage, waste stabilization pond sludge, agro-food industry).

15.4.2 *Treatment sequence*

For systems treating human generated wastewater or other effluents, the stage that the TW occupies in the treatment sequence should be identified as: primary (preceded by preliminary screening and grit removal only), secondary (prior removal of the majority of particulate and organic matter content) or tertiary (with the purpose of further polishing, disinfection or nutrient removal). Hence, the type of preceding treatment technologies should be specified.

15.4.3 Systems with multiple TW Units

Systems incorporating several treatment steps are increasingly being used in order to satisfy various source-water and end-use requirements. Many TW systems also incorporate numerous TW units that are operated in parallel for maintenance reasons. Thus, it is important that the number and type of TW units that make up a treatment system are clearly identified and whether they are organised in series or parallel. The first step is to distinguish between whether the system is a "Single-stage" or "Multi-stage" system. A Single-stage system is composed of one single wetland unit, whereas a Multi-stage system includes a succession of several wetland units operated in series (and potentially other treatment technologies).

15.4.3.1 Multiple Units in Series

A multi-stage TW system can be classified as "Mono-typic" (e.g. a succession of several Horizontal Flow systems), or "Poly-typic" (commonly referred to as a hybrid or combined system), which is a succession of different TW types (e.g. a Down Flow unit followed by a Horizontal Flow unit). Poly-typic systems should be defined stage by stage, using the terminology described previously in this chapter, and identifying the location of each TW unit within the treatment sequence.

15.4.3.2 Multiple Units in Parallel

Full scale TW systems often include multiple wetland units operated in parallel, even if consisting of only one stage in series. All of the parallel units may be operated simultaneously, or the flow may be alternated between the parallel units on a cyclical basis. The number, size and method of splitting and managing the parallel systems are important characteristics which can significantly affect the performance and longevity of a system and should therefore be clearly defined. If flow is alternated between parallel units on a cyclical rotation, then the operational details of this rotation and resting should be specified.

15.4.4 Scale and Dimensions

The scale of the system should be clearly identified, such as full-scale, pilot-scale (with its ratio to full scale), mesocosm (intermediate between bench and pilot scales) and microcosm (lab or bench scale). It is also important to identify whether the system is a full-scale operational system exposed to all the perturbations experienced in the "real world" or a more strictly controlled research facility.

The actual operating dimensions (length, width, depth) of the TW are some of the most important fundamental details that should be provided. This should include the normal water depth as well as the depth of substrate if applicable.

15.4.5 Recirculation

Any recirculation of part of the flow should be identified, including the location from where and to where the recycled portion is directed. The recycling stage should be defined, as well as the part (percentage) of recycled flow or recirculation ratio.

15.4.6 Other Important Design and Operational Information

Any other design and operational details that are necessary for the reader to understand how the system functions and behaves, or to interpret the data presented, should also be provided. These may include details about the:

- System age (commissioning date) and any relevant abnormalities in the operational history (such as periods of resting or overloading);
- Media type and size specifications for HF and DF systems (nominal size, d_{10}, uniformity Coefficient, washed or unwashed, crushed rock or alluvial material), or the type of soil substrate used for plant establishment in SF systems;
- Plant species used, date of planting (and planting technique and density if relevant) and any routine harvesting practices;
- Hydraulic loading rates. These should be reported as areal rates in units such as L m^{-2} d^{-1} or mm d^{-1}. For systems where a residence time is clearly definable, the Hydraulic Retention Time should be given, at least as a nominal estimation if tracer studies have not been conducted. It should be clearly stated whether these parameters are based on actual measured inflow and outflow rates, or assumed design loadings. In some situations it is also informative to comment on what the original design loading rates were and whether the system is over- or underloaded.
- Treatment performance data and influent loading rates. Wherever pollutant removal data is summarised for a TW, the influent and effluent concentration for the contaminants of interest should be reported at the very least. If possible, influent loading rates, exit loads and mass removal rates should be reported, either on an areal basis (such as g m^{-2} d^{-1} or tonnes per hectare per year), or volumetric basis (e.g. g m^{-3} d^{-1}). If the system dimensions are provided, then the reader can inter-change between areal and volumetric units via simple calculations. While it is sometimes convenient to summarise performance data into average concentration percent reductions in order to highlight trends or simplify the situation, such data is rarely very useful or informative on its own. It can often be useful to summarise data into reaction rate coefficients, such as *k* rates using a kinetic

model. In such cases, it is important to define the model and its parameters, and the assumptions made in calculating the model parameters.

- Inlet loading rates for SSF systems. Clogging of the substrate can be a major operational limitation in SSF systems. The loading of total suspended solids (TSS) and biochemical oxygen demand (BOD) at the inlet of the wetland can play a significant role in clogging. Hence, the TSS and BOD loading at the inlet zone of SF TWs should be reported along with a description of the clogging related experiences for the system. The inlet TSS and BOD loading rates should be reported in a standardised and relevant way, such as grams per square metre per day (g m^{-2}.d), with the surface area being the cross-sectional surface area at the point where the wastewater is introduced to the bed, typically in the plane that is perpendicular to the predominant flow direction. For example, for a DF TW the surface area will be the upper surface of the bed, whereas for a HF TW it will be the cross-sectional surface area as defined by the bed width multiplied by the wetted depth at the inlet end.
- Influent loading conditions, such as continuous gravity flow, or intermittent pumped loading. Where the influent is pumped intermittently into the TW, the loading frequency (e.g. daily, hourly) and duration should be reported.
- Effluent quality target and national standards can also be relevant, as they can explain and justify the type of TW and its design parameters.

Acknowledgments We sincerely want to thank here all colleagues who have contributed to this paper with their support, reflections, comments, criticisms, ideas, graphs, and references. In alphabetical order, special thanks to V. Dias, J. Nivala, C. Tanner, F. Van Dien, J. Vymazal, S. Wallace, and C. Weedon.

References

Austin D. C. (2006). Influence of cation exchange capacity (CEC) in a tidal flow, flood and drain wastewater treatment wetland. *Ecological Engineering, 28*, 35–43.

Brix, H. (1993). Wastewater treatment in constructed wetlands: system design, removal processes, and treatment performance. In A. G. Moshiri (Ed.), *Constructed wetlands for water quality improvement* (pp. 9–22). Boca Raton: CRC Press.

Brix, H., & Schierup, H.-H. (1989). The use of aquatic macrophytes in water-pollution control. *Ambio, 18*, 100–107.

Cooper, P. F., Job, G. D., Green, M. B., & Shutes, R. B. E. (1996). *Reed beds and constructed wetlands for wastewater treatment*, Medmenham, Marlow: WRc Publications.

Greenway, M. (2003). Suitability of macrophytes for nutrient removal from surface flow constructed wetlands receiving secondary treated sewage effluent in Queensland, Australia. *Water Science and Technology*, *48* (2), 121–128.

Kadlec, R. H., & Knight, R. L. (1996). *Treatment wetlands* (893 p). New York/Boca Raton, FL: CRC Press/Lewis Publishers.

Kadlec, R. H., & Wallace, S. D. (2008). *Treatment Wetlands* (2nd ed.). Boca Raton: CRC Press.

Liénard, A., Guellaf H., & Boutin, C. (2001). Choice of sand for filters used for secondary treatment of wastewater. *Water Science and Technology, 44*(2–3), 189–196.

Mitsch, W. J., & Gosselink, J. G. (2007). *Wetlands* (4th ed.). New York: Wiley.

Nielsen, S. (2003). Sludge drying reed beds. *Water Science and Technology*, *48* (5), 101–109.

Vymazal, J. (2009). The use of constructed wetlands with horizontal sub-surface flow for various type of waste water. *Ecological Engineering*, *35*, 1–17.

Vymazal, J., & Kröpfelova, L. (2008). *Wastewater treatment in constructed wetlands with horizontal sub-surface flow*. Dordrecht: Springer.

Wallace, S. D., Higgins, J. P., Crolla, A. M., Bachand, A., & Verkuijl, S. (2006). High-rate ammonia removal in aerated engineered wetlands. In V. Dias & J. Vymazal (Eds.), *Proceedings of 10th International Conference on Wetland Systems for Water Pollution Control*. (pp. 255–264). Lisbon: Ministério de Ambiente, do Ordenamento do Território e do Desenvolvimento Regional (MAOTDR) and IWA.

Wetzel, R. G. (2001). *Limnology: Lake and river ecosystems* (3rd ed.). San Diego, CA: Academic.

Xanthoulis X., Tilly J., Fonder N., Wauthelet M., Bergeron, Ph., Brix, H., et al. (2008). *Curriculum on low cost wastewater treatment*. Europe Aid Cooperation office. Contract VN/Asia-Link/012(113128) 2005–2008.

Younger, P. L., Banwart, S. A., & Hedin, R. S. (2002). *Mine Water: Hydrology, pollution, remediation*. Environmental Pollution 5. Dordrecht: Kluwer.

Chapter 16
Comparison of Temperature Regimes of Two Temperate Herbaceous Wetlands in the Course of the Growing Season

Jakub Brom, Alžběta Rejšková, and Jan Procházka

Abstract Wetlands are habitats with highly variable species composition, hydrology and functioning in terms of energy exchange and forming of local climate. We compared two graminoid temperate wetlands situated nearby and at similar latitude to see how much their temperature characteristics differed with changing conditions during a growing season. The results showed that the two habitats, a littoral zone of a fishpond and a wet meadow, differed markedly. The daily means at 2 m were on average higher in the littoral as a consequence of lower temperature minima in the wet meadow. However, the daily maxima at 2 m were higher in the wet meadow due to advection of warm air from the surrounding habitats. We suppose that the temperature of the air at 2 m was predominantly influenced by temperature characteristics of the surrounding habitats (i.e. local meteorology). The temperature within vegetation and at the soil surface was dependent on the soil water content. This relation was stronger in the wet meadow than in the littoral due to the complex microtopography of the latter *Carex* dominated habitat. Vegetation played an important role in influencing temperature at the soil surface and within the vegetation cover, i.e. in the very proximity of the land surface. Because of the higher LAI, this trend was stronger in the wet meadow.

Keywords Wetland · Temperature regime · Vegetation season · Global radiation · Soil water content · Albedo · Leaf area index · Vegetation height

16.1 Introduction

Ambient temperature represents one of the most important environmental factors influencing presence and physiological functioning of plants and other organisms (e.g. Körner, 2006). On the other hand plants through their physiological processes

J. Brom (✉)
Department of Landscape Management, Faculty of Agriculture, University of South Bohemia in České Budějovice, České Budějovice, Czech Republic
e-mail: jbrom@zf.jcu.cz

J. Vymazal (ed.), *Water and Nutrient Management in Natural and Constructed Wetlands*, DOI 10.1007/978-90-481-9585-5_16,

influence temperature of their environment (Nobel, 1999; Fitter & Hay 2002; Kabat et al., 2004). Even though the area of wetlands in the cultural landscape is very limited (approx. 2–3% of the global land area, Linacre (1976)), these habitats with high biomass production and sufficient water supply play an important role in the landscape functioning because they have the capacity to influence humidity and temperature status of their environment by intensifying evaporation and condensation processes (Ripl, 1995, 2003; Pokorný, 2001; Kravčík, Pokorný, Kohutiar, Kováč, & Tóth, 2007; Brom & Pokorný, 2009). Therefore it is important to consider not only the biodiversity impact but also environmental functions of wetlands in landscape planning including recultivation. Although the potential role of wetlands in landscape functioning is large, the relations to and interconnections between the water cycle and temperature regime of a wetland and its surroundings are complex and poorly understood. Moreover, there is little information about temperature variability among wetlands with different species composition, situation and environmental conditions. Also the importance of different biophysical factors for local climate are rarely studied (Jackson et al., 2008).

The aims of this study were (1) to compare seasonal temperature regime of two temperate grass wetlands and (2) to estimate the importance of selected biophysical factors (incoming solar radiation, soil humidity and albedo) for temperature regimes of the wetlands.

16.2 Material and Methods

16.2.1 Site Description

The studied area is located in the Třeboň Biosphere Reserve (TBR), South Bohemia, Czech Republic. The average altitude of the flat TBR is 440 m a.s.l. The average annual precipitation is 618 mm (Pokorný & Kučerová, 2000). Annual average temperature is 8.2°C. More than 50% of the TBR is covered by forest, 30% by farmland and about 10% by fishponds and wetlands. The rest of TBR is covered by built-up areas, roads and anthropogenic and ruderal areas.

We studied the temperature regimes of two wetland localities, a littoral of a pond and a wet meadow near the town of Třeboň (Fig. 16.1). The littoral is situated at an inflow part of the Ruda fishpond and covers about 18 ha. It is surrounded by a coniferous forest. The habitat is dominated by *Carex acuta* L., which forms tussocks up to 0.6 m high. A substantial part of the littoral is covered by gaps between the tussocks. They are either without any vegetation or occupied by small water flora e.g. *Lemna* sp. The habitat is not managed. During our measurements the water level fluctuated near the base of the tussocks.

The wet meadow is located in the floodplain of the Rožmberk fishpond. The area of wet meadows complex covers approximately 190 ha. The homogenous part which we studied is approx. 12 ha large. The habitat hydrology is interconnected with the fishpond. The herbaceous wetland is dominated by *Carex acuta* L. and *Urtica dioica* L. *Carex acuta* forms approximately 0.3 m tall tussocks in this locality.

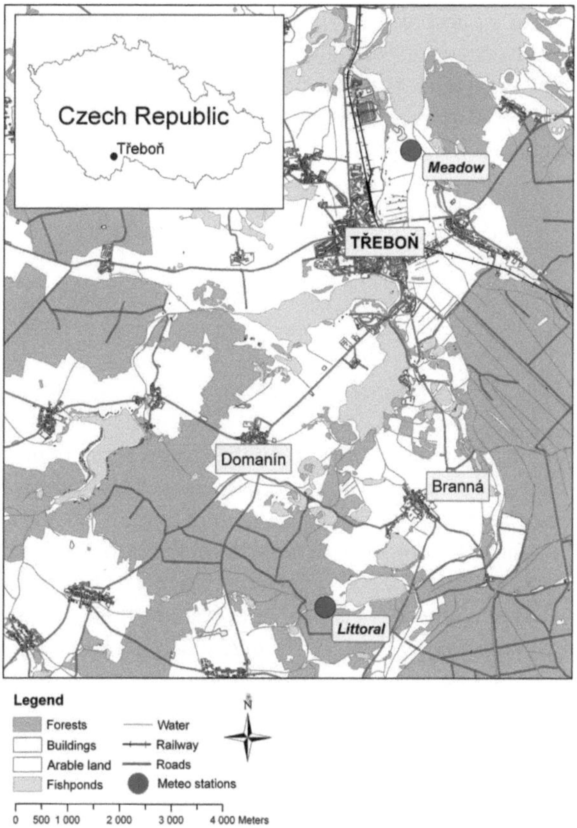

Fig. 16.1 Map of the area of interest. Studied localities are marked with a red sign

Urtica dioica L. and *Persicaria hydropiper* (L.) Dalarbre are dominant in the gaps. The habitat is not being managed. More details about the wet meadow vegetation characteristics were reported by Prach (2008).

16.2.2 Data Description

The meteorological data were recorded at 10-min intervals from May 1st to August 29th 2008. We measured air temperature 2 m above the soil surface and at the vegetation surface, (Ta and Tc, °C; T+Rh probes, accuracy ± 0.1°C), temperature at the soil surface (Ts, °C; Pt 100, accuracy ± 0.1°C), incoming and reflected shortwave (global) radiation ($R_{S\downarrow}$ and $R_{S\uparrow}$, respectively, W m^{-2}; CM3 pyranometers, Kipp and Zonen, the Netherlands, spectral range from 310 to 2800 nm) and volumetric content of liquid water in the soil at 0.05 m under the ground (θ, %; Wirrib, AMET, Czech Republic, accuracy ± 0.01 m^3 m^{-3}).

The vegetation cover height (H in meters) was measured manually. Leaf area index was measured using plant canopy analyser Li 2000, Li-COR, USA (LAI,

$m^2\ m^{-2}$; accuracy $\pm$ 15% in optimal conditions, for more information see Welles and Norman (1991). The parameters of vegetation cover were recorded approximately once in 3 weeks.

The daily mean albedo (α; %) was computed from the daily means of incoming and reflected shortwave global radiation as follows.

$$\alpha = 100\frac{R_{s\uparrow}}{R_{s\downarrow}} \quad (1)$$

16.2.3 Statistical Analysis

For comparison of the monthly temperature regimes of the two localities and for regression analyses of relations between temperatures and the environmental parameters and daily means of temperature, maxima and minima were used.

Multiple comparisons were developed using the pair *t*-tests with Bonferroni corrections (Salkind, 2007). If Bonferroni correction was used *p*-level was marked p_{cor}.

The relations between temperature time series (daily means and maxima of temperature) and environmental variables ($R_{s\downarrow}$ – incoming global radiation, α – albedo, θ – water content in soil) were tested using simple linear regressions.

All tests were tested at 5% probability level ($p < 0.05$).

16.3 Results

A summary of monthly temperature values and temperature maxima and minima at three heights as well as statistical evaluations of differences between the studied habitats are given in Table 16.1. With the exception of two cases the differences between the studied localities were always significant. The exceptions were insignificant differences between the localities (1) in average air temperature at 2 m and average temperature of the soil surface in July and (2) temperature maxima at the vegetation surface in June.

Daily temperature means Tc and Ta in the littoral were similar and their courses were tightly correlated. In May, July and August Tc was slightly higher than Ta (pair *t*-test: $t = -6.67, \text{df} = 30, p_{cor} < 0.001, t = -4.5, \text{df} = 30, p_{cor} < 0.001$ and $t = -4.23, \text{df} = 28, p_{cor} < 0.01$, respectively). In June the difference was not significant. Ts differed substantially from Ta and Tc during the whole measuring period ($p_{cor} < 0.001$).

Ta and Tc significantly differed also in the wet meadow (pair *t*-tests, $p_{cor} << 0.05$). In this habitat, however, Tc was always lower than Ta (Fig. 16.2). Whereas in May there was no significant difference between Ts and both Ta and Tc, the difference was significant during other studied months (pair *t*-test: $p_{cor} < 0.001$).

Table 16.1 Monthly means of daily temperatures (T), daily temperature maxima (T_{max}) and minima (T_{min}) at 2 m above ground (Ta), at the vegetation surface (Tc) and at the soil surface (Ts) in the littoral and in the meadow and results of statistical tests (t, pair t-tests) evaluating differences between localities. d.f., degrees of freedom

	Month	Mean littoral	Mean meadow	t	d.f.	p-level
Ta	May	13.9	13.5	5.52	30	***
	June	18.0	17.8	2.57	29	*
	July	18.1	18.0	0.99	30	n.s.
	August	18.0	17.7	4.68	28	***
Tc	May	14.3	13.3	8.02	30	***
	June	18.0	16.9	6.94	29	***
	July	18.3	16.3	8.14	30	***
	August	18.3	15.8	11.82	28	***
Ts	May	11.3	13.0	−13.80	30	***
	June	14.6	15.5	−12.83	29	***
	July	15.7	15.5	1.05	30	n.s.
	August	15.6	15.1	3.61	28	**
Ta_{max}	May	21.4	21.9	−3.95	30	***
	June	25.6	26.4	−5.06	29	***
	July	25.2	26.1	−5.61	30	***
	August	25.2	25.2	−6.10	28	***
Tc_{max}	May	24.2	25.5	−6.08	30	***
	June	28.5	27.8	1.62	29	n.s.
	July	27.9	24.3	9.50	30	***
	August	28.3	27.3	11.98	28	***
Ts_{max}	May	12.2	18.6	−17.61	30	***
	June	15.7	18.6	−15.33	29	***
	July	16.2	17.9	−6.19	30	***
	August	15.8	17.8	−8.01	28	***
Ta_{min}	May	6.4	4.3	8.9	30	***
	June	11.3	9.7	9.0	29	***
	July	12.2	10.6	7.2	30	***
	August	11.3	9.2	11.0	28	***
Tc_{min}	May	5.5	1.4	12.3	30	***
	June	9.6	6.8	13.2	29	***
	July	11.6	8.6	9.8	30	***
	August	10.8	7.3	14.1	28	***
Ts_{min}	May	10.5	8.3	9.5	30	***
	June	13.4	12.6	7.7	29	***
	July	15.3	13.1	9.4	30	***
	August	15.4	12.1	16.6	28	***

Significance: *** significant at $p < 0.001$, ** significant at $p < 0.01$, * significant at $p < 0.05$, n.s. – not significant

The temperature maxima Ta_{max} and Tc_{max} differed significantly in all cases (pair t-tests: $p_{cor} << 0.05$). Whereas in the littoral the mean Tc_{max} was always higher than Ta_{max}, in the wet meadow this was true only during May and June (Fig. 16.3). In July and August the trend in the wet meadow reversed and the mean Ta_{max} was higher than Tc_{max}.

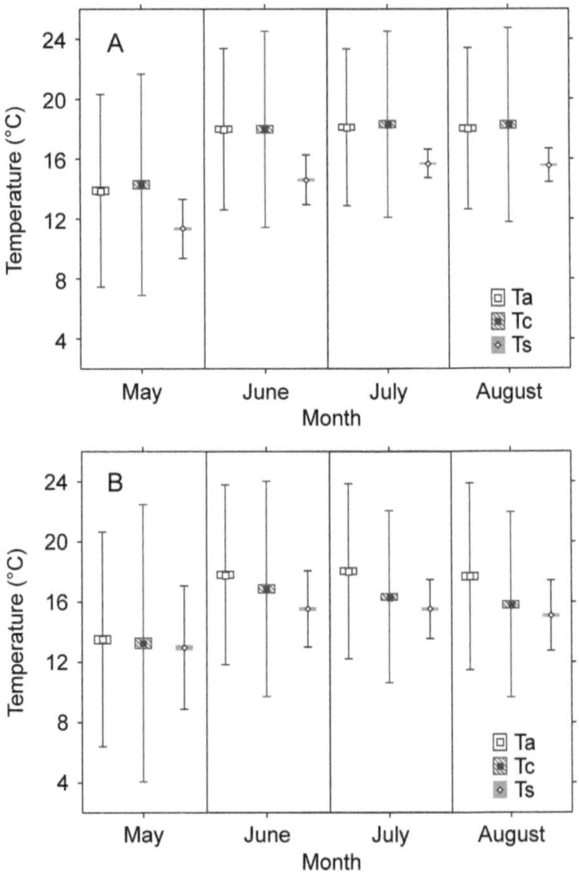

Fig. 16.2 Hourly means of temperature at 2 m, at vegetation surface and at the soil surface from May to August 2008 in the littoral (**a**) and in the wet meadow (**b**). Centres represent means, boxes standard errors of means and antennae standard deviations of means

Ts_{max} was always significantly lower than Ta_{max} and Ts_{max} (pair *t*-test: $p_{cor} < 0.001$) (Fig. 16.3). In the littoral the daily maximum temperature was in May at all heights lower than in other months. In the wet meadow Ts_{max} was higher in May and June than in July and August. The daily minimum temperatures at all heights were always lower in the wet meadow than in the littoral (Table 16.1).

We evaluated the impact of incoming radiation, albedo, soil humidity, vegetation cover height and LAI on temperature profile in the two wetlands. The parameters and their importance for impacting temperature profile of the habitats differed during the growing season. Monthly means of incoming shortwave radiation, volumetric content of water in soil and albedo are presented in Table 16.2.

The incoming shortwave radiation was comparable for the two habitats. The divergence was due to different cloud movement at the sites (Fig. 16.4a).

The two studied wetlands were characterized by different soil water content dynamics. Whereas in the littoral the soil profile was fully saturated for most of the time of the measurement, in the wet meadow the soil water content was more

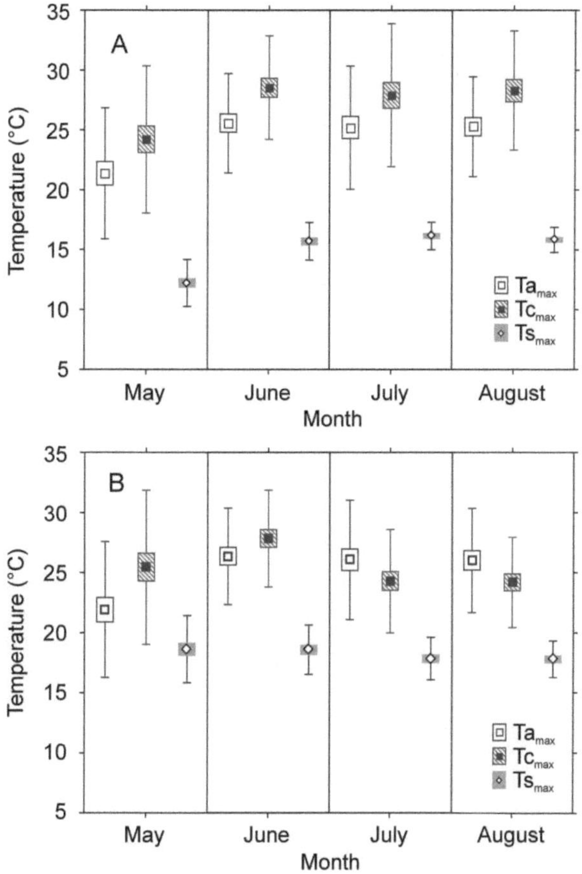

Fig. 16.3 Daily maximum temperature from May to August 2008 in the littoral (**a**) and in the wet meadow (**b**) at 2 m, in the canopy layer and at the soil surface. Centres represent means, boxes standard errors of means and antennae standard deviations of means

Table 16.2 Mean monthly values of incoming shortwave radiation ($R_{s\downarrow}$), volumetric content of water in soil (θ) and albedo (α) from May to August 2008

Locality	Month	$R_{s\downarrow}$ (W m^{-2})	θ (% vol.)	α (%)
Littoral	May	233.50	52.88	14.80
	June	234.06	49.54	17.85
	July	216.43	51.95	17.47
	August	202.59	53.12	15.80
Wet meadow	May	227.67	43.28	19.02
	June	236.30	43.55	22.67
	July	210.37	39.78	21.44
	August	191.99	37.31	18.75

dependent on the actual precipitation and therefore the water content in the soil fluctuated throughout the measurement period although it was rather high all the time (Fig. 16.4b). In May and in June the soil water content in the wet meadow was thus higher than in July and August.

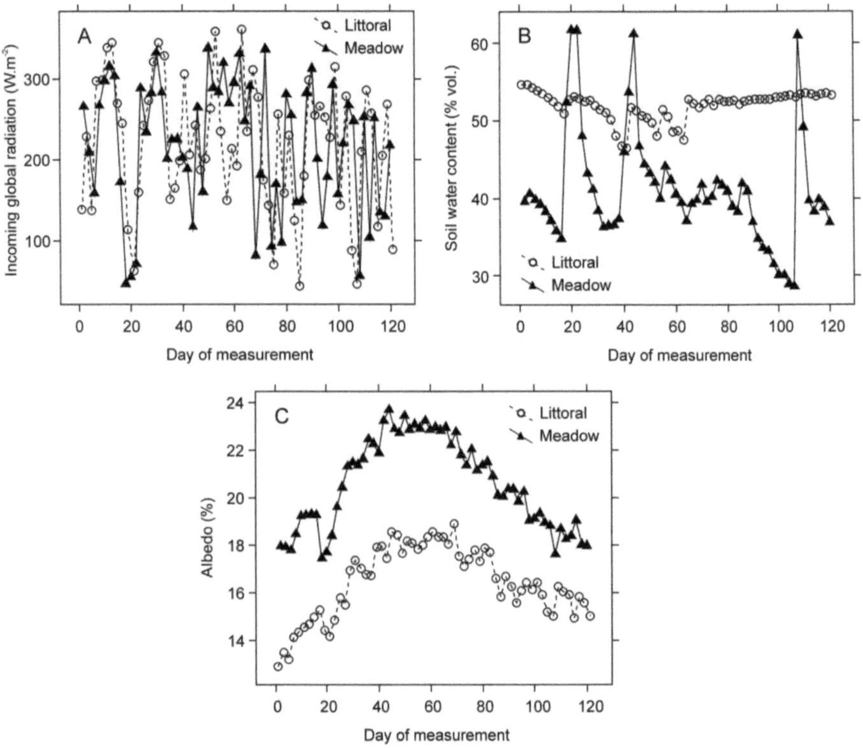

Fig. 16.4 (**a**) Incoming global radiation ($R_{s\downarrow}$), (**b**) volumetric content of water in soil and (**c**) albedo in the littoral and in the wet meadow during the observed period. The x axis represents number of days from the beginning of the measurement

Albedo in the littoral was generally lower than albedo in the wet meadow (Fig. 16.4c). The values, however, were not constant throughout the growing season. In May and August albedo was lower than in June and July in both habitats.

In general LAI in the wet meadow was approximately twice as high as LAI in the littoral. The habitats differed also in LAI and H dynamics throughout the growing season (Fig. 16.5). Maximum LAI of 4.36 was reached in the wet meadow as early as in mid May. The values began to decrease in mid June. In the littoral the maximum LAI of 2.53 was reached in mid June and from the end of July LAI gradually decreased. In the littoral, maximum H was reached sooner than in the wet meadow but it began to decrease gradually already in mid June. In the wet meadow it took longer at the beginning of the growing season to reach maximum, however then H remained rather stable throughout the whole monitored period.

In the littoral the incoming shortwave radiation turned out to be the most important environmental factor influencing Ta, Tc, Ta_{max} and Tc_{max}. Ts was not correlated with any of the monitored environmental factors. Ts_{max} was correlated with incoming shortwave radiation in May, June and July.

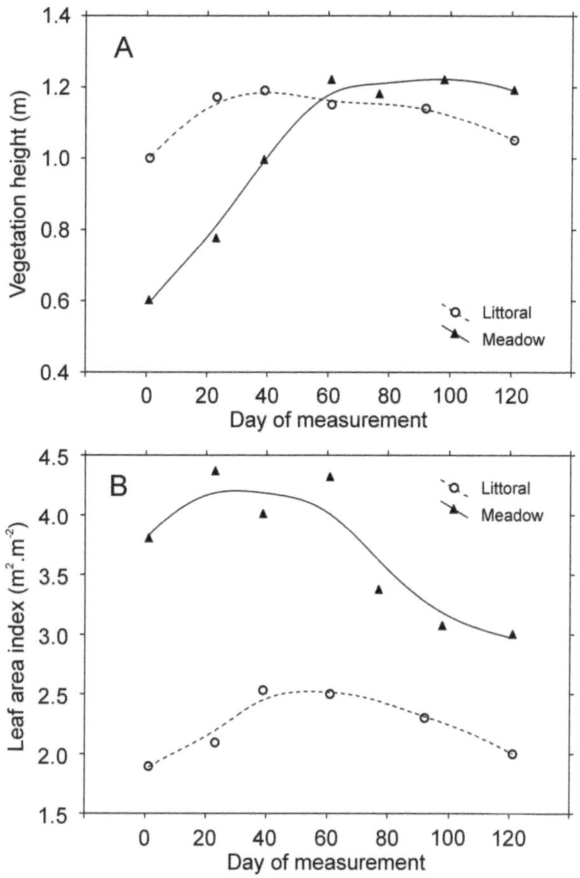

Fig. 16.5 Vegetation cover height (**a**) and leaf area index (LAI) (**b**) in the littoral and in the wet meadow. The x axis represents number of days from the beginning of the measurement. The lines represent trend of the parameters change

In the wet meadow both Ta, Tc and Ta_{max} Tc_{max} were correlated with incoming shortwave radiation during all four monitored months. Ts was in the wet meadow correlated with incoming shortwave radiation only in May and July, although the relation with Ts_{max} was significant for all months. All significant relations between temperature means and maxima in both localities manifested a positive trend, i.e. the temperature grew with the increasing amount of incoming shortwave radiation.

Another environmental factor important for temperature regime of an ecosystem was water content in the soil. Daily temperature means as well as temperature maxima in the wet meadow were significantly negatively correlated with soil water content in all cases with the exception of mean temperatures in May. In the littoral, however, only the relation between Ta and Tc and soil water content in May was significant. For temperature maxima the relation was significant in May and in June. The amount of water in the soil in the littoral was correlated with the daily mean temperature and maximum temperature at the soil surface in May and August and in May, July and August, respectively.

Table 16.3. Summary of correlations between daily means of temperature and maximum daily temperature (measured at 2 m above ground, at vegetation surface and at the soil surface) and environmental variables (incoming shortwave radiation, $R_{s\downarrow}$; water content in soil, θ; albedo, α) for observed months. Values significant at 5% probability level are marked with bold letters

			Daily temperature means			Daily temperature maxima		
Locality	Measuring level	Month	$R_{s\downarrow}$ (W m^{-2})	θ (% vol.)	α (%)	$R_{s\downarrow}$ (W m^{-2})	θ (% vol.)	α (%)
Littoral	2 m	May	**0.62**	**−0.72**	**0.86**	**0.82**	**−0.54**	**0.65**
		June	**0.69**	−0.25	−0.11	**0.77**	**−0.38**	−0.08
		July	**0.74**	−0.26	−0.18	**0.75**	−0.25	−0.17
		August	**0.64**	−0.13	0.27	**0.86**	0.1	**0.44**
	Vegetation surface	May	**0.67**	**−0.7**	**0.85**	**0.84**	**−0.48**	**0.61**
		June	**0.7**	−0.24	−0.15	**0.8**	**−0.44**	−0.11
		July	**0.76**	−0.23	−0.18	**0.79**	−0.24	−0.14
		August	**0.7**	−0.12	0.29	**0.88**	0.17	**0.48**
	Soil surface	May	0.35	**−0.75**	**0.89**	**0.45**	**−0.78**	**0.9**
		June	0.32	−0.14	−0.25	**0.53**	−0.27	−0.12
		July	0.31	−0.14	0.26	**0.47**	**−0.51**	**0.43**
		August	−0.06	**−0.83**	0.04	−0.06	**−0.86**	0.06
Meadow	2 m	May	**0.58**	**−0.4**	**0.9**	**0.84**	**−0.71**	**0.82**
		June	**0.6**	**−0.59**	−0.18	**0.74**	**−0.56**	−0.11
		July	**0.67**	**−0.43**	−0.11	**0.74**	**−0.46**	−0.13
		August	**0.56**	**−0.7**	**0.45**	**0.85**	**−0.55**	**0.38**
	Vegetation surface	May	**0.58**	**−0.36**	**0.9**	**0.88**	**−0.72**	**0.82**
		June	**0.56**	**−0.65**	**−0.38**	**0.66**	**−0.65**	**−0.55**
		July	**0.5**	**−0.38**	−0.16	**0.8**	**−0.55**	0.03
		August	**0.41**	**−0.67**	**0.39**	**0.83**	**−0.48**	0.25
	Soil surface	May	**0.44**	−0.21	**0.87**	**0.87**	**−0.71**	**0.77**
		June	0.28	**−0.59**	**−0.4**	**0.5**	**−0.63**	**−0.47**
		July	**0.47**	**−0.38**	−0.14	**0.74**	**−0.51**	−0.05
		August	0.15	**−0.67**	**0.54**	**0.65**	**−0.65**	**0.56**

The relation between albedo and temperature at all three levels throughout the growing season was not clear and thus impossible to interpret. The statistically significant correlations of albedo and temperature in May and August were most probably a consequence of concurrence of seasonal changes of albedo and other studied factors. All correlations between environmental factors and temperature in the studied wetlands are presented in Table 16.3.

16.4 Discussion

The two studied graminoid wetlands differed in many aspects in the measured profile of height both in the daily mean temperature and in the temperature extremes. The daily temperature means at 2 m (Ta) and at the vegetation surface (Tc) were always higher in the littoral than in the wet meadow, with the exception of Ta in July when the difference was not significant. Whereas in May and June Ts was higher in the wet meadow than in the littoral, in July the relation was not significant and in August Ts was higher in the littoral than in the wet meadow. The differences in temperature characteristics of the two habitats were caused by a whole range of factors whose importance differed for every height level. As the studied habitats were not remote and were situated in similar altitude, the generally lower Ta and Tc in the wet meadow were most probably due to local conditions determined by micro-topography. Long term meteorological measurements had previously shown that slight depression of this locality causes in-flowing of cool air (Květ, 2009, personal communication). This trend was confirmed also by our data, as daily minimum temperature was lower in the wet meadow than in the littoral (Table 16.1). The temperature differences between the localities were more marked when the sky was clear and atmosphere stable (data not shown). Air temperature is determined also by current humidity of the air, which influences temperature at which condensation occurs, i.e. the dew point (Running, Nemani, & Hungerford, 1987; Hayden, 1998; Lhomme & Guilioni, 2004). It is the evaporation-condensation process that connects plant physiology (transpiration) with air temperature and under suitable conditions represents a mechanism of temperature stabilisation-high transpiration cools and condensation warms the air near the surface (Brutsaert, 1982; Gates, 1980).

Besides meteorological conditions transpiration rate and temperature distribution in the stand are dependent on biomass distribution (Busch, 2000), physiological characteristics and architecture of the stand (Kabat et al., 2004; Graser, Verma, & Rosenberg, 1987; Fliervoet & Werger 1984; Baldocchi, Verma, Rosenberg, Bald, & Specht, 1985). These parameters influence humidity in the stand and light as well as wind penetration into the stand etc. (Ross, 1975; Přibáň, Jeník, Ondok, & Popela, 1992). The littoral stand of *Carex acuta* is more open to the surroundings than the wet meadow stand.

The role of transpiration in temperature regulation was apparent in our measurements when Tc_{max} was compared. Apart from May, Tc_{max} was always higher in the littoral, which reached approximately only half the LAI of the wet meadow (Fig. 16.5), than in the wet meadow, even though Ta_{max} was higher in the wet meadow. The higher Ta_{max} in the wet meadow was most probably due to the advection of warm air from the drained fields and meadows in the proximity of the site (our preliminary results, data not shown). On the contrary, the littoral is directly surrounded by a fishpond and by a forest with rather cool and moist air. The correlation between temperature maxima and incoming shortwave radiation was stronger in May and August than in June and July, during the period with the highest LAI when the temperatures near the ground were more intensely down-regulated by transpiration from the vegetation.

Functioning of a stand is influenced also by root architecture (Yadav, Mathur, & Siebel, 2009; Schymanski, Sivapalan, Roderick, Beringer, & Hutley, 2008). The tussocks of *Carex acuta* in the littoral may have a specific influence on the habitat hydrology. The 0.6 m high tussocks are covered with vegetation only in their upper parts. Between the tussocks there are mostly shaded gaps covered by dead straw. Dead biomass is an insulation material, especially when it is dry, and its presence can lead to soil surface heating while the deeper soil profile is not much influenced (Rouse, 2000; Królikowska, Přibáň, & Šmíd, 1998; Brom & Pokorný, 2009). When dry, the emerging parts of the tussocks may cause water stress to the vegetation even though the base of the tussock is still covered with water. Touchette, Iannacone, Turner, and Frank (2007) showed a quick decrease of xylem water potential and transpiration rate when related species *Carex alata* was exposed to drought. Water stress as well as warming up of the dry parts of the tussocks contribute to temperature increase within the littoral stand (Idso, Jackson, Pinter, Reginato, & Hatfield, 1981; Jackson, Idso, Reginato, & Pinter, 1981). A thick layer of organic detritus is present also in the wet meadow. As it is shaded with vegetation throughout the growing season, its isolating and warming features are decisive for soil surface temperature only in May, when vegetation is still thin (Table 16.1).

The impact of incoming radiation on Ts was also found to be related to the vegetation growth. This was most obvious in particular when assessing Ts_{max} which correlated with incoming radiation more strongly during the periods when vegetation was still sparse. While in the littoral the relation between Ts and incoming radiation was not significant, this relation was significantly correlated in the wet meadow. Here the relation was weaker in May due to still thin vegetation cover and in July, probably due to a spell of drought which might have caused transpiration reduction. The insignificant relations in the littoral were probably due to heterogeneous vegetation cover, as micrometeorological conditions which exist between the tussocks can be largely decoupled from the incoming radiation. The temperature of the soil can be also influenced by water flowing to the lake from its inflow.

The impact of different environmental factors on temperature regimes of the two sites changed dynamically throughout the growing season (Table 16.3). The daily mean temperature (Ta and Tc) was predominantly influenced by incoming shortwave radiation. The correlations were stronger in the littoral than in the wet meadow. The differences between the correlations of the two habitats was most probably related to the characteristics of the vegetation cover (LAI, amount of biomass, photosynthesis rate etc.). With increasing biomass and increasing transpiration rate the vegetation cover influenced the temperature of the air near the surface more strongly and thus decoupled partly the temperature from the straight influence of the incoming radiation.

In the short-term (couple of days) the changes of albedo were small due only to changing ratio of direct and diffuse radiation (Přibáň et al., 1992), with changing humidity (Idso, Jackson, Reginato, Kimball, & Nakayama, 1975) and short term changes of the stand architecture. Changes of albedo throughout the season showed a clear, and in both habitats, similar trend given both by the changing angle of incidence of the solar radiation and by evolution of stand architecture during the growing season (Ross, 1975). The actual values differed in the habitats with those in the littoral being markedly lower due to low LAI and large gaps in the vegetation stand. The vegetation cover on the wet meadow was more homogenous with higher abundance of broadleaf species. The importance of vegetation for albedo values is reflected in low albedo values during May when the dark organic detritus was still not covered by vegetation. Maximum albedo values correspond with the peak amount of biomass in the habitats. The follow-up decline of albedo values in August is related to starting senescence.

Another factor important for temperature regime of the habitats was the water content in the soil. In the wet meadow the water content in the soil negatively correlated with both temperature means and temperature maxima, with the only exception of daily mean soil surface temperature in May. The decreasing water content in the soil probably caused impairment of water connectivity between water in the soil and roots thus lowering transpiration and supporting decoupling of vegetation from the temperature of the surrounding air. In the littoral the daily temperature means correlated with soil water content at all three height levels only in May. As already mentioned the specific architecture of *Carex acuta* stand makes the hydrology of this habitat complicated as the tussocks do not dry out linearly. The significant correlations in May (Table 16.3) are probably related to the fact that the surface of the tussocks is still wet and does not isolate soil from the air. Water can thus rise from the soil through the detritus or evaporate straight from the water surface between the tussocks which are not yet covered by vegetation. Later during the growing season the hydrology of the tussocks is probably too complicated to be borne out by the presented measurements and the relations thus remained unclear.

16.5 Conclusions

Temperature characteristics of the two studied graminoid wetlands, a littoral zone of a fishpond and a wet meadow, differed markedly. The incoming global irradiance was shown to be the most influential environmental variable determining temperature of the studied wetlands, in particular at the 2 m height above ground. Also the water content in the soil played an important role, though not always easy to interpret. The daily means at 2 m were on average higher in the littoral as a consequence of lower temperature minima in the wet meadow. The daily maxima at 2 m were higher in the wet meadow due to advection of warm air from the surrounding habitats. Mean and maximum temperatures within the vegetation cover were lower in the wet meadow. This trend was related to markedly higher LAI in the wet meadow. *Phalaris arundinacea*, one of the dominant species in the studied wet meadow, has very high transpiration rate also under hot weather conditions (unpublished data). According to our results vegetation played an important role in influencing temperature at the soil surface and within the vegetation cover, i.e. in the very proximity of the land surface. However, the temperature of the air at 2 m was predominantly influenced by temperature characteristics of the surrounding habitats (i.e. local meteorology) and vegetation played much weaker role.

Acknowledgement This work was supported by the projects of the Ministry of Education, Youth and Sports of the Czech Republic (NPV 2B06023 and MSM 6007665806) and The National Agency for Agriculture Research of the Ministry of Agriculture of the Czech Republic (QH92034). We would like to thank the colleagues from ENKI o.p.s. and the Department of Landscape Management for their excellent technical help and personal support.

References

Baldocchi, D. D., Verma, S. B., Rosenberg, N. J., Bald, B. L., & Specht, J. E. (1985). Microclimate-plant architectural interactions: Influence of leaf width on the mass and energy exchange of a soybean canopy. *Agricultural and Forest Meteorology, 35*, 1–20.

Brom, J., & Pokorný, J. (2009). Temperature and humidity characteristics of two willow stands, a peaty meadow and a drained pasture and their impact on landscape functioning. *Boreal Environment Research, 14*, 389–403.

Brutsaert, W. (1982). *Evapotranspiration into the atmosphere. Theory, history and applications*. Dortrecht,: D. Reidel Publishing Company.

Busch, J. (2000). Canopy transpiration rates in eutrophic wetlands dominated by sedges (*Carex* spp.) differ in a species specific way. *Physics and Chemistry of the Earth*, Part B: *Hydrology, Oceans and Atmosphere, 25*, 605–610.

Fitter, A. H., & Hay, R. K. M. (2002). *Environmental physiology of plants* (3rd ed.). London: Academic Press.

Fliervoet, L. M., & Werger, M. J. A. (1984). Canopy structure and microclimate of two wet grassland communities. *New Phytologist, 96*, 115–130.

Gates, D. M. (1980). *Biophysical ecology*. New York: Dover Publications, Inc.

Graser, E. A., Verma, S. B., & Rosenberg, N. J. (1987). Within-canopy temperature patterns of sorghum at two row spacings. *Agricultural and Forest Meteorology, 41*, 87–205.

Hayden, B. P. (1998). Ecosystems feedback on climate at the land scale. *Philosophical Transaction of the Royal Society London, B353*, 5–18.

Idso, S. B., Jackson, R. D., Pinter, P. J., Jr., Reginato, R. J., & Hatfield, J. L. (1981). Normalizing the stress degree day for environmental variability. *Agricultural Meteorology, 24*, 45–55.

Idso, S. B., Jackson, R. D., Reginato, R. J., Kimball, B. A., & Nakayama, F. S. (1975). The dependence of bare soil albedo on soil water content. *Journal of Applied Meteorology, 14*, 109–113.

Jackson, R. D., Idso, S. B., Reginato, R. J., & Pinter, P. J., Jr. (1981). Canopy temperature as a crop water stress indicator. *Water Resources Research, 17*, 1133–1138.

Jackson, R. B., Randerson, J. T., Canadell, J. G., Anderson, R. G., Avissar, R., Baldocchi, D. D., et al. (2008). Protecting climate with forests. *Environmental Research Letters, 3*, 044006.

Kabat, P., Claussen, M., Dirmeyer, P. A., Gash, J. H. C., Bravo de Guenni, L., Meybeck, M., et al. (Eds.). (2004). *Vegetation, water, humans and the climate. A new perspective on an interactive system*. Berlin: Springer.

Körner, C. (2006). Significance of temperature in plant life. In: J. I. L. Morison & M. D. Morecroft (Eds.), *Plant growth and climate change*. Oxford: Blackwell Publishing.

Kravčík, M., Pokorný, J., Kohutiar, J., Kováč, M., & Tóth, E. (2007). *Water for the recovery of the climate – A new water paradigm*. Bratislava: Municipalia.

Królikowska, J., Přibáň, K., & Šmíd, P. (1998). Micro-climatic conditions and water economy of wetland vegetation. In D. F. Westlake, J. Květ, & A. Szczepański (Eds.), *The production ecology of wetlands* (pp. 367–404). Cambridge: Cambridge University Press.

Lhomme, J. P., & Guilioni, L. (2004). A simple model for minimum crop temperature forecasting during nocturnal cooling. *Agricultural and Forest Meteorology, 123*, 55–68.

Linacre, E. T. (1976). Swamp. In: J. L. Monteith (Ed.), *Vegetation and atmosphere* (Vol. II., pp. 329–347). London: Academic Press.

Nobel, P. S. (1999). *Physicochemical and environmental plant physiology.* New York: Academic.

Pokorný, J. (2001). Dissipation of solar energy in landscape – controlled by management of water and vegetation. *Renewable Energy, 24*, 641–645.

Pokorný, J., & Kučerová, A. (2000). Monitoring klimatu a atmosférických depozic v CHKO Třeboňsko [Climate and atmospheric deposition monitoring in Biosphere Reserve Třeboňsko]. In Třeboňsko 2000, Ekologie a ekonomika po dvaceti letech [Třeboň Region 2000. Ecology and economy after 20 years], (pp. 87–99). Třeboň, Czech Repubic: ENKI o.p.s., Správa CHKO Třeboňska, National Committee MaB UNESCO.

Prach, K. (2008). Vegetation changes in a wet meadow complex during the past half-century. *Folia Geobotanica, 43*, 119–130.

Přibáň, K., Jeník, J., Ondok, J. P., & Popela, P. (1992): *Analysis and modeling of wetland microclimate*. Studie ČSAV 2–92. Praha: Academia.

Ripl ,W. (1995). Management of water cycle and energy flow for ecosystem control – the Energy-Transport-Reaction (ETR) model. *Ecological Modelling, 78*, 61–76.

Ripl, W. (2003). Water: the bloodstream of the biosphere. *Philosophical Transaction. The Royal Society of London, B358*, 1921–1934.

Ross, J. (1975). Radiative transfer in plant communities. In: J. L. Monteith (Ed.), *Vegetation and the atmosphere. Vol. 1. Principles* (pp. 13–55). London: Academic Press.

Rouse, W. R. (2000). The energy and water balance of high-latitude wetlands: controls and extrapolation. *Global Change Biology, 6*(Suppl. 1), 59–68.

Running, S. W., Nemani, R. R., & Hungerford, R. D. (1987). Extrapolation of synoptic meteorological data in montainous terrain and its use for simulating forest evapotranspiration and photosynthesis. *Canadian Journal of Forest Research, 17*, 472–483.

Salkind, N. J. (Ed.). (2007). *Encyclopedia of measurement and statistics.* Thousand Oaks, CA: Sage.

Schymanski, S. J., Sivapalan, M., Roderick, M. L., Beringer, J., & Hutley, L. B. (2008). An optimality-based model of the coupled soil moisture and root dynamics. *Hydrology and Earth System Sciences, 12*, 913–932.

Touchette, B. W., Iannacone, L. R., Turner, G. E., & Frank, A. R. (2007). Drought tolerance versus drought avoidance: a comparison of water drought relation in herbaceous wetland plants subjected to water withdrawal and repletion. *Wetlands, 27*, 656–667.

Welles, J. M., & Norman, J. M. (1991). Instrument for indirect measurement of canopy architecture. *Agronomy Journal, 83*, 818–825.

Yadav, B. K., Mathur, S., & Siebel, M. A. (2009). Soil moisture dynamics modeling considering the root compensation mechanism for water uptake by plants. *Journal of Hydrologic Engineering, 14*, 913–922.

Chapter 17
Chemical Properties of the Sediment Interstitial Water at Lake Fertő/Neusiedler See

Edit Ágoston-Szabó and Mária Dinka

Abstract The chemical parameters of the surface and sediment interstitial water from the 0–20 and 20–40 cm sediment layer were investigated in different reed stands of Lake Fertő/Neusiedler See. The temperature, pH, redox potential, SO_4^{2-}, NO_2^--N and NO_3^--N concentrations decreased, while the electrical conductivity and the concentrations of $Na^+, Ca^{2+}, K^+, Mg^{2+}, Cl^-, PO_4^{3-}$-P and S^{2-} increased in the function of sediment depth. As comparing the vegetated and unvegetated areas of the sampling sites: PO_4^{3-}-P, S^{2-} and NO_2^--N concentrations of the interstitial water were lower and the SO_4^{2-} and NO_3^--N concentrations were higher at vegetated areas, which demonstrated the effect of internal oxygen transport of the *Phragmies australis* rhizosphere on the chemical characteristics of interstitial water. The degraded reed stand was characterised by low redox potential and high S^{2-} concentrations. The trace element concentrations of the surface water decreased in the order $B^{3+} > Fe^{2+} > Al^{3+} > Pb^{2+} > Zn^{2+} > Mn^{2+}$ and increased with the depth of the sediment; their concentrations were higher in the sediment interstitial water of unvegetated than of the vegetated areas.

Keywords Chemical parameters · Nutrients · Reed stand · Sediment interstitial water

17.1 Introduction

The reed degradation processes which were observed in Middle-Europe in the last few decades (review in Ostendorp, 1989) had stimulation effect on the investigation of *Phragmites australis* die-back. According to this aligned researches were carried out in the frame of the European Research Program on Reed Die-Back and

---

E. Ágoston-Szabó (✉)
Hungarian Danube Research Station of the Hungarian Academy of Sciences, Vácrátót, Hungary
e-mail: edit@botanika.hu

J. Vymazal (ed.), *Water and Nutrient Management in Natural and Constructed Wetlands*, DOI 10.1007/978-90-481-9585-5_17,

Progression (EUREED) between 1993 and 1999, in which has been pointed out among others that the physico-chemical parameters of the surface and interstitial water are important controlling factors of the reed stand conditions (Van der Putten, 1997; Armstrong & Armstrong, 1998; Brix, 1999; Dinka, 1998, 2001; Čížková et al., 2001; Urban-Berčič & Gaberščik, 2004; Ágoston-Szabó, 2004, 2007). Van der Putten (1997) revealed that at Lake Fertő/Neusiedler See the high sulphide concentration represents the principal phytotoxin for the reed stand.

In previous works conducted also in the Hungarian part of the lake Fertő/Neusiedler See (Armstrong & Armstrong, 1998; Dinka, 1998, 2001; Ágoston-Szabó, 2004, 2007) significant differences were found between the chemical composition of the sediment interstitial water of the healthy and degraded reed stands and between the reed covered and reedless zones of the lake and significant changes were found in the sediment interstitial water chemistry in the function of depth, which was also observed by Gunatilaka (1984) on the Austrian part of the lake.

The study of the chemical composition of the sediment interstitial water in shallow lakes gives a better understanding of the nutrient dynamics and the interaction between the surface water and the associated sediment (de Vincente, Amores, & Cruz-Pizzaro, 2006). The chemistry of interstitial water, the concentration profiles of various chemical parameters can indicate ongoing biogeochemical processes in the sediment, which are interrelated to degradation of organic matter (Berner, 1980). Organic matter reaching the sediment surface is mineralised by micro-organisms with different metabolic pathways; aerobic processes have been considered to play the main role, however in the suboxic and anoxic zones of the sediment anaerobic processes also have an important role in the oxidation of organic matter (Jones, 1985). The redox related cycling of elements in *Phragmites* dominated ecosystems has been reported by Čížková et al. (2001), Urbanc-Berčič and Gaberščik (2004) or Armstrong and Armstrong (1998).

The aim of our study was to investigate the changes in the chemical parameters of the sediment interstitial water as a function of sediment depth, health condition of the reed stands and reed cover.

17.2 Study Site

Lake Fertő/Neusiedler See, situated on the Hungarian-Austrian border (47 42′ N, 16 46′ E, altitude: 115.5 m, Fig. 17.1) has a surface area of 309 km^2 (Hungarian part 75 km^2), with sodic water and a mean depth of 1.1 m (see detailed description of the lake in Löffler, 1979). 54% of the whole lake and 85% of the Hungarian part is covered by reed. Sampling sites were selected in the Hungarian part of the lake.

Site 2 and 5 are healthy reed stands: site 2 is a homogenous reed area with shallow water (20–40 cm), while site 5 is a thinning reed area with deep water (80–120 cm). Site 3 is a degraded reed area, with clumped distribution of the culms, shallow water (30–50 cm) and organic matter rich sediment.

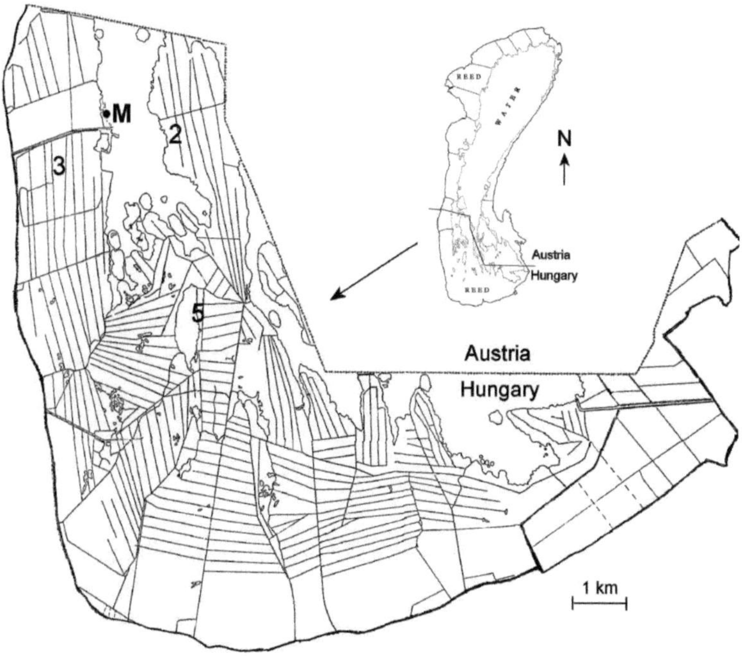

Fig. 17.1 Sampling sites at Lake Fertő/Neusiedler See

Samples were taken both from the water among reed shoots and the unvegetated areas of the thinned reed area. The different character of these plots will be referred as "in" (in reed tussocks, among reed shoots) and "out" (unvegetated area around the reed tussocks) in the figures.

17.3 Material and Methods

17.3.1 Methods of Measurement

Interstitial water was taken from 0–20 and 20–40 cm sediment layers (because 80% of the reed rhizome system is found in these layers) in three replicates. Interstitial water was taken from previously set plastic sampling tubes (with 5 cm diameter and with 1 mm wide and 10–25 mm long gaps on the lower 20 cm part) at the sampling sites. The bottom of the tubes was previously sealed with cotton material in order to prevent sediment particles getting into the sampling tubes. Before sampling, stagnant water was removed from the tubes, and the time of water samples touching air was reduced to the minimum.

The temperature, electric conductivity, pH and redox potential of the water were determined on the spot with a Hydrolog 2100 type field equipment. The Na^{+}, K^{+}, Mg^{2+}, Ca^{2+}, Fe^{2+}, Mn^{2+}, Zn^{2+}, Pb^{2+}, B^{3+}, Al^{3+} concentrations of the interstitial water and surface water were determined with ICP, the Cl^-, $SO_4{}^{2-}$, $NO_3{}^-$-N, $NO_2{}^-$-N, $PO_4{}^{3-}$-P , $NH_4{}^+$-N concentrations of the interstitial water and surface water were determined with standard chemical analysis (Felföldy, 1980; Golterman, Clymo, & Ohnstad, 1978) while the S^{2-} concentration was measured with iodometric titration, after $ZnCl_2$ precipitation (Golterman et al., 1978).

17.3.2 Data Analysis and Statistical Methods

An analysis of variance, Pearson's correlation and cluster analysis were performed (Statgraf 1.0 for Windows and Past programme package, Hammer, Harper, & Ryan, 2001). Only the results of the statistical analyses are given in this study.

17.3.3 Date of Sampling

Samples were collected on 06.08.1996 and on 29.07.1997, in both years the water level was high (1.28 m, which corresponds to 115.79 m above the Adriatic See level). There were no significant differences ($p<0.01$) between the results of interstitial water from 1996 and 1997, therefore only the results obtained in 1997 are presented in this study.

17.4 Results

17.4.1 Changes in the Chemical Characteristics of the Interstitial Water as a Function of Depth, Reed Cover and Health Condition of the Reed Stands

17.4.1.1 In situ Measured Parameters

The *pH* values in the surface water ranged from 7.7 to 8.6 (Table 17.1). The highest pH was measured at site 5. The pH values were significantly ($p<0.05$) lower in the sediment interstitial water as compared to the surface water and higher values were recorded in the interstitial water of the vegetated sites than in the corresponding unvegetated areas.

The *temperature* of the surface water varied between 19.7 and 22.5°C (Table 17.1), and decreased with the sediment depth. The highest temperature values were measured in the surface and interstitial water of site 3 (shallow water degraded reed stand). In unvegetated areas of site 3 and 5 the temperature of the sediment interstitial water was significantly ($p<0.05$) lower than that of the surface water; in vegetated areas of these sites and also at site 2 only the temperature of the interstitial water from the 20–40 cm sediment layer differed significantly ($p<0.05$) from the corresponding surface water and from the 0–20 cm sediment interstitial water.

Table 17.1 Changes in the in situ measured chemical parameters

Sites	pH			T (ºC)			Conductivity (μS cm^{-1})			Redox potential (mV)		
cm	0	20	40	0	20	40	0	20	40	0	20	40
2	7.7^{c}	7.2^{b}	6.9^{a}	19.7^{b}	19.2^{b}	17.8^{a}	1705^{a}	1943^{a}	2405^{b}	142^{b}	–75^{a}	–97^{a}
3 in	8.5^{b}	7.5^{a}	7.4^{a}	22.5^{b}	21.8^{b}	20.3^{a}	1674^{a}	1667^{a}	1761^{a}	136^{b}	–72^{a}	–83^{a}
3 out	8.5^{b}	7.0^{a}	7.0^{a}	22.5^{b}	20.0^{a}	19.7^{a}	1674^{a}	2620^{b}	3863^{c}	136^{b}	–108^{a}	–121^{a}
5 in	8.6^{b}	8.4^{b}	7.5^{a}	21.3^{b}	21.6^{b}	18.5^{a}	1642^{a}	1632^{a}	3621^{b}	102^{c}	–6^{b}	–109^{a}
5 out	8.6^{b}	7.0^{a}	7.2^{a}	21.3^{b}	19.5^{a}	19.2^{a}	1642^{a}	2763^{b}	3155^{b}	102^{b}	–127^{a}	–123^{a}

values followed by different letters are significantly different at the 0.05 probability level.
0: surface water, 20: 0–20 cm interstitial water, 40: 20–40 cm interstitial water

The *electrical conductivity* of the surface water varied between 1642 and 1705 μS cm^{-1} and significantly ($p<0.05$) increased with the sediment depth (Table 17.1) except site 3 in. The conductivity of the interstitial water in unvegetated sites was higher than in the corresponding vegetated areas (except with the 20–40 cm sediment layer of site 5).

The *redox potential* of the surface water varied between 102 and 142 mV (Table 17.1). The redox potentials were sensitive to depth and reed cover (lower values were obtained in the unvegetated zones) and it was significantly ($p<0.05$) lower in the sediment interstitial water than in the surface water at all sampling sites.

17.4.1.2 Nutrient Concentrations

In the surface water the NH_4^+*-N* concentrations varied between 0.23 and 1.62 μg L^{-1}, the NO_2^-*-N* concentrations between 8.46 and 12.83 μg L^{-1} and the NO_3^-*-N* concentrations between 27.44 and 38.41 μg L^{-1} (Fig. 17.2). The NO_3^--N and NO_2^--N concentrations decreased with the sediment depth at all sampling sites. Significant ($p<0.05$) differences were found between the $NO_3{}^-$-N concentrations of the interstitial water of both layers and the corresponding surface water at site 3 (in and out) and between the 20–40 cm interstitial water and corresponding surface water at site 5 (in and out), respectively. The NH_4^+*- N* concentrations increased with the sediment depth at sites 3 in and 5 (in and out), but significantly ($p<0.05$) only at site 5.

Higher NO_2^--N and lower NO_3^--N concentrations were obtained in the sediment interstitial water of unvegetated areas at site 3 and 5 than in the corresponding vegetated areas. The highest NO_3^--N concentrations were measured in surface water of the degraded reed stand (Fig. 17.2).

The N:P molar ratio in the surface water ranged from 313 to 824, in the 0–20 cm interstitial water from 104 to 540 and it was higher than in the 20–40 cm interstitial water, where it ranged from 28 to 181.

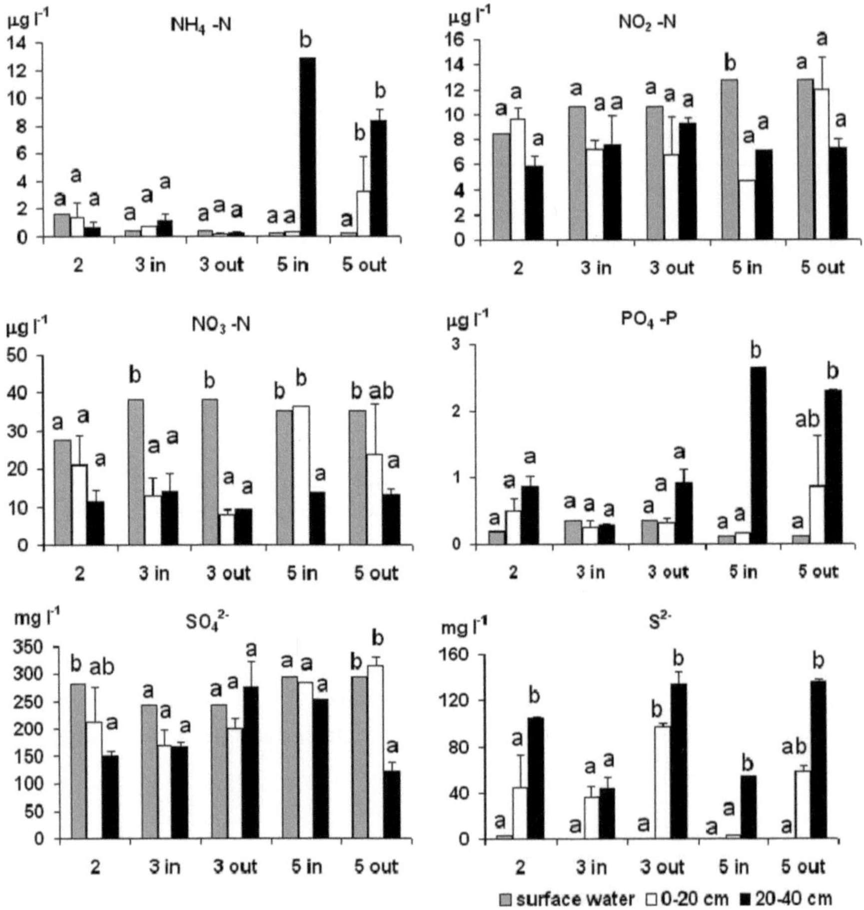

Fig. 17.2 Changes in the nutrient and sulphate, sulphide concentrations of the surface water and 0–20 cm, 20–40 cm sediment interstitial water. Significantly different (p<0.05) values are indicated by different letters

The PO_4^{3-}*-P* concentrations of the surface water ranged from 0.13 to 0.35 µg L^{-1} (Fig. 17.2), it increased with the sediment depth but significantly (p<0.05) only at site 5. The PO_4^{3-}-P concentrations were higher in the unvegetated than in the corresponding vegetated sites in except with 20–40 cm sediment layer of site 5. In the surface water the highest PO_4^{3-}-P concentrations were obtained in the degraded reed stands.

17.4.1.3 Changes in Concentrations of Main Ions

The $SO_4{}^{2-}$ concentrations of the surface water varied between 241.3 and 294.8 mg L^{-1}, it decreased in the function of the sediment depth but significant (p<0.05)

differences were found only between the surface and 20–40 cm interstitial water of sites 2 and 5 out. The SO_4^{2-} concentrations were higher in unvegetated than in the vegetated sites (except with site 5 out 20–40 cm layer). Surface water of degraded reed (site 3) had lower SO_4^{2-} concentrations as compared to healthy reeds (Fig. 17.2).

Concentrations of S^{2-} in the surface water were only detected at site 2 (0–2.72 mg L^{-1}). The S^{2-} concentrations increased with the sediment depth (Fig. 17.2); but significantly ($p<0.05$) only in the interstitial water of site 3 out and the 20–40 cm interstitial water of site 2 and 5 out. Lower S^{2-} values were measured in the vegetated than in the unvegetated areas and higher values in the degraded than in the healthy reed stands.

The changes in the concentrations of the Na^+, K^+, Ca^{2+}, Mg^{2+} and Cl^- are presented in Table 17.2.

Significant ($p<0.05$) differences were observed between the Na^+, K^+, Mg^{2+}, Cl^- concentrations of the surface and of the sediment interstitial waters from both layers at site 3 out, and in the Ca^{2+} concentrations at site 2. The K^+, Mg^+ concentrations of the 20–40 cm sediment interstitial water significantly ($p<0.05$) differed from that of the surface water at sites 2, 3 out, 5 in, 5 out. The concentrations of the Na^+, K^+, Ca^{2+}, Mg^{2+}, Cl^- were higher in the unvegetated as compared with the vegetated areas.

A close correlation was found between the electrical conductivity and the Na^+, Mg^{2+}, Cl^- concentrations in the surface and interstitial water of 0–20 cm and 20–40 cm sediment layer ($R^2_{\text{el. conduct.}-\text{Na}} = 0.96$ and 0.99, $R^2_{\text{el. conduct.}-\text{Mg}} = 0.34$ and 0.94, $R^2_{\text{el. conduct.}-\text{Cl}} = 0.74$ and 0.93).

17.4.1.4 Changes in Concentrations of Trace Elements

The surface water trace element concentrations decreased in the order B^{3+}(401 – 467 μg L^{-1}) > Fe^{2+}(12–39 μg L^{-1}) > Al^{3+}(1–24 μ g L^{-1}) > Pb^{2+}(1–18 μ g L^{-1}) > Zn^{2+}(1–6 μg L^{-1}) > Mn^{2+}(1–2 μg L^{-1}), and increased with the depth of the sediment (Fig. 17.3). Their concentrations in the interstitial water decreased in the order: $B^{3+} > Al^{3+} > Fe^{2+} > Mn^{2+} > Pb^{2+} > Zn^{2+}$ and were

Table 17.2 Changes in the concentrations of chloride and the main cations

Sites	Cl^- (mg L^{-1})			Na^+ (mg L^{-1})			Mg^{2+} (mg L^{-1})			K^+ (mg L^{-1})			Ca^{2+} (mg L^{-1})		
cm	0	20	40	0	20	40	0	20	40	0	20	40	0	20	40
2	228^a	294^a	737^b	254^a	284^a	349^b	107^a	127^a	165^b	29^a	35^a	47^b	27^a	45^b	51^b
3 in	165^a	211^a	202^a	248^a	244^a	264^a	113^a	111^a	118^a	22^a	21^a	23^a	34^a	34^a	34^a
3 out	165^a	638^b	1080^c	248^a	411^b	684^c	113^a	170^b	283^c	22^a	36^b	46^b	34^a	43^a	59^b
5 in	146^a	169^a	545^a	244^a	248^a	626^b	105^a	107^a	254^b	27^a	27^a	71^b	32^a	33^a	48^b
5 out	146^a	390^a	830^b	244^a	375^b	570^c	105^a	131^a	173^b	27^a	38^a	54^b	32^a	37^a	36^a

values followed by different letters are significantly different at the 0.05 probability level.
0: surface water, 20: 0–20 cm interstitial water, 40: 20–40 cm interstitial water

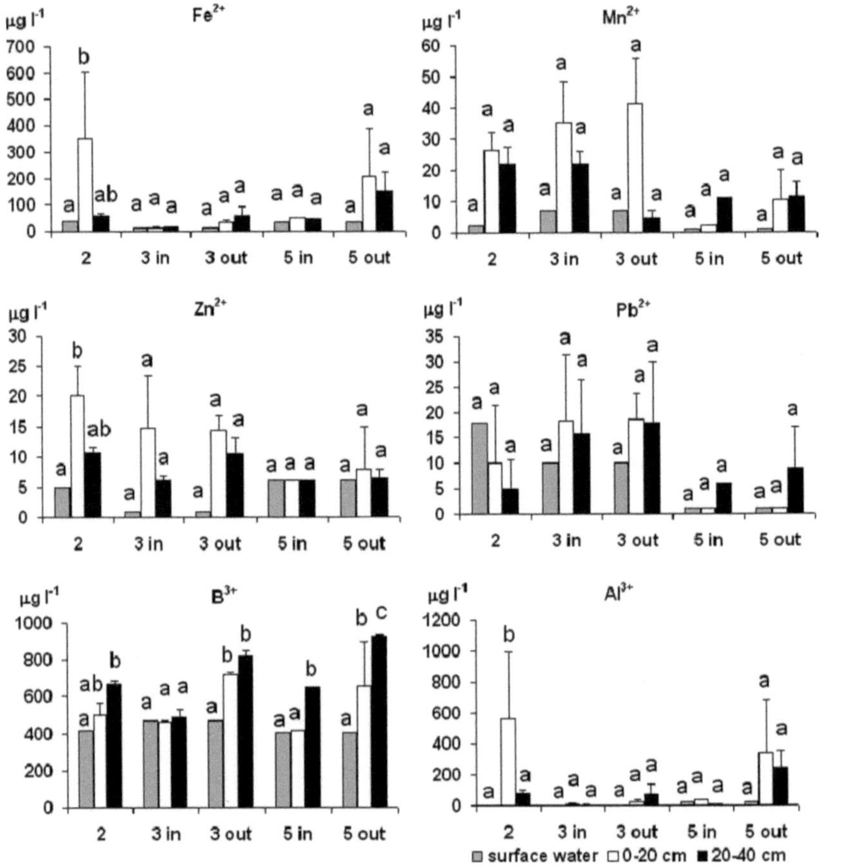

Fig. 17.3 Changes in the concentrations of trace elements in the surface water and 0–20, 20–40 cm sediment interstitial water. Significantly different (p<0.05) values are indicated by different letters

higher in the unvegetated than in the vegetated areas. Higher concentrations of Mn^{2+} and Pb^{2+} were measured in the surface and interstitial water of site 3 and 2 than in that of site 5. The lowest Fe^{2+} and Zn^{2+} concentrations of the surface water were observed at site 3.

17.4.2 Statistical Evaluations

For the illustration of the similarity between the sampling sites the results of the cluster analysis are presented on the Fig. 17.4. In the 0–20 cm sediment interstitial layer, the vegetated and the unvegetated areas were involved in different similarity groups: the first similarity group was represented by the sites: 3 out, 5 out (unvegetated areas) and the second similarity group by the sites 3 in, 5 in and 2 (vegetated

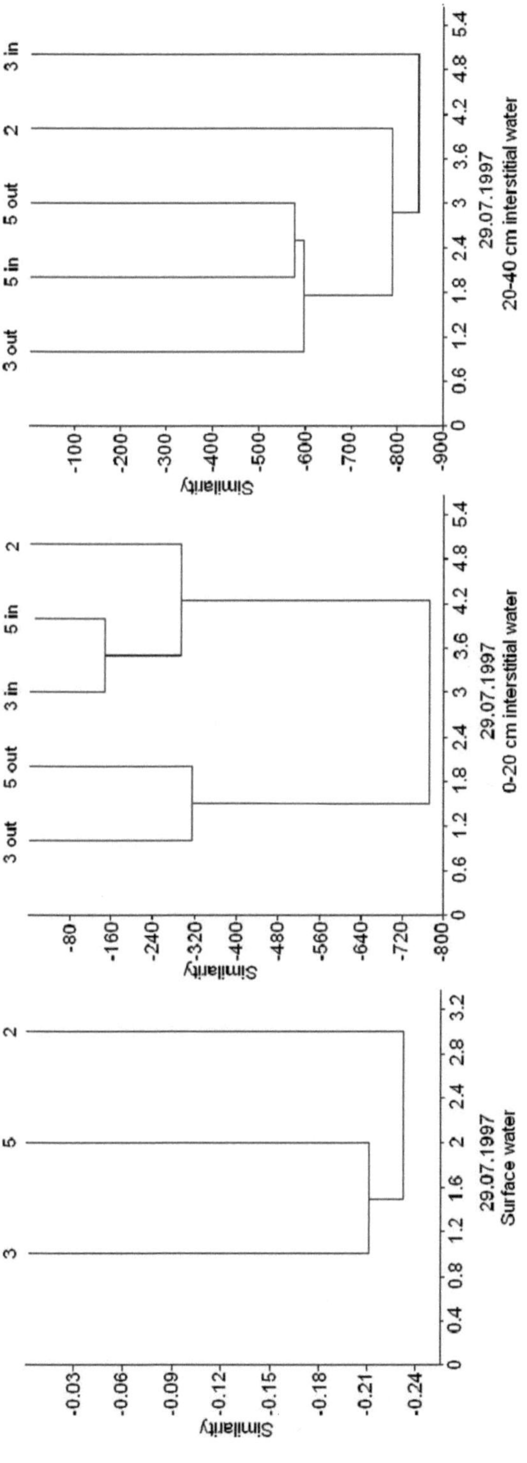

Fig. 17.4 Cluster analysis of the sampling sites

areas). In the 20–40 cm sediment layer the first group involved the sites: 3 out, 5 in, 5 out, 2 and the second group only the site 3 in.

17.5 Discussion

Chemistry of sediment interstitial water was analyzed in the function of sediment depth, reed cover and health condition of the reed stands at Lake Fertő/Neusiedler See. The results of these comparative investigations suggested that the reed stands influence the cycling of elements in the shallow lake ecosystems.

17.5.1 Nutrients

Usually the increase in PO_4^{3-} concentrations as a function of sediment depth is the reason for the release of the iron bound phosphate and the decomposition of organic matter in reductive conditions (Istvánovics, 1988).

The sediment of the Lake Fertő/Neusiedler See is dolomitic clayey marl type, which has a high capacity to absorb phosphate (Kitano, Okumura, & Idogaki, 1978). According to Dinka (1989a), Dinka (1994) 60.0–78.5% of the sediment TP is bound to the Ca and 13–28% to the organic matter and only 14% is NaOH hydrolysable. The co-precipitation of the PO_4^{3-} with carbonate can decrease the interstitial water P concentrations (Otsuki & Wetzel, 1972).

Close correlations were found between the aboveground reed biomass (Dinka, Ágoston-Szabó, & Szeglet, 2010) and PO_4^{3-}-P concentrations of the 20–40 cm interstitial water ($y = 1.6659x + 0.5705, R^2 = 0.64$), while there were no correlations with the PO_4^{3-}-P concentrations of the surface and 0–20 cm interstitial water.

The lower PO_4^{3-}-P concentrations in vegetated areas can be explained by the fact that aquatic macrophytes can lower the P concentration of the interstitial water by the uptake of the dissolved P (James, Barko, & Field, 1996; Debbie, Moss, & Phillips, 1997) and by creating favourable aerobic conditions in their rhizosphere to the precipitation of P with Fe oxyhydroxides (Andersen & Ring, 1999).

The elevated NH_4^+ concentrations in the interstitial water may have been the result of degradation of N-containing organic matter (Van Luijn, Boers, Lijklema, & Sweerts, 1999). The analysis of nitrogen speciation in our study indicated a NO_3^--N dominance in the surface and interstitial water, which suggest that the nitrification and coupled denitrification may have been the dominant processes in the sediment N transformation, but according to McClung and Frankenberger (1985) the anoxic conditions, high pH and salinity inhibit nitrification. In marine sediments the anoxic denitrification has been observed by Mortimer et al. (2004) and the coupled anoxic nitrification-manganese reduction processes by Freitag and Posser (2003). The runoff contribution from agriculture, the possible infiltration of the NO_3^- from the catchments area through the underground water, during this rainy year (1997), can not be excluded. Variation in N:P molar ratio and the lack of the correlations between the nutrients and aboveground biomass suggested a P limitation in the surface and 0–20 cm sediment layer.

17.5.2 Main Ions

Similarly to the results of Dinka (2001) and Ágoston-Szabó (2004, 2007) the pH, redox-potential, SO_4^{2-} concentrations and the temperature decreased, while the electrical conductivity, cations, S^{2-} and Cl^- concentrations increased with the depth of the sediment. When comparing our electrical conductivity, Cl^- concentrations values with the results of the above mentioned authors, similar differences were found between the vegetated and unvegetated sites.

The SO_4^{2-} concentration decrease was paralleled with an increase in S^{2-} concentration as a function of the sediment depth. Based on our results and the results of previous studies of Armstrong and Armstrong (1998), Dinka (2001), Ágoston-Szabó (2004, 2007) one of the dominant processes in the sediment of Lake Fertő/Neusiedler See is the sulphate reduction, which induce high sulphide concentrations. The sulphide rich interstitial water correlated with the low redox values (0–20 cm: $R^2_{\text{redox-S}^{2-}} = -0.84$, 20–40 cm: $R^2_{\text{redox-S}^{2-}} = -0.77$).

The differences in the redox potential values and sulphide concentrations in the interstitial water of the unvegetated and vegetated areas may be due to the *Phragmites* mediated gas transport. According to the result of Armstrong and Armstrong (1998) the efficiency of the oxygen transport in the *P. australis* rhizosphere influences the redox-potential and the concentrations of the redox sensitive compounds in the sediment. The oxygen transported through the *Phragmites* roots into the surrounding sediment creates favourable conditions to the oxidation of reduced compounds and to the detoxification of the sediment (Azzoni, Giordani, Bartoli, Welsh, & Viaroli, 2001).

The increase in the concentrations of Na^+, Ca^{2+}, K^+, Mg^{2+} and Cl^- with the sediment depth suggests that these ions diffuse from the deeper sediment layers toward the sediment surface. Diffusion across the sediment water interface is an important process controlling water quality of shallow lakes (Ma & Liu, 1999). The differences in the concentrations of the Na^+, Ca^{2+}, K^+, Mg^{2+}, and Cl^- between the vegetated and corresponding unvegetated areas can be explained by the salt uptake of the reed. Batty and Younger (2004) found that the depth profile of the Ca^{2+} is influenced by the degree of Ca^{2+} uptake by plants. Mann and Wetzel (2000) also found a significant difference in the interstitial water chemistry between the vegetated and unvegetated sites in the *Juncus effusus* dominated Talladega Wetland Ecosystem.

17.5.3 Trace Elements

In the surface water the measured concentrations of the trace elements were mostly below the recommended drinking water standards by the EU with the exception of boron.

The variation of the redox potential values and the concentrations of the redox-active elements: Fe^{2+}, Mn^{2+} with depth showed the reductive mobilization of these elements in the sediment. The increasing reduction of Fe (III) and Mn (IV) oxy-hydroxides results in increasing concentrations of aqueous Fe^{2+} and Mn^{2+} (Vink, 2002). The concentration profile of the trace elements suggests that they are moving from the sediment toward the surface water. Increased dissolved trace metal concentrations were found in the sulphidic sediment zones and a correlation was observed between $Mn^{2+} - S^{2-}(R^2 = 0.69 - 0.51), Zn^{2+} - S^{2-}(R^2 = 0.37 - 0.59), Pb^{2+} - S^{2-}(R^2 = 0.54–0.09)$ concentrations of the 0–20 cm and 20–40 cm interstitial water, respectively, which suggests that these elements probably associate the sulphides and compete for binding opportunities between the reactive sulphide phase, carbonates and relatively large amounts of organic matter (Vink, 2002; Yu, Tsai, Chen, & Ho, 2001).

The correlation found between the redox active trace elements (Fe^{2+}, Mn^{2+}) and heavy metals (Zn^{2+}, Pb^{2+}) in the 0–20 cm sediment layer: $Mn^{2+}-Zn^{2+}(R^2 = 0.74)$, $Fe^{2+} - Zn^{2+}(R^2 = 0.43)$, and $Mn^{2+} - Pb^{2+}(R^2 = 0.97)$ suggest the associated occurrence of these elements. According to Yu et al. (2001) the levels of various heavy metal binding phases decrease with the depth of the sediment, which was also observed by Dinka (1989b) in the sediments of Lake Fertő/Neusiedler See. It could be the reasons for increased concentrations of these elements in the sediment interstitial water.

17.6 Conclusions

The investigation of the chemical parameters of surface and sediment interstitial water in the reed stands of Lake Fertő/Neusiedler See contributes to a better understanding of the biogeochemical cycle of the elements and of the nutrient cycling in the lake.

Knowing the pH, redox potential, NH_4^+, NO_3^-, S^{2-}, SO_4^{2-} concentration profiles, information can be obtained about the quality and direction of the processes taking place in the sediment (e.g. sulphate reduction, nitrification, denitrification).

The differences in interstitial water chemistry between vegetated and unvegetated sites and between the degraded and healthy reed stands demonstrated that the changes in the chemical parameters are strongly interrelated with the presence and the health condition of the reed stand.

Acknowledgments This work was supported by EC-EUREED (IC-CT96-0020/CT960020 A2) and AKP, KVM/MTA.

References

Ágoston-Szabó, E. (2004). Chemical characteristics of sediment interstitial water at Lake Fertő/Neusiedler See. *Ecohydrology & Hydrobiology*, *4*(1), 67–76.

Ágoston-Szabó, E. (2007). *Physical and chemical characteristics of the sediment interstitial water at Lake Fertő/Neusiedler See regarding to the reed stands as well as to the interactions with the sediment and related waters*. Ph.D. Dissertation, Budapest.

Andersen F. O., & Ring, P. (1999). Comparison of phosphorus release from littoral and profundal sediments in a shallow eutrophic lake. *Hydrobiologia, 408/409*, 175–183.

Armstrong, J., & Armstrong, W. (1998). *Phragmites australis* – A preliminary study of soil oxidizing sites and internal gas transport pathways. *New Phytologist, 108*, 373–382.

Azzoni, R., Giordani, G., Bartoli, M., Welsh, D. T., & Viaroli, P. (2001). Iron, sulphur and phosphorus cycling in the rhizosphere sediments of a eutrophic *Ruppia cirrhosa* meadow (Valle Smarlacha, Italy). *Journal of Sea Research, 45*, 15–26.

Batty, L. C., & Younger, P. L. (2004). Growth of *Pragmites australis* (Cav) Trin ex. Steudel in mine water treatment wetlands: effects of metal and nutrient uptake. *Environtal Pollution, 132*, 85–93.

Berner, R. A. (1980). *Early diagenesis, A Theoretical approach*. Princeton: Princeton University Press.

Brix, H. (1999). The European research project on reed die-back and progression (EUREED). *Limnologica, 29*, 5–10.

Čížková, H., Pechar, L., Husák, š., Květ, J., Bauer, V., Radová, J. et al. (2001). Chemical characteristics of soils and pore waters of three wetland sites dominated by *Phragmites australis*: relation to vegetation composition and reed performance. *Aquatic Botany, 69*, 235–249.

Debbie, S., Moss, B., & Phillips, G. (1997). Do rooted macrophytes increase sediment phosphorus release? *Hydrobiologia, 342/343*, 27–34.

de Vincente, I., Amores, V., & Cruz-Pizzaro, L. (2006). Instability of shallow lakes: A matter of the complexity of factors involved in sediment and water interaction. *Limnetica, 25*, 253–270.

Dinka, M. (1989a). Über den P-Gehalt des Sediments im Neusiedler See. *BFB Bericht, 71*, 89–93.

Dinka, M. (1989b). Über den Zn-, Cu-, Pb-, Mn-, Fe- und Sr- Gehalt des Sediments im Neusiedler See. *BFB Bericht, 71*, 95–99.

Dinka, M. (1994). *Spatial changes of Lake Fertő/Neusiedler See based on sediment chemistry studies* [in Hungarian: A Fertő térbeli eltérései üledékkémiai tanulmányok alapján]. Dissertation, Vácrátót, Hungary.

Dinka, M. (1998). Über die Chemischen Verhältnisse des interstitialen Wassers in Schilfbestanden des Neusiedler Sees/Fertő. *4. Arbeitstagung des Bereiches Umwelt "Erdwissenschaftliche Aspekte des Umweltschutzes"* (pp. 343–349). Vienna: Tagungsband.

Dinka, M. (2001). Differences in interstitial water conditions in a degraded reed stand area. *Verhand-lungen Internationale Vereinigung Limnologie, 27*, 3385–3388.

Dinka, M., Ágoston-Szabó, E., & Szeglet P. (2010). Comparison between biomass and C, N, P and S contents of vigourous and die-back reed stands of Lake Fertő/Neusiedler See, *Biologia, 65/2*, 237–247.

Felföldy, L. (1980). *A biológiai vízminősítés*. [Biological water qualification]. Budapest: VIZDOK.

Freitag, T. E., & Prosser J. I. (2003). Community structure of ammonia-oxidizing bacteria within anoxic marine sediments. *Applied Environmental Microb*iology, *69*, 1359–1371.

Golterman, H. L., Clymo, R. S., & Ohnstad, M. A. M. (1978). *Method for Physical and Chemical Analysis of Freshwaters* (IBP Handbook No. 8). London: Blackwell Scientific Publications.

Gunatilaka, A. (1984). Nahrstoffkreislauf im Schilfgürtel des Neusiedler-Sees-Auswirkungen des Grünschnittes. *AGN Forschungsbericht*, 1981–1984. Herausgeben von den Bundesministerien für Wissenschaft und Forschung und Gesundheit und Umweltschutz und dem Land Burgenland – Landesmuseum als Sonderband 72 der Wissenschaftlichen Arbeiten aus dem, Burgenland: 223–311.

Hammer, R., Harper, D. A. T., & Ryan P. D. (2001). Past Paleontological statistics software package for education and data analysis. *Paleontologia electronica, 4*, 1–9.

Istvánovics, V. (1988). Seasonal variation of phosphorus release from the sediments of shallow Lake Balaton (Hungary). *Water Research, 22*, 1473–1481.

James, W. F., Barko, J. W., & Field, S. J. (1996). Phosphorus mobilization from littoral sediments of an inlet region in Lake Delavan, Wisconsin. *Archiv für Hydrobiologie, 138*, 245–257.

Jones, J. G. (1985). Microbes and microbial processes in sediments. *Philosophical Transaction. The Royal Society of London B, 315*, 3–17.

Kitano, Y., Okumura, M., & Idogaki, M. (1978). Uptake of phosphate ions by calcium carbonate. *Journal of Geochemistry, 12*, 29–37.

Löffler, H. (Ed.). (1979). *Neusiedler See: The limnology of a shallow lake in Central Europe*. Boston, MA: The Hague.

Ma, Y. J., & Liu, C. Q. (1999). Seasonal migration of redox boundary and its influence on water quality in Lake Lugu, Yunnan, China. *Ninth Annual V. M. Goldschmidt Conference*. Aug. 22–27, 1999, Cambridge, Massachusetts, Abstract No. 7026.

Mann, C. J., & Wetzel, R. G. (2000). Effects of the emergent macrophyte *Juncus effusus L.* on the chemical composition of interstitial water and bacterial productivity. *Biogeochemistry, 48*, 307–322.

McClung, G., & Frankenberger, W. T. (1985). Soil nitrogen transformations as affected by salinity. *Soil Science, 139*, 405–411.

Mortimer, R. J. G., Harris, S. J., Krom, M. D., Freitag, T. E., Prosser, J. I., Barnes, J., et al. (2004). Anoxic nitrification in marine sediments. *Marine Ecology Progress Series, 276*, 37–51.

Ostendorp, W. (1989). "Die-back" of reeds in Europe – a critical review of literature. *Aquatic Botany, 35*, 5–26.

Otsuki, A., & Wetzel, R. G. (1972). Coprecipitation of phosphate with carbonates in marl lake. *Limnology and Oceanography, 17*, 763–767.

Urbanc-Berčič, O., & Gaberščik, A. (2004). The Relationship of the processes in the rhizosphere of common reed *Phragmites australis* (Cav.) TRIN. EX STEUDEL to water fluctuation. *International Revue für Hydrobiologie, 89*, 500–507.

Van der Putten (1997). Die-back of *Phragmites australis* in European wetlands: an overview of the European Research Programme on Reed Die-back and Progression (1993 – 1994). *Aquatic Botany, 59*, 263–275.

Van Luijn, F., Boers, P. C. M., Lijklema, L., & Sweerts, J.-P. R. A. (1999). Nitrogen fluxes and processes in sandy and muddy sediments from a shallow eutrophic lake. *Water Research, 33*, 33–42.

Vink, J. P. M. (2002). Measurement of heavy metal speciation over the redox gradients in natural water– Sediment interfaces and implication for uptake by benthic organisms. *Environtal Science and Technology, 36*, 5130–5138.

Yu, K. C., Tsai, L. J., Chen, S. H., & Ho, S. T. (2001). Chemical binding of heavy metals in anoxic river sediments. *Water Research, 35*, 4086–4094.

Chapter 18
Water Regime Changes and the Function of an Intermittent Wetland

Nataša Dolinar, Nina Šraj, and Alenka Gaberščik

Abstract The intermittent Lake Cerknica comprises a variety of wetland habitats, which support numerous species, many of them included on the Red List. The exchange of wet and dry periods influences the through-flow of energy and the turnover of matter in the lake. In the last five decades, the water level fluctuations were highly variable. Since 1961 the May and June water levels have shown a decreasing trend line, in October the trend line revealed an increase, while from November to February no changes of trend could be observed. High water levels in winter and during the peak vegetation season negatively affected reed productivity. The decomposition of reed leaves increased with the wetness of the survey location, while the soil mineralisation rate was only slightly related to soil wetness. In comparison to the tributaries, the lake water contained relatively low amount of nutrients, which was a consequence of a densely vegetated area.

Keywords Decomposition · Evapotranspiration · Intermittent lake · *Phragmites australis* · Wetland habitats

18.1 Introduction

Lake Cerknica is an intermittent wetland appearing at the bottom of the Cerknica polje karst depression. It is characterised by extreme inter-annual water level fluctuations, which are the result of a high precipitation rate and water runoff into an extensive underground drainage system in carbonate rocks. The catchment area of the lake extends over 475 km^2. The majority of the lake's inflow consists of karstic waters (80%), while only one tributary brings surface water (15%). The amount of water in the lake depends on precipitation and potential evapotranspiration of the

A. Gaberščik (✉)
Department of Biology, Biotechnical Faculty, University of Ljubljana, Ljubljana, Slovenia
e-mail: alenka.gaberscik@bf.uni-lj.si

J. Vymazal (ed.), *Water and Nutrient Management in Natural and Constructed Wetlands*, DOI 10.1007/978-90-481-9585-5_18,

polje and its surroundings (Kranjc, 2002b). During the rainy period, water is collected from the upstream-lying karst poljes, from the surrounding karst plateaus and directly from rainfall.

The studies of processes in intermittent wetlands revealed that water levels during the vegetation period, together with the intensity, timing and the extent of floods and droughts, affect primary production, life cycles of animals, i.e. spawning of fish and nesting of birds, as well as mineralisation and decomposition (Boulton & Brock, 1999; Gaberščik, Urbanc-Berčič, Kržič, Kosi, & Brancelj, 2003; Dinka et al., 2008). Water level fluctuations create a variety of habitats which support diverse communities. Habitats are delineated by the range of changes in water level and soil properties. Clear zoning of different vegetation types was described for the lake (Martinčič, 2002) as was also observed in other wetlands (Finlayson, 2005). In the past, life cycles of the majority of organisms in the lake were intimately coupled to more or less regular water level fluctuations and the primary producers were well adapted to water level changes. In the last few decades, flood and drought events have became scattered and irregular (Dolinar, Rudolf, Šraj, & Gaberščik, 2010), possibly due to climate change. As a consequence, conditions for plant growth as well as for other processes are changing.

The present contribution is a critical review of the processes studied in the Lake Cerknica intermittent wetland summarising the trends of changes.

18.2 Water Regime

An intermittent water regime has made Lake Cerknica an attractive research subject throughout history. In the 17th century, Slovenian polymath J. V. Valvasor, a member of the English Royal Society, tried to decipher the mechanism of water level fluctuations by a set of underground lakes appearing at different levels (Fig. 18.1) (Rupel, 1951). Only a century later, a simplified explanation was offered by Nagel and Gruber (Kranjc, 2002a). They claimed that the lake appeared at the bottom of the polje when the inflow was bigger than the outflow, and as such the lake was the result of a surplus of water in the area.

The present understanding of the lake's hydrology is much more complex. The main characteristics of the Lake Cerknica area are presented in Table 18.1. A large difference between the inflow and the outflow indicates a high importance of evapotranspiration rate in the area. The lake can fill up within 1 week, when the water level reaches a normal level of 550 m a.s.l., covering about 53% of the polje (Kranjc, 2002b). When the water exceeds this level, extreme floods occur, the last one being in November 2000 (552.2 m a.s.l.). The lake begins to disappear in late spring and the water dries up in 2–4 weeks (Kranjc, 1986).

Monthly water level fluctuations in the last five decades were highly variable (Fig. 18.2). The year with the biggest extremes was 1985, while the maximum water level at the measuring location (at Stržen Dolenje jezero) was measured in 2000 (654 cm). The average water level from November to February has remained more or less the same, in May and June a trend of decrease has been observed, while in October the trend line has shown an increase (the trend lines are not shown).

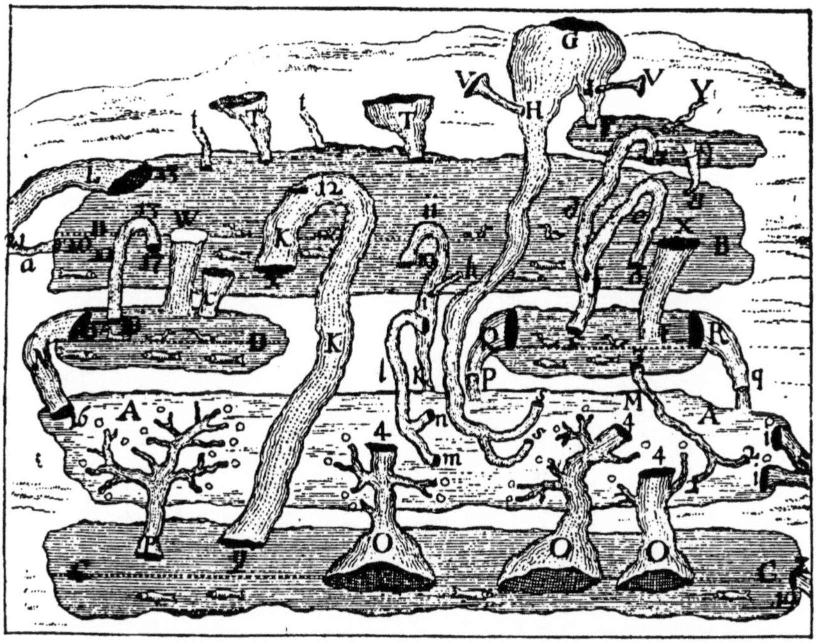

Fig. 18.1 The system of lakes beneath Lake Cerknica as seen by polymath J. V. Valvasor in 17th century (source: Rupel, 1951). Reprinted with permission of Mladinska knjiga

Table 18.1 The main characteristics of the Lake Cerknica area (Zupančič, 2002; Kranjc, 2002a, 2002b)

Northing	45°45′
Easting	14°20′
Average yearly precipitation	1700–1800 mm
Potential yearly Evapotranspiration	750 mm
Catchment area	475 km^2
Area of polje	38 km^2
Elevation of polje	549 m a.s.l.
Area of normal lake	20 km^2
Elevation of normal lake	550 m a.s.l.
Average yearly duration of floods	260 days
Maximum inflow	210–240 $m^3\ s^{-1}$
Maximum outflow	40–90 $m^3\ s^{-1}$

The results also revealed large differences in water level within the vegetation periods when the availability of water is essential for plant activity. An early onset of the dry season might be a problem for primary producers due to more frequent heavy rains which occasionally flood the area. Such events, even over a short time, could affect plant success and consequently the survival of other species, resulting in poorer plant communities around the lake. The year with the biggest extremes was 1985, while the maximum water level at the measuring location (at Stržen Dolenje jezero) was measured in 2000 (654 cm).

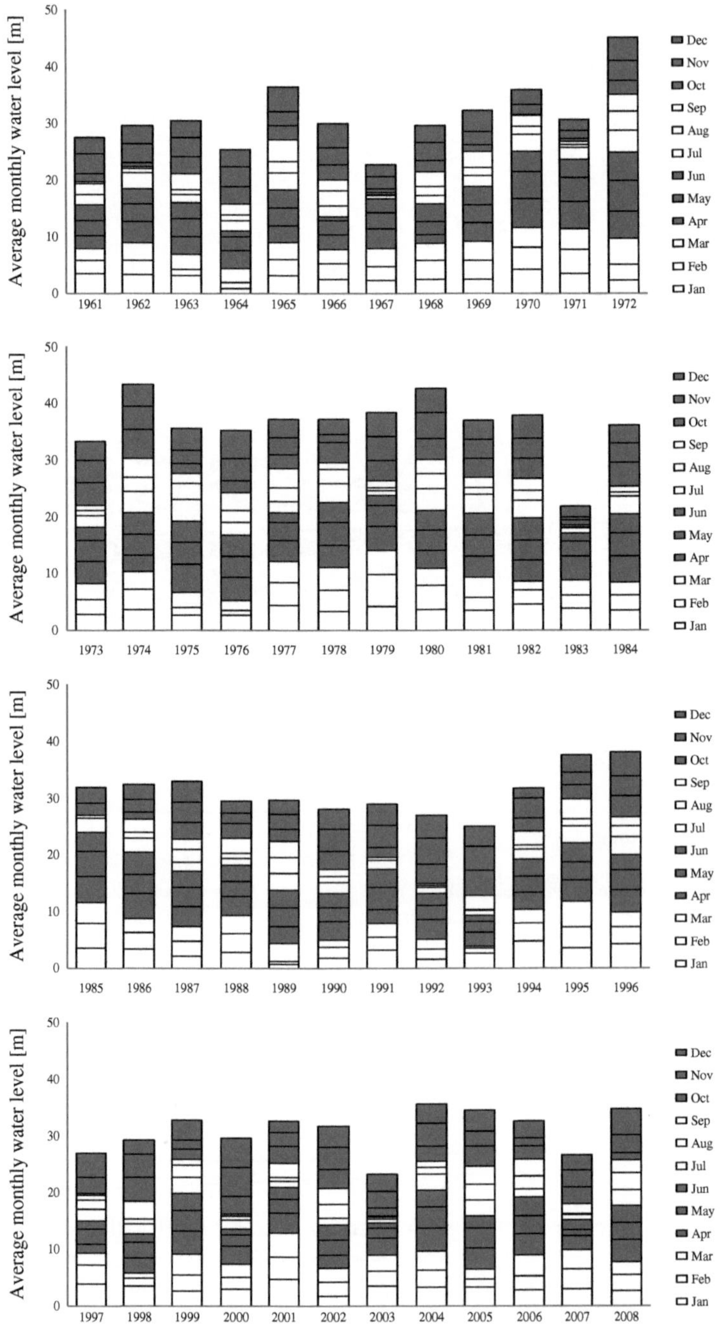

Fig. 18.2 Monthly changes of water levels at Lake Cerknica for the periods of 1961–1972, 1973–1984, 1985–1996 and 1997–2008 (data set provided by the Environmental Agency of the Republic of Slovenia)

18.3 Primary Production

Reed stands (*Phragmites australis*) present an efficient trap for solar energy, contributing the most to the primary production of Lake Cerknica, influencing the water cycle (Dolinar et al., 2010) and functioning as a sink for nutrients (Gaberščik et al., 2003). We distinguish between three different types of reed stands at Lake Cerknica, namely littoral stands thriving in shallower parts of the lake which are flooded for about 9 months per year, ecotonal stands thriving along the tributaries and the main stream Stržen, and terrestrial stands colonising the soil, which is often water-saturated but seldom flooded. The reed primary production ranged from 130 to 1600 g m^{-2}, being the highest in the ecotonal stands. Vis, Hudon, Carignan, and Gagnon (2007) report that variations in the water level lead to shifts in the distribution and biomass of emergent and submerged macrophytes, as was also the case at Lake Cerknica. In general, it is stated that the higher the water level fluctuations, the greater the production potential of the ecosystem (Boulton & Brock, 1999). Our recent and previous measurements revealed that the variability in primary production and abundance of reed plants could be more than 50% (Gaberščik, Urbanc-Berčič, & Martinčič, 2000). Nonetheless, this species contributes the most to the overall primary production of the lake.

Long-term data analysis has shown that the timing and length of floods also dictate the development and success of reeds. High water levels, especially in winter time, correspond with lower biomass production of reed. This could be attributed to the occurrence of waves and formation of ice cover, followed by a decrease in water level. Waves and broken ice can damage reed culms and prevent the transfer of oxygen into the underground parts, affecting early growth in spring (Dolinar et al., 2010). The relation between reed biomass production and the average water level in December and January is shown on a graph (Fig. 18.3). The variability in reed biomass production between different years could indicate a presence of stress

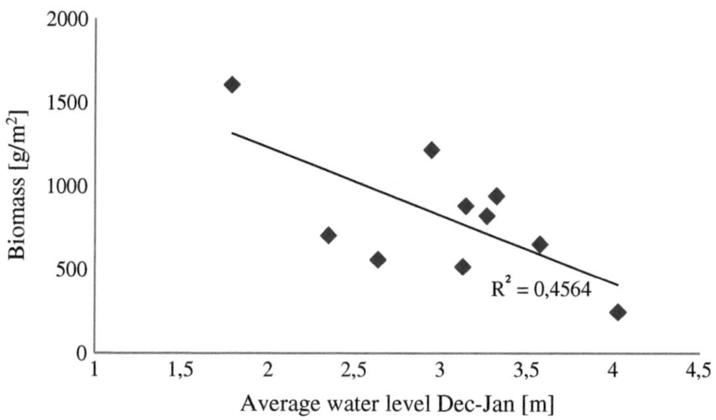

Fig. 18.3 Relation between average water level in winter and reed biomass production in the following year

due to a high water level during the peak vegetation period, which was shown by measurements of actual photochemical efficiency and amino acid analysis in littoral stands (Dinka et al., 2008).

Tall sedges colonising wet or only occasionally flooded areas are also widespread in the area of Lake Cerknica (Martinčič, 2002). Their primary production was less variable, ranging from 330 to 480 g m^{-2}.

18.4 Organic Matter Decomposition

The extent of floods and droughts influenced the decomposition of plant biomass at various locations of the lake. The decomposition of reed leaves at dry, occasionally flooded and permanently flooded locations was significantly different and increased with the wetness of the location. However, this was not the case in the decomposition of reed culms. The decomposition of sedges was comparable to the decomposition of reed culms (Fig. 18.4). The results of Pinna and Basset (2004) were in agreement with our results, showing higher decomposition rates at sites with normal water level compared to sites exposed to complete summer desiccation. The opposite results were found in a reed decomposition study of a temporary Mediterranean stream, where the overall decomposition was higher when leaves emerged as opposed to being submerged (Menendez, 2005).

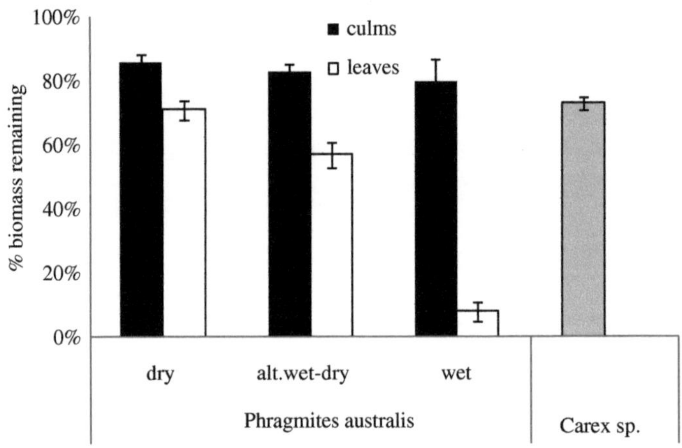

Fig. 18.4 Decomposition rates for *Phragmites australis* culms and leaves in different habitats (dry, alternating wet and dry, wet) and *Carex* sp

18.5 Other Processes

Water level fluctuations are known to increase the soil mineralisation rate (Boulton & Brock, 1999). To get an insight into the mineralisation processes of Lake Cerknica, the relation to soil water content and soil respiration potential measured as

ETS(electron transport system)-activity were monitored at various survey locations throughout the year. The highest ETS-activity was detected at Dujice (0.12–0.23 mg O_2 gDM^{-1} h^{-1}), while at Drvošec and Zadnji kraj, ETS-activities were lower than 0.008 mg O_2 gDM^{-1} h^{-1} (Urbanc-Berčič & Gaberščik, 2001). The organic structure of the soil enabled high field-water capacity, supporting undisturbed mineralisation processes even in dryer periods (Urbanc-Berčič & Gaberščik, 2004). It was expected that the ETS-activity of soil would be positively correlated with soil water content, however the correlation was rather weak. It was also expected that ETS-activity would correlate to the amounts of soluble nutrients released into the pore water, but the correlation was not significant.

The densely vegetated lake also functions as a sink for nutrients discharging into the lake (Jørgensen, 1990; Pieczynska, 1990; Wetzel, 1990, Mitsch et al., 2001, Gaberščik et al., 2003). In line with the above, the water quality monitoring indicated the lowest amounts of nutrients in the lake water, while relatively high content was measured in the surface tributary Cerkniščica and moderate in the karst tributaries, even though they could occasionally bring nutrients from distant areas (Gaberščik, Kosi, Krušnik, Urbanc-Berčič, & Bricelj, 1994; Gaberščik et al., 2003).

18.6 Biodiversity

In the period from 1772 to 1996, about 100,000 ha of wetlands were destroyed in Slovenia. The remaining areas total up to only 36,000 ha, with the 3800 ha of Lake Cerknica presenting a precious island in the landscape, supporting high biodiversity. Even though the importance of biodiversity for system functioning is well recognised (Tilman & Downing, 1994, Schulze, Beck, & Müller-Hohenstein, 2005), its role in intermittent wetland systems is not sufficiently documented and understood. The exchange of dry and wet periods influences the distribution, survival and development of organisms in the area (Gaberščik et al., 2003). The majority of animal species is able to avoid extreme conditions employing various strategies of overcoming drought. They either have a mobile adult stage (dragonflies, water-beetles, mayflies...), desiccation resistant eggs (zooplankton, snails...) or migrate to other wetlands (birds) (Boulton & Brock, 1999). This is not the case with plants, which are rooted and need to overcome changes in water level. Their occurrence, reproduction, and distribution, and the structure of communities in intermittent wetland habitats, are determined by short term changes such as duration of inundation, water depth and water retention capacity (Barrat-Segretain, Henry, & Bornette, 1999; Riis & Hawes, 2002; Capers, 2003; Warwick & Brock, 2003). Under such conditions amphibious plants have a competitive advantage over truly aquatic and wetland plant species, which are less tolerant to the variable water regime (Fernández-Aláez, Fernández-Aláez, & Bécarez, 1999; Cronk & Fennessy, 2001).

Water level fluctuations and heterogeneous hydrologic patterns result in a variety of habitats in time and space, offering habitat to numerous species (Table 18.2) among which many are protected (Pravilnik o uvrstitvi ogroženih rastlinskih in živalskih vrst v rdeči seznam, 2002). According to the vegetation survey of the

Table 18.2 Biodiversity and protected species of Lake Cerknica and its surroundings

	Number of species		An example of endangered species	References from Gaberščik A (ed) Monograph on Lake Cerknica 2002
	recorded	protected		
Higher plants	?	50	*Drosera intermedia*	Martinčič (2002)
Algae	329	?	?	Kosi (2002)
Mammals (Mammalia)	45	25	*Lutra lutra*	Polak (2002b)
Birds (Aves)	256	73	*Crex crex*	Polak (2002a)
Reptiles (Reptilia)	11	11	*Emys orbicularis*	Polak (2002c)
Amphibians (Amphibia)	13	13	/	Veenvliet and Poboljšaj (2002)
Fish (Pisces)	8	4	*Lota lota*	Povž (2002)
Butterflies (Lepidoptera)	120	40	*Maculiena alcon*	Čelik (2002)
Caddisiflies (Trichoptera)	33	2	*Wormaldia subnigra*	Urbanič and Krušnik (2002)
Beetles (Coleoptera)	690[a]	?	*Lucanus cervus*	Drovenik (2002)
Bees (Apoidea)	58	12	/	Gogala (2002)
Grasshoppers (Orthoptera)	38[b]	2	/	Veenvliet (2002)
Mayflies (Ephemeroptera)	11	6	/	Zabric (2002)
Dragonflies (Odonata)	36	4	/	Bedjanič (2002)
Lower crustaceans	37	3	/	Brancelj (2002)
Mollusks (Mollusca)	142	35	*Zospeum exiguum*	Slapnik (2002)

Species estimated: [a]1300–1600 in wider area, [b] 50; ? – no reliable data exist

Cerknica polje, the distribution of plant community follows the hydrologic gradient defined by the extent and duration of floods. In line with the above, significant effects of water level fluctuations on the shore vegetation pattern were reported for Lake Michigan (Lyon, Drobney, & Olson, 1986). The edge of the Cerknica polje, where the soil is wet or only occasionally flooded, is dominated by regularly mown wet grassland communities (*Molinietum caeruleae* and *Deschampsio-Plantaginetum altissimae*) (Martinčič, 2002). Where the water level never exceeds 2 m, wetland vegetation developed. The largest part of the polje is covered by reed stands (*Phragmites australis*), while different sedge communities (*Carex* sp.) which are used for strewing are common as well. Truly aquatic plants are confined to permanent water bodies, such as some deeper depressions and the Stržen stream with tributaries (Kržič, Gaberščik, & Germ, 2004; Šraj-Kržič & Gaberščik, 2005). The most common plants are stoneworts (Characeae) and pondweeds (*Potamogeton* sp.) (Martinčič, 2002).

The alternation of floods and dry periods results in a flora, dominated by plant species exhibiting amphibious character that enables their survival under varying water conditions (Warwick & Brock, 2003; Williams et al., 2003). Morphological and biochemical features and reproduction strategies enable them to have a continuous physiological functioning over the hydrologic gradient from water to dry land (Boulton & Brock, 1999; Šraj-Kržič & Gaberščik, 2005; Germ & Gaberščik, 2003). One of the most flexible species is *Polygonum amphibium,* which is usually very abundant in Lake Cerknica (Gaberščik & Martinčič, 1992; Gaberščik, 1993). If the water level in spring is high, tall amphibious plants prevail (*Senecio paludosus* and *Sium latifolium*), while at lower water levels shorter plants dominate (*Mentha aquatica, Gratiola officinalis, Rorippa amphibia, Teucrium scordium* and *Myosotis palustris*) (Martinčič, 2002).

The lake area provides temporary shelter and foraging habitat for numerous mammal species, inhabiting the surrounding forests. The most important are the three protected large carnivores – the brown bear (*Ursus arctos*), wolf (*Canis lupus*) and lynx (*Lynx lynx*). Some other mammals are quite frequent as well: i.e. the wild cat (*Felis silvestris*), red fox (*Vulpes vulpes*), red deer (*Cervus elaphus*) and roe deer (*Capreolus capreolus*) (Polak, 2002b). One of the most endangered mammals in Slovenia, the otter (*Lutra lutra*), was believed to be extinct from the area, but it has been seen again in one of the tributaries. In the area of Lake Cerknica, 256 bird species were detected and out of them, more than 100 are considered permanent or occasional breeders with 73 species listed as protected. The area is one of the most important breeding sites for the corn crake (*Crex crex*) in the country (Polak, 2002a). The majority of reptiles, amphibians, and fish are protected, because of a rapid decline in the number of some species during the last decades. As an example, the pond terrapin (*Emys orbicularis*) is likely to be extinct (Polak, 2002c). The main threats to reptiles are abandonment of the traditional land use and the forestation of dry meadows, while amphibians are threatened by the loss of small water-bodies, introduction of fish, pollution and new infrastructure building (Veenvliet & Poboljšaj, 2002). Lake Cerknica is also characterised by a high diversity of invertebrates. High diversity of snails is characteristic of the subterranean watercourses of the lake and 10 species are partly endemic to this area.

For a long time the intermittent character of Lake Cerknica has presented a natural protection against the invasion of alien plant species that have spread across Slovenia in the last few decades (Martinčič & Leskovar, 2002). However, due to an untypical water regime in recent years, small stands of the invasive species *Solidago canadensis* has colonised dryer locations near Vodonos and Rešeto.

References

Barrat-Segretain, M.-H., Henry, C. P., & Bornette, G. (1999). Regeneration and colonization of aquatic plant fragments in relation to the disturbance frequency of their habitats. *Archiv für Hydrobiologie, 145*, 111–127.

Bedjanič, M. (2002). On the dragonflies. In A. Gaberščik (Ed.), *Jezero, ki izginja – Monografija o Cerkniškem jezeru* [The Vanishing Lake – Monograph on Lake Cerknica] (pp. 138–147). Ljubljana, Slovenia: Društvo ekologov Slovenije (in Slovenian with English summary).

Boulton, A. J., & Brock, M. A. (1999). *Australian freshwater ecology. Processes and management*. Glen Osmond, Australia: Gleneagles Publishing.

Brancelj, A. (2002). Anostracans, cladocerans and copepods. In A. Gaberščik (Ed.), *Jezero, ki izginja – Monografija o Cerkniškem jezeru* [The Vanishing Lake – Monograph on Lake Cerknica] (pp. 130–137). Ljubljana, Slovenia: Društvo ekologov Slovenije (in Slovenian with English summary).

Capers, R. S. (2003). Six years of submerged plant community dynamics in a freshwater tidal wetland. *Freshwater Biology, 48*, 1640–1651.

Cronk, J. K., & Fennessy, M. S. (2001). *Wetland plants: biology and ecology*. London: Lewis.

Čelik, T. (2002). Butterflies – a long hidden treasure of Cerkniško polje. In A. Gaberščik (Ed.) *Jezero, ki izginja – Monografija o Cerkniškem jezeru* [The Vanishing Lake – Monograph on Lake Cerknica] (pp. 186–199). Ljubljana: Društvo ekologov Slovenije (in Slovenian with English summary).

Dinka, M., Agoston-Szabo, E., Urbanc-Berčič, O., Germ, M., Kržič, N., & Gaberščik, A. (2008). Reed stands conditions at selected wetlands in Slovenia and Hungary. In J. Vymazal (Ed.), Wastewater treatment, plant dynamics and management in constructed and natural wetlands (pp. 1–12). Dordrecht: Springer.

Dolinar, N., Rudolf, M., Šraj, N., & Gaberščik, A. (2010). Environmental changes affect ecosystem services of the intermittent Lake Cerknica. *Ecological Complexity, 7*, 403–409. doi:10.1016/j.ecocom.2009.09.004

Drovenik, B. (2002). Beetles. In A. Gaberščik (Ed.), *Jezero, ki izginja – Monografija o Cerkniškem jezeru* [The Vanishing Lake – Monograph on Lake Cerknica] (pp. 166–179). Ljubljana: Društvo ekologov Slovenije (in Slovenian with English summary).

Fernández-Aláez, C., Fernández-Aláez, M., & Bécarez, E. (1999). Influence of water level fluctuation on the structure and composition of the macrophyte vegetation in two small temporary lakes in the northwest of Spain. *Hydrobiologia, 415*, 155–162.

Finlayson, C. M. (2005). Plant ecology of Australia's tropical floodplain wetlands: A review. *Annals of Botany, 96*, 541–555.

Gaberščik, A. (1993). Measurements of apparent CO_2 flux in amphibious plant *Polygonum amphibium* L. growing over environmental gradient. *Photosyntetica, 29*, 473–476.

Gaberščik, A., Kosi, G., Krušnik, C., Urbanc-Berčič, O., & Bricelj, M. (1994). Water quality in Lake Cerknica and its tributaries. *Acta Carsologica, 23*, 269–287.

Gaberščik, A., & Martinčič, A. (1992). Spreminjanje lastnosti listov vodne dresni (*Polygonum amphibium* L.) v gradientu kopno/voda. *Biološki vestnik, 40*, 1–11.

Gaberščik, A., Urbanc-Berčič, O., Kržič, N., Kosi, G., & Brancelj, A. (2003). The intermittent lake Cerknica: Various faces of the same ecosystem. *Lakes and Reservoirs: Research and Management, 8*, 159–168

Gaberščik, A, Urbanc-Berčič, O., & Martinčič, A (2000) The influence of water level fluctuations on the production of reed stands (*Phragmites australis*) on intermittent lake Cerknica. In S. Christofor, A. Sarbu, & M. Adamescu (Eds.), *Proceedings of. International Workshop and 10th Macrophyte Group Meeting IAD-SIL* (pp. 29–34). Danube Delta, Romania: University of Bucarest.

Germ, M., & Gaberščik, A. (2003). Comparison of aerial and submerged leaves in two amphibious species, *Myosotis scorpioides* and *Ranunculus trichophyllus*. *Photosyntetica, 41*, 91–96

Gogala, M. (2002). Bees. In A. Gaberščik (Ed.), *Jezero, ki izginja – Monografija o Cerkniškem jezeru* [The Vanishing Lake – Monograph on Lake Cerknica] (pp. 160–165). Ljubljana: Društvo ekologov Slovenije (in Slovenian with English summary).

Jørgensen, S. E. (1990). Erosion and filtration. In S. E. Jørgensen & H. Löffler (Eds.), *Guidelines of shore management* (Vol 3, pp. 13–20). Japan: International Lake Environmental Comittee. UNEP.

Kosi, G (2002) Algae in the lake and its tributaries. In Gaberščik A (Ed.), *Jezero, ki izginja – Monografija o Cerkniškem jezeru* [The Vanishing Lake – Monograph on Lake Cerknica]. Ljubljana: Društvo ekologov Slovenije, pp. 66–71 (in Slov. with Engl. summary)

Kranjc, A. (1986). The lake of Cerknica and its floods. *Geografski zbornik, 25*, 75–116.

Kranjc, A. (2002a). Geology and geomorphology. In A. Gaberščik (Ed.), *Jezero, ki izginja – Monografija o Cerkniškem jezeru* [The Vanishing Lake – monograph on Lake Cerknica] (pp. 18–26). Ljubljana: Društvo ekologov Slovenije (in Slovenian with English summary).

Kranjc, A. (2002b). Hydrology. In A. Gaberščik (Ed.) *Jezero, ki izginja – Monografija o Cerkniškem jezeru* [The Vanishing Lake – monograph on Lake Cerknica] (pp. 26–38). Ljubljana: Društvo ekologov Slovenije (in Slovenian with English summary).

Kržič, N., Gaberščik, A., & Germ, M. (2004). The phenotypic plasticity of *Glyceria fluitans* growing over the water/land gradient. *Acta biologica slovenica, 47*, 65–73.

Lyon, J. G., Drobney, R. D., & Olson, C. E., Jr. (1986). Effects of lake Michigan water level fluctuations on wetland soil chemistry and distribution of plants in the straits of Mackinac. *Journal of Great Lakes Research, 12*, 175–183

Martinčič, A. (2002). Vegetation. In A. Gaberščik (Ed.), *Jezero, ki izginja – Monografija o Cerkniškem jezeru* [The Vanishing Lake – monograph on Lake Cerknica] (pp. 80–96). Ljubljana: Društvo ekologov Slovenije (in Slovenian with English summary).

Martinčič, A., & Leskovar, I. (2002). Pteridophytes and spermatophytes. In A. Gaberščik (Ed.), *Jezero, ki izginja – Monografija o Cerkniškem jezeru* [The Vanishing Lake – monograph on Lake Cerknica] (pp. 72–79). Ljubljana, Slovenia: Društvo ekologov Slovenije (in Slovenian with English summary).

Menendez, M. (2005). Decomposition of the common reed *Phragmites australis* in a Mediterranean stream. *Archiv für Hydrobiologie, 163*, 101–115

Mitsch, W. J., Day, J. W., Jr., Gilliam, J. W., Groffman, P. M., Hey, D. L., Randall, G. W., et al. (2001). Reducing N loading to the Gulf of Mexico from the Mississippi River Basin: Strategies to counter a persistent ecological problem. *BioScience, 51*, 373–388

Pieczynska, E. (1990). Littoral habitats and communities. In S. E. Jørgensen & H. Löffler (Eds.), *Guidelines of shore management* (Vol 3, pp. 39–72). Japan: International Lake Environmental Comittee. UNEP.

Pinna, M., & Basset, A. (2004). Summer drought disturbance on plant detritus decomposition processes in three River Tirso (Sardinia, Italy) sub-basins. *Hydrobiologia, 522*, 311–319.

Polak, S. (2002a). Lake Cerknica – an internationally important bird area (IBA). In A. Gaberščik (Ed.), *Jezero, ki izginja – Monografija o Cerkniškem jezeru* [The Vanishing Lake – Monograph on Lake Cerknica] (pp. 236–247). Ljubljana: Društvo ekologov Slovenije (in Slovenian with English summary).

Polak, S. (2002b). Mammalian fauna of Cerkniško polje and its surrounding. In A. Gaberščik (Ed.), *Jezero, ki izginja – Monografija o Cerkniškem jezeru* [The Vanishing Lake – Monograph on Lake Cerknica] (pp. 248–257). Ljubljana: Društvo ekologov Slovenije (in Slovenian with English summary).

Polak, S. (2002c). The reptile fauna of Lake Cerknica and its surrounding. In A. Gaberščik (Ed.), *Jezero, ki izginja – Monografija o Cerkniškem jezeru* [The Vanishing Lake – Monograph on Lake Cerknica] (pp. 231–235). Ljubljana: Društvo ekologov Slovenije (in Slovenian with English summary).

Povž, M. (2002). Freshwater fish and fishery management. In A. Gaberščik (Ed.), *Jezero, ki izginja – Monografija o Cerkniškem jezeru* [The Vanishing Lake – Monograph on Lake Cerknica] (pp. 200–207). Ljubljana: Društvo ekologov Slovenije (in Slovenian with English summary).

Pravilnik o uvrstitvi ogroženih rastlinskih in živalskih vrst v rdeči seznam (Rules on the inclusion of endangered plant and animal species in the Red List). (2002). Praprotnice in semenke (Pteridophyta & Spermatophyta), Priloga (Supplement) 1, Uradni list RS, št. 82/2002, Uradni list RS, št. 42/2010, Ministrstvo za okolje, prostor in energijo.

Riis, T., & Hawes, I. (2002). Relationships between water level fluctuations and vegetation diversity in shallow water of New Zealand lakes. *Aquatic Botany, 74*, 133–148.

Rupel, M. (1951). *Valvazorjevo berilo*. Ljubljana: Mladinska knjiga.

Schulze, E. D., Beck, E., & Müller-Hohenstein, K. (2005). *Plant ecology*. Heidelberg: Springer.

Slapnik, R. (2002). Mollusks. In A. Gaberščik (Ed.), *Jezero, ki izginja – Monografija o Cerkniškem jezeru* [The Vanishing Lake – Monograph on Lake Cerknica] (pp. 116–129). Ljubljana: Društvo ekologov Slovenije (in Slovenian with English summary).

Šraj-Kržič, N., & Gaberščik, A. (2005). Photochemical efficiency of amphibious plants in an intermittent lake. *Aquatic Botany, 83*, 281–288.

Tilman, D., & Downing, J. A. (1994). Biodiversity and stability in grasslands. *Nature, 367*, 363–365.

Urbanc-Berčič, O., & Gaberščik, A. (2001). The influence of water table fluctuations on nutrient dynamics in the rhizosphere of common reed (*Phragmites australis*). *Water Science and Technology, 44*, 245–250.

Urbanc-Berčič, O., & Gaberščik, A. (2004). The relationship of the processes in the rhizosphere of common reed *Phragmites australis*, (Cav.) Trin. ex Steudel to water fluctuation. *International Revue für Hydrobiologie, 89*, 500–507.

Urbanič, G., & Krušnik, C. (2002). Caddisifleis of the lake and its tributaries. In A. Gaberščik (Ed.), *Jezero, ki izginja – Monografija o Cerkniškem jezeru* [The Vanishing Lake – Monograph on Lake Cerknica] (pp. 180–185). Ljubljana: Društvo ekologov Slovenije (in Slovenian with English summary).

Veenvliet, P. (2002). Grashoppers. In A. Gaberščik (Ed.), *Jezero, ki izginja – Monografija o Cerkniškem jezeru* [The Vanishing Lake – Monograph on Lake Cerknica] (pp. 154–159). Ljubljana: Društvo ekologov Slovenije (in Slovenian with English summary).

Veenvliet, P., & Poboljšaj, K. (2002). Amphibia. In A. Gaberščik (Ed.), *Jezero, ki izginja – Monografija o Cerkniškem jezeru* [The Vanishing Lake – Monograph on Lake Cerknica] (pp. 208–217). Ljubljana: Društvo ekologov Slovenije (in Slovenian with English summary).

Vis, C., Hudon, C., Carignan, R., & Gagnon, P. (2007). Spatial analysis of production by macrophytes, phytoplankton and epiphyton in a large river system under different water-level conditions. *Ecosystems, 10*, 293–310.

Warwick, N. W. M., & Brock, M. A. (2003). Plant reproduction in temporary wetlands: the effects of seasonal timing, depth, and duration of flooding. *Aquatic Botany, 77*, 153–167.

Wetzel, R. G. (1990). Land-water interfaces: Metabolic a limnological regulators. *Verh Internat Verein Limnol, 24*, 6–24.

Williams, P., Whitfield, M., Biggs, J., Bray, S., Fox, G., Nicolet, P., et al. (2003). Comparative biodiversity of rivers, streams, ditches and ponds in an agricultural landscape in Southern England. *Biological Conservation, 115*, 329–341.

Zabric, D. (2002). The swarming mayflies. In A. Gaberščik (Ed.), *Jezero, ki izginja – Monografija o Cerkniškem jezeru* [The Vanishing Lake – Monograph on Lake Cerknica] (pp. 148–153). Ljubljana: Društvo ekologov Slovenije (in Slovenian with English summary).

Zupančič, B. (2002). Climate. In A. Gaberščik (Ed.), *Jezero, ki izginja – Monografija o Cerkniškem jezeru* [The Vanishing Lake – Monograph on Lake Cerknica] (pp. 4–17). Ljubljana: Društvo ekologov Slovenije (in Slovenian with English summary).

Chapter 19
Effect of Mau Forest Clear Cut on Temperature Distribution and Hydrology of Catchment of Lakes Nakuru and Naivasha: Preliminary Study

Petra Hesslerová and Jan Pokorný

Abstract A decrease in total area of tropical forests is considered as a significant factor that influences landscape functioning, including the hydrological cycle; it contributes to climate change, and has many other consequences. This paper presents the extent of deforestation in Nakuru and Naivasha region (Rift Valley, Central Kenya) between the years 1986 and 2005 and its effects upon thermal characteristics of the landscape. Such changes have immense impact on hydrological regimes of the Rift Valley lakes in central Kenya. Multispectral data from Landsat TM, ETM^+ and Terra Aster were used to determinate dense and humid forests as well as their changes over time. A field observation, realized during the "dry" rainy season in October 2008, confirmed evident decline of precipitation and consequent low water discharge from the deforested catchment in the rivers and water level in lakes.

Keywords Deforestation · Regional climate · Remote sensing · Water balance

19.1 Introduction

Lakes Nakuru, Elmenteita and Naivasha are examples of several lakes lying on the floor in eastern African Rift Valley. The first two are shallow alkaline-saline, the third is freshwater. Because of the physical-chemical parameters of water, these lakes provide habitats to large populations of water birds and support highly specialised organisms. High rates of solar radiation and sufficient nutrients supply lead to abundance of phytoplankton. Blue green algae (cyanobacteria) and benthic algae

P. Hesslerová (✉)
ENKI, o.p.s., Třeboň, Czech Republic; Department of Applied Geoinformatics and Spatial Planning, Faculty of Environmental Sciences, Czech University of Life Sciences Prague, Praha, Czech Republic
e-mail: M42barta@seznam.cz

J. Vymazal (ed.), *Water and Nutrient Management in Natural and Constructed Wetlands*, DOI 10.1007/978-90-481-9585-5_19,

are the base of the food chain, influencing fish population and consequently supporting piscivorous bird populations. These lakes provide habitats for several species falling into different conservation categories as threatened, vulnerable and endangered at different levels, e.g. Lesser Flamingo (*Phoenicopterus minor*) and Greater Flamingo (*Phoenicopterus ruber roseus*). Because of their international importance as wetlands and habitats, all lakes have been proclaimed as Ramsar sites. In addition they recharge groundwater, supply drinking water to a substantial part of the population, are used for irrigation, power generation, and are abused for horticulture (Becht & Harper, 2002). Lake ecosystems are very vulnerable and sensitive, depending upon water level fluctuations and precipitation regime. Variations in salinity invoke changes in phytoplankton abundance and consequently in bird populations (Odada, Raini, & Ndetei, 2006).

The water budget of these lakes is maintained by recharge from different sources – hot springs, inflowing rivers, underground recharge, surface runoff and direct rainfall. Increasing human population and stress on the environment leads to large destruction and excessive exploitation of the landscape. Deforestation is considered as the most significant factor influencing ecosystem functioning and has a wide range of other consequences (Foley et al., 2005; Shukla, Nobre, & Sellers, 1990; Lawton, Nair, Pielke, & Welch, 2001; Bruijnzeel, 2005). The Rift Valley has been undergoing large land cover change that has accelerated mainly in the last 20 years and has a huge impact on hydrological regime and climate. One of the most important water towers that recharge several rivers and lake basins in the valley is the Mau Forest. Unfortunately, it belongs to one of the most affected and endangered ecosystems, with serious consequences on water balance of the lakes and surrounding ecosystems.

The Mau Forest Complex (4000 km^2) is referred as one of the largest remaining continuous blocks of indigenous forest in East Africa. A high amount of precipitation is an abundant water source that feeds twelve rivers (Njoro, Sondu, Ng'iro etc.) and a number of systems including not only Lakes Nakuru, Elementaita, Naivasha, but also Lakes Natron, Mara and Victoria. Water from the Mau Forest is estimated to serve more than three million people in Kenya and Tanzania, as well as supporting the surrounding ecosystems. The rural lifestyle of people traditionally inhabiting the area did not contribute to large forest destruction. However recently, imigration has caused large parts of the forest area to be cleared for settlement, agriculture land and teagardens. The stress on the ecosystem was supported by politics, mainly in election periods during the 1980s and 1990s when the promise of "give vote, get land" was popular. Some 39% of the officially gazetted forest has been taken illegally. Such large deforestation has dramatically affected surrounding ecosystems, local climate and the hydrology of the catchment, reducing water flows. The government subsequently took action against further destruction of the forest; between 2004 and 2006 more than a hundred thousand persons were forcibly evicted from the Mau Forest Complex, but no alternative place to live was offered (AI, 2007). A new hydropower plant at Sondu-Miriu constructed on the river of the same name, was not able to produce its 60 MW as planned due to a water shortage. The Japanese Bank for International Cooperation stopped construction of the second hydropower station in the region. In July 2008, the Kenyan Prime Minister Odinga declared that, during the last 10 years, the Mau Forest had lost 100,000 ha of area due to agriculture and illegal cutting; the losses were estimated to be valued at 300 million USD. The Kenyan Government decided to evict 200,000 people from the Mau Forest, and to fence the Forest in order to prevent illegal logging, and restore the natural hydrology of the catchment.

There are several studies (Shivoga et al., 2007; Kitaka, Harper, Mavuti, & Pacini, 2002; Raini, 2009a, 2009b; Mogaka, Gichere, Davis, & Hirji, 2006) dealing with the hydrological response of catchments, river outflows, water quality and hydrological regimes of lakes. A common feature of these papers is the monitoring of land cover/use changes (Baldyga, Miller, Driese, & Gichaba, 2007) and its impact on water quality. There is an agreement that human-induced disturbances resulting from land cover/use changes lead to enhanced erosion, nutrient losses, degradation of water resources, and deterioration of water quality in rivers. Much attention has been paid to Njoro River, which is the main freshwater source for Lake Nakuru (Mathooko, 2001; Karanja, China, & Kundu, 1986; Lelo, Chiuri, & Jenkins, 2005). These studies are mostly targeted on water channels flows. However there are other factors influencing many processes in the landscape. The land cover temperature and its amplitudes are indicators, determining the balance between water cycling, energy and matter flows within the landscape. There are both direct and indirect effects of temperature on the water cycle (rainfall distribution and intensity, its source, water quality, aquifer level), local and global climate (circulation systems, wind, convection flows, and air humidity), and matter loss. Land cover type, its structure and condition (e.g. chlorophyll content and wetness) affect the temperature and its distribution. Depending on these parameters, the functional landscape dissipates solar energy mainly via evapotranspiration – ideally 80% is transformed into latent heat that does not contribute to rises in temperature. The daily temperature amplitude is low. Permanent vegetation (forest) keeps water in the soil, prevents erosion and slows down the decomposition rate of organic matter in soil. Matter losses (both erosion particles and dissolved ions) from forested catchment are lower than from agricultural and drained land (Ripl, 2003). Remote sensing data and techniques enable monitoring of land cover and its changes in time. Processing of Landsat and Terra Aster multispectral satellite images provide additional assessments of distribution of radiation temperature and surface wetness that are results of feedback landscape functioning (Procházka, Včelák, Wotavová, Štíchová, & Pechar, 2006).

Distribution of surface temperature can be observed by the thermal infrared sensor that scans the radiation in wavelengths 3–14 μm, approximately. This temperature is actually recoded thermal radiation of the Earth's surface. These wavelengths identify the emitted radiation of objects within reflected solar radiation. The objects radiate energy as a function of their temperature. This emitted energy is an external manifestation of an object's energy state and that is remotely sensed using thermal scanners and satellite sensors. The emitted energy is used to determine radiation temperature of the surface and there is a relationship between kinetic (real) and radiation temperature (see basic laws of thermodynamics – Wien's displacement law, Stefan-Boltzmann law, Kirkhof's law). To determine energetic efficiency of landscape and the way in which it is dissipated, multispectral and thermal remote sensing data from Landsat TM, ETM+ or Terra Aster can be used. Landsat channel TM6 (10.4–12.5 μm) scans its thermal radiance of the Earth's surface, with a spatial resolution 120, 60 m eventually. In case of Terra Aster data, the spatial resolution of five thermal channels (ranging from 8,125–11,65 μm) is 90 m. The main significance is that there is no need for data interpolation and in addition the real (kinetic) temperature of the surface can be derived.

The area of Mau Forest has been visited several times within the projects Fingerponds (INCO-DEV-CT-2001-10037) and BOMOSA (INCO-CT-2006-032 103). The changes of land cover and rapid decrease of water level in the lakes and rivers has been very significant for the last 10 years. These circumstances adversely affect the life style of rural community insomuch that the government has been forced to evince an interest and adopt very unpopular steps to solve the problem. These have resulted in high social tensions and riots.

The aim of this paper is to target the analyses and relationships of vegetation cover (forest occurrence) and temperature that are the results of evapotranspiration, and thus the water cycle. We assume such a vast deforestation will have an effect in a rise of temperature of these areas. Satellite images, available since the 1980s, have been analyzed to get information about these significant environmental changes.

19.2 Data and Methods

Landsat satellite data are available from the 1980s and enable evaluation of thermal changes of landscape as consequences of land cover changes. Unfortunately because of the discontinuity of the Landsat mission in 2001 for the African region, substitution of other data (e.g. Terra Aster) to get up-to-date information is essential. Landsat images to which we do have access cover the area of Eastern, Western, South West, Southern Mau, Transmara, Maasai Mau, Mau Narok, West Molo and Molo and others. Deforestation impacts upon the landscape were observed on broader surroundings of the Mau Forest, delineated as a subset of 124 × 125 km (15,500 km^2) from the scenes mosaic 169–060 and 169–061. Time of acquisition was on 28 January 1986 and 27 January 2000. These dates correspond to the pre-green period. To complete the assessment of as much up-to-date data as possible, Terra Aster data were used to analyze the situation on 5 August 2005 in the most exploited Eastern Mau Forest region. The scene size is 60 × 60 km (3600 km^2).

Supervised classification (algorithm maximum likelihood) was used to discriminate dense and humid forest in case of both data types. The classifier considers both the variances and covariances of the class signatures. The assumption is Gaussian distribution of a training class, and classes are characterized by the mean vector and the covariance matrix. Each pixel in the image is valued by statistical probability to determine the membership of the pixels in the class. Each pixel is assigned to the class to which it has the highest probability of being a member. The maximum likelihood classifier is considered to give relatively accurate results.

Dense forest, detected in 1986, was created by step-by-step classification. The class was divided into two sub-categories. A maximum likelihood algorithm was used, threshold value 10%, channels used TM 4-5-2. The same procedure was applied to the scene from 2000. To get final results, three sub-classes of forest were discriminated. The first two sub-classes were classified by the same parameters, for the third, channels ETM 4-5-7 were used, threshold by 15%. Terra Aster image classification from year 2005 was done by two forest sub-classes and the following parameters: maximum likelihood algorithm, threshold 10%, channels 2-3-4-5 and 1-3-4-5. The major problem was a thick cloudiness in the bottom part of the image that could not be removed.

The maps of the relative temperatures were acquired from the cumulated histogram of band TM6 (ETM+ 6 and AST 15 respectively) values using the equal-area method. Five temperature classes were defined (each containing ca 20% of pixels): lowest temperature – low – middle – high – highest temperature. The temperature evaluation did not use absolute values (in degrees Celsius) since they are not suitable for comparisons between different dates. Uses of relative temperatures enable comparisons of scenes acquired under different conditions.

19.3 Results and Image Interpretation

The information provided in Fig. 19.1a–c shows main land cover types in the selected area. RGB synthesis of channels TM, ETM+ 4-5-3 and Aster channels 3-4-2 was used. These combinations involve red visible, near and mid-infrared parts of electromagnetic spectra, and therefore they enhance most information about land cover. The Mau Forest complex is displayed in reddish brown, lakes as black, dark or navy blue. Green tones indicate pastures or grassland with dry vegetation, cyan and represent bare grounds.

The results of satellite image classification (Fig. 19.2a, b) show the extent of the Mau Forest in the years 1986 and 2000. Visual comparison of the images enables detection of time changes. The decrease of total area of dense and humid forests is evident. The most affected regions are Eastern, South West and Maasai Mau complexes. During a 14 year period, 55,000 ha were excised (from the total area assessed). In 1986 the forests covered 400,000 ha in 2000 it was 345,000 ha.

The Terra Aster image (Fig. 19.2c) displays mainly the Eastern Mau Forest complex and part of the Maasai Mau that was unfortunately covered by cloudiness. However it was possible to extract the total area changes of Eastern Mau. From 48,00 ha in 1986, in 2000 only 36,000 ha remained. Five years later the total area sharply decreased by 19,500 ha (16,500 ha remained). There is an opposite trend in a few areas, where the expansion of forest vegetation was detected. This could be caused by the fact that the satellite is able to detect only forests that are more than 10 years old. Slow reforestation of these areas (probably establishment of plantation forests) could be assumed.

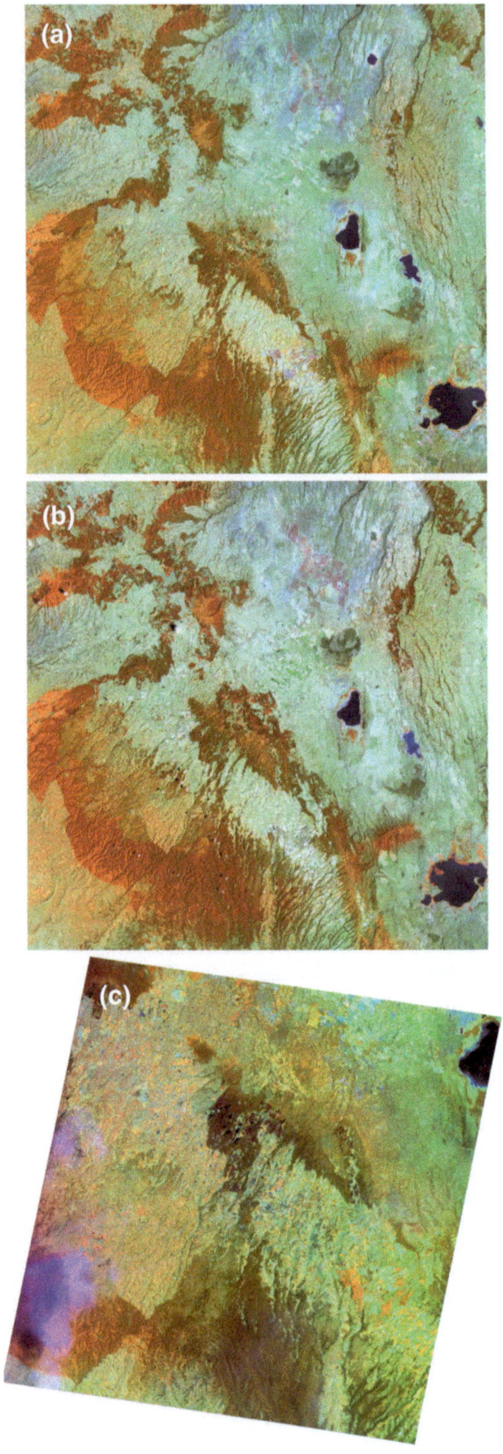

Fig. 19.1 Land cover of the Mau Forest region in (**a**) 1986, (**b**) 2000 and (**c**) 2005. Forests appear in this RGB syntheses *dark red* (nearly brown), non-forest vegetation as *red*, dry non-forest vegetation *green*, by cyan bare grounds are displayed and water bodies nearly *black* or *dark blue* hues. *Violet* on Terra Aster image (**c**) indicates clouds

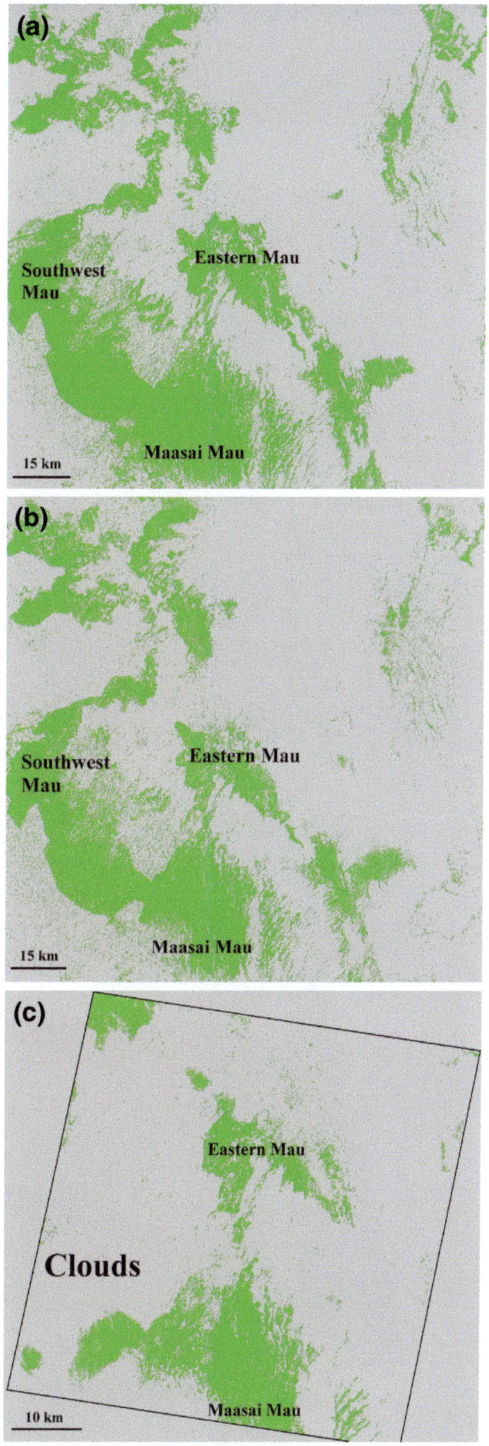

Fig. 19.2 Total area of dense and humid forests within the Mau Forest region in the years (**a**) 1986, (**b**) 2000 and (**c**) 2005. *Green colour* indicates the forest, *grey* is the background

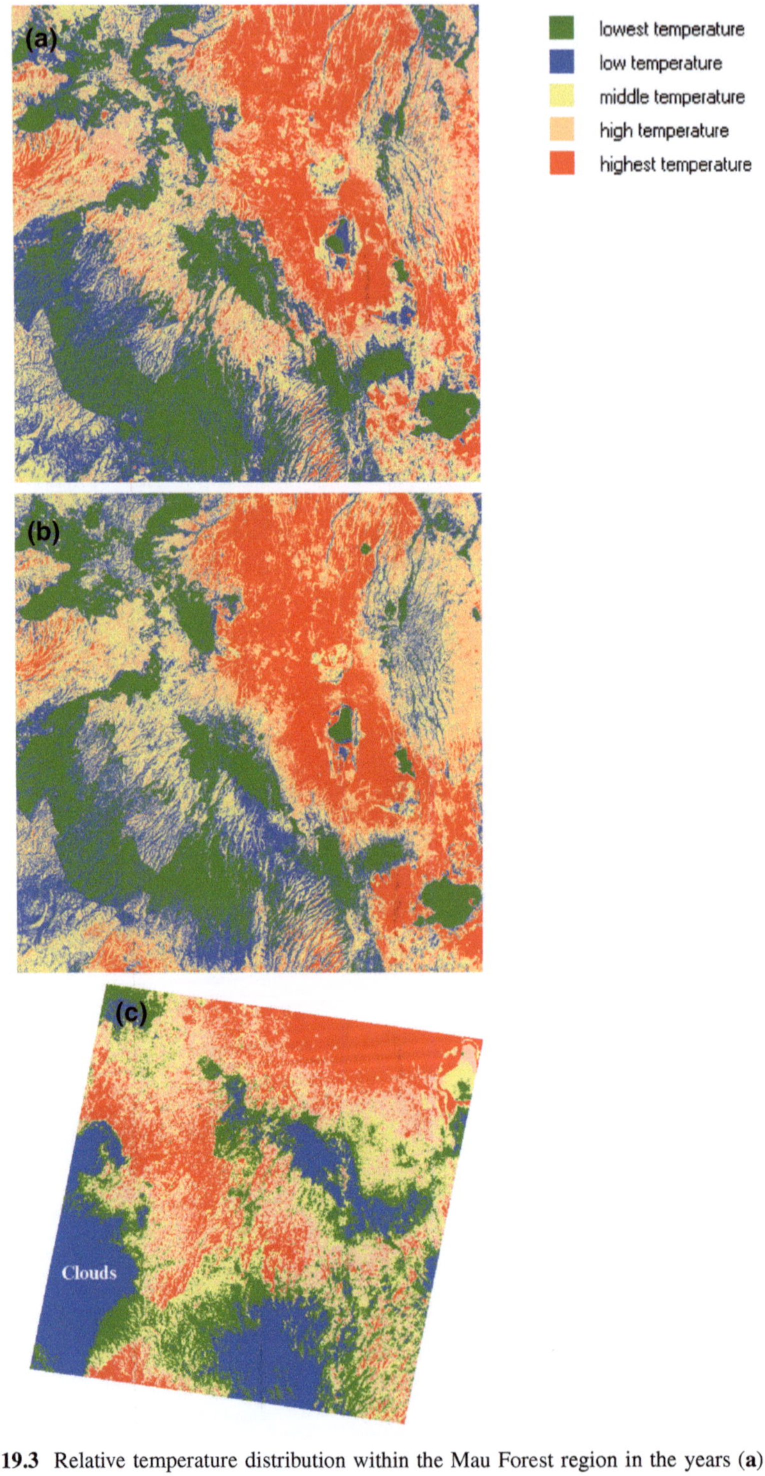

Fig. 19.3 Relative temperature distribution within the Mau Forest region in the years (**a**) 1986, (**b**) 2000 and (**c**) 2005

Such extensive changes of forest cover significantly influence thermal characteristics of the landscape. The temperature distribution (Fig. 19.3a, b) is affected by land cover (Procházka et al., 2006). The lowest temperature categories are bounded with dense and humid forests, or with lakes. The same is evident from thermal Terra Aster data (Fig. 19.3c), where in addition the lowest temperature values cover the area shaded by cloudiness. In this scene, the temperature contrast between forested and non-forested areas is very sharp. Because of total removal of scattered trees in the neighbouring Eastern Mau complex, the zone of highest temperature class is directly adjacent. Unfortunately the Aster scene does not cover all three lakes (Nakuru, Naivasha and Ementaita). Lake Nakuru is surrounded by an extremely large "heat island", without any fragmentation. The tiny fringes of forest cannot cool the landscape; in addition there is a sharp increase of water temperature in the lake in comparison with 1986. Disappearance and rarefaction of vegetation is accompanied by temperature increases. It can be observed in areas with sparse tree cover where indigenous dense forest has been cleared. If the surface is cleared of well-watered vegetation, this has an effect in rapid temperature increase. Clearing of well-watered vegetation results in a rapid temperature increase. The main causes of this are low evapotranspiration rates. As a result incident solar radiation is converted mainly into sensible heat; latent heat (that cools the surface) is in the minority. The Rift Valley area with large lakes is affected especially by this effect. Lack of functional vegetation (forests, wetlands) is evident from Fig. 19.1a–c in this area, resulting in vast areas of overheated landscape.

A significant rise of temperature is observed mainly in the areas of deforestation (Eastern Mau and south of Maasai Mau). The same trend is typical for the Rift Valley, the most affected region is the area between lakes Nakuru and Naivasha. Especially the north shore of Lake Naivasha is affected by a rapid rise of temperature. Positive trends of slight temperature decrease are typical for those areas, when compared to year 1986, but still remains non-forest vegetation. It has proved that even fractional and scarce non-forest vegetation has a cooling effect on landscape, compared to bare grounds, e.g. in the Rift Valley.

19.4 Conclusion

There are no arguments in defence of what has happened. The changes of local climate and the hydrological regime in Nakuru and the Rift Valley region cannot result only from a rain shadow of the Mau Escarpment, but from global climate change and population growth accompanied by rising exploitation of water resources (Mogaka et al., 2006). As shown by satellite images, forest excision on such a large scale as it has been happening in the case of the Mau Forest has induced a rapid decrease of humidity, guided by a consequent rise of temperature. And not only does this appear on clear-cut areas, where it is most evident, but also in whole catchments where the changes have been happening.

A remarkable decrease of water discharge and temperature rise in the deforested areas of the Mau Forest provides evidence that proves the positive effect of forests in the water balance of larger catchments. Consequently it is shown that human activities and, to be concrete, water and vegetation management, directly causes climate change on a regional scale.

Acknowledgement Many thanks to our African partners involved in the BOMOSA project for their support during ground truthing, for comments and reviews – Kenya Marine & Fisheries Research Institute, Sangoro Aquaculture Station, Kenya Ethiopian Institute of Agricultural Research, NFLARR, Sebetha, Ethiopia, Egerton University, Egerton, Kenya,Ministry of Agriculture, Animal Industry and Fisheries, Dept of Aquaculture, Entebbe, Uganda, Moi University, Dept of Zoology, Eldoret, Kenya, Kenyan Ministry of Livestock and Fisheries Development, Sagana Fish Farm, Kenya, as well as our European partners from University of Natural Resources & Applied Life Sciences, Vienna, Austria, University of Bologna, Italy, Austrian Academy of Sciences, Institute for Limnology, IPGL-Office, Mondsee Austria. Financial support was provided by BOMOSA (INCO-CT-2006-032103) and by the Ministry of Education, Youth and Sports of the Czech Republic NPV 2B06023 projects. Special thanks to Prof. Patrick Denny for valuable comments on the issue and for correction of the paper. The source for a free Landsat data set was the Global Land Cover Facility (GLCF), www.landcover.org.

References

Amnesty International. (2007). Kenya nowhere to go: forced evictions in mau forest briefing paper, May 2007.

Baldyga, T. J., Miller, S. N., Driese, K. L., & Gichaba Ch. M. (2007). Assessing land cover change in Kenya's Mau forest region using remotely sensed data. *African Journal of Ecology, 46*, 46–54.

Becht, R., & Harper, D. M. (2002). Towards an understanding of human impact upon the hydrology of Lake Naivasha, Kenya. *Hydrobiologie*, *488*, 1–11.

Bruijnzeel, L. A. (2005). Tropical montane cloud forest: a unique hydrological case. In M. Bonell, & L. A. Bruijnzeel (Eds.), *Forests, Water and People in the Humid Tropics* (pp. 462–483). Cambridge: UNESCO, Cambridge University Press.

Foley, A. J., DeFries, R., Asner, G. P., Barford, C., Bonan, G., Carpenter, S. R., et al. (2005). Global consequences of land use. *Science, 309*, 570–574.

Karanja A. K., China S. S., & Kundu P. (1986). The influence of land use on Njoro River catchment between 1975 and 1985. *Proceedings of the 3rd National Workshop on Soil and Water Conservation*, Nairobi, Kenya, September, 1986, Department of Agricultural Engineering, University of Nairobi, and Swedish International Development Cooperation Agency.

Kitaka, N., Harper, D. N., Mavuti, K. M., & Pacini, N. (2002). Chemical characteristics, with particular reference to phosphorus, of the rivers draining into Lake Naivasha, Kenya. *Hydrobiologia, 488*, 57–71.

Lawton, R. O., Nair, U. S., Pielke Sr., R. A., & Welch, R. M. (2001). Climatic Impact of Tropical Lowland Deforestation on Nearby Montane Cloud Forests. *Science, 294*, 584–587.

Lelo, F. K., Chiuri, W., & Jenkins, M. W. (2005). Managing the River Njoro Watershed, Kenya: Conflicting laws, policies, and community priorities. *International workshop African Water Laws: Plural Legislative Frameworks for Rural Water Management in Africa*, 26–28 January 2005, Johannesburg, South Africa.

Mathooko, J. M. (2001). Disturbance of a Kenyan Rift valley stream by the daily activities of local people and their livestock. *Hydrobiologia, 458*, 131–139.

Mogaka, J., Gichere, S., Davis, R., & Hirji, R. (2006). *Climate variability and water resources degradation in Kenya.* Washington, DC: The World Bank. Available online from http://books.google.cz/books?id=jyNHmgONtk0C&printsec=frontcover [accessed online on 5 August 2009].

Odada, E. O., Raini, J., & Ndetei, R. (2006). Lake Nakuru. experience and lessons learned brief. Available online from http://www.ilec.or.jp/eg/lbmi/pdf/18_Lake_Nakuru_27 February2006.pdf [accessed online on 1 September 2009].

Procházka, J., Včelák, V., Wotavová, K., Štíchová, J., & Pechar, L. (2006). Holistic concept of landscape asessment: Case study of three small catchments in the Šumava mountains. *Ekológia, 25*, 5–17.

Raini, J. A. (2009a). Impact of land use changes on water resources and biodiversity of Lake Nakuru catchment basin, Kenya. *African Journal of Ecology, 47*(1), 39–45

Raini, J. A. (2009b). *Long-term trends in water quality, water quantity and biodiversity at Lake Nakuru, Kenya*. Fmr Project Ecologist, World Wide Fund for Nature, Lake Nakuru Project, (P.O Box 33 Nakuru, Kenya) Available from http://iodeweb1.vliz.be/odin/bitstream/1834/2071/1/WLCK-57-62.pdf [accessed on 24 August 2009].

Ripl, W. (2003). Water: Bloodstream of the biosphere. Philosophical transaction of the royal society of London (B). *Biological Sciences, 358*, 1921–1934.

Shivoga, W. A., Muchiri, M., Kibichi, S., Odanga, J., Miller, S. N., Baldyga, T. J., et al. (2007). Influences of land use/cover on water quality in the upper land middle reaches of River Njoro, Kenya. *Lakes & Reservoirs: Research and Management, 12*, 97–105.

Shukla, J., Nobre, C., & Sellers, P. (1990). Amazon deforestation and climate change. *Science, 247*, 1322–1325.

Waki Commission of Inquiry into Post-Election Violence final report, 15 October 2008.

Chapter 20
Wetland Analyses in Lake Kyoga Region and Kamuli District in Uganda

Petra Hesslerová and Jan Pokorný

Abstract Fluctuations of water level and the hydrological regime of Lake Kyoga influence the dynamics and occurrence of wetlands and water reservoirs within its surroundings. During the last 20 years many of them have become nearly entirely overgrown and afflicted by siltation. The aim of the Ministry of Agriculture, Animal Industry and Fisheries in Uganda is to use these waters and wetlands for extensive fish production. Landsat and Terra Aster satellite data were used to analyze the occurrence of free water and wetlands between the years 1986 and 2006. Multispectral classification of these data provides information on how to detect localities potentially suitable for free water restoration and establishing of the BOMOSA fish farming system. The results have shown significant decrease of wetlands and water bodies within the analyzed time horizon.

Keywords BOMOSA fish farming · Landsat · Macrophytes overgrowing · Reservoirs

20.1 Introduction

One of the major issues of many African countries is to ensure a sufficient amount of nutritious food, especially proteins for the population. Expansion of the fishery industry is one of the possible solutions, as it can provide large amounts of animal proteins. One way in which this issue is being addressed in the East African region is by integrating the BOMOSA cage fish farming system in reservoirs, ponds and temporary water bodies. In this way, fish farming can provide a great amount of nutritious food as well as self-employment for local communities. The BOMOSA

P. Hesslerová (✉)
ENKI, o.p.s., Třeboň, Czech Republic; Department of Applied Geoinformatics and Spatial Planning, Faculty of Environmental Sciences, Czech University of Life Sciences Prague, Praha, Czech Republic
e-mail: M42barta@seznam.cz

J. Vymazal (ed.), *Water and Nutrient Management in Natural and Constructed Wetlands*, DOI 10.1007/978-90-481-9585-5_20,

project aims to validate this farming method in selected localities, and later to promote it in similar aquatic ecosystem locations in Eastern Africa. The cage fish farming system was originally proposed to be established in open-water reservoirs. However, Africa faces a lack of eligible water bodies. Even though Uganda is an inland country, in comparison with the other states of Eastern Africa it could be considered as well-watered country. Due to the fact that the central part of the country is formed by the East Africa plateau, nearly one-fifth (40,000 km^2) of this territory is swampy. Lake Kyoga and the surrounding basin dominate central Uganda. The Kamuli district, the region where the first two model BOMOSA plots were established, lies within the swampy area of Lake Kyoga. These model plots are situated in Ndolwa and Busoga High School near district town Kamuli that has replaced a Kasolwe plotendangered by high-siltation. This process is not only natural; human activities play a substantial role. Especially during dry seasons the wetlands are cultivated, consequently eutrophication appears, siltation and overgrowing speed up.

Because of positive acceptance of the BOMOSA project in local communities, the idea has spread to bring the cage fish farming system to more communities within the Kamuli district. Interviews with local people revealed that there used to be many open water bodies, however in the last 20 years many have become overgrown with aquatic vegetation. The aim, supported by the Ugandan Ministry of Agriculture, is to use these overgrowing water bodies and wetlands for fish farming and restore free water. In order to get more water bodies for BOMOSA cage fish farming, macrophyte vegetation must be harvested and recent sediment partly removed.

The question is how to identify appropriate locations in which to restore water bodies for cage farming. Remote sensing methods and archive satellite images offer an objective method of looking back on a prospective location's history. The main task of the project described in this paper was to detect free water from Landsat satellite images and wetlands, in which water still prevails over macrophytes. Comparison of two Landsat scenes from 1986 to 2001 enabled us to detect the extent of free water and wetlands and to make change-over-time analyses. It is anticipated that through archive images it will be possible to detect water reservoirs or wetlands that are afflicted with overgrowth. The removal of macrophytes from such reservoirs could provide new water resources for fish farming, without any negative effect on the water regime of surrounding wetlands. Although the image from the Terra Aster satellite with which we began covers only part of the Kamuli district (vicinity of Kamuli town), we were able to acquire the full information we needed from data acquired in the year 2006.

20.2 Geography of Lake Kyoga Region

The primary inflows into Lake Kyoga are the Victoria Nile and water courses originating in the Mount Elgon area, and the lake is drained through the Victoria Nile to Lake Albert. Lake Kyoga is a shallow basin lake (depth 3–5.7 m) with several extensions, including Lake Kwania, Lake Bugondo, and Lake Opeta (which in the dry season could be considered as a separate lake). A bimodal rainfall regime brings precipitation in two seasons. The first starts usually at the turn of February and March, the second starts in July and ends in October. Precipitation rates in the central part of Uganda vary between 1100 and 1500 mm (CIA, 2007). The amount of precipitation received during the rainy season influences the total area of Lake Kyoga that could range from 1600 km^2 to 4100 km^2, as well as the area of the Finger Lakes and their extension to marshy areas (ILEC, 2001). The Kamuli district is situated south of the Lake Kyoga swamp zone, and is bordered by Kayunga District to the west, Kaberamaido District and Soroti District to the north, across Lake Kyoga, and on the eastern part borders on Kaliro, Iganga and Jinja District.

As well as other African Lakes, Lake Kyoga and its tributaries are affected by overgrowing of macrophytes. Areas less than 3 m deep, especially swampy shorelines are continuously covered by emerged macrophytes such as Papyrus (*Cyperus papyrus*), Hippo grass (*Vossia cuspidata*), Cattails (*Typha* sp.), and Water hyacinth (*Eichhornia crassipes*); floating mats of Water lilies (*Nymphaeaceae*) are more than common. Excessive rains in 1997–1998, accredited to the El Niño effect, led to increased water-level. As a consequence, papyrus, water lilies and hyacinth mats have been spread in forms of floating islands over the lake. In addition macrophytes have accumulated by the outlet of the lake and caused a rapid increase of water level and floods. (ILM, 2004)

20.3 Data and Methods

The analyses of free water and wetlands were done from Landsat TM and ETM+ data. Comparison of two scenes from 1986 to 2001 enabled us to detect the extent of free water and wetlands and to make change-over-time analyses. The dates of acquisition were 10 January 1986 and 27 January 2001, both representing the dry season. The scenes 171-069 (Lake Kyoga) have been processed and cover 35721 km^2 (189 $\times$ 189 km). The area involves nearly all of Lake Kyoga (up to Busembatia town, except the most eastern part), reaches the town Lira on the north, has a lake outlet on the west and nearly touches Jinja on the south. The 2001 scene is partly covered by cumulus clouds and their shades are incorrectly classified as water or wetlands. This effect could not been removed from the data, nor could haze in the1986 scene.

To complete as much as possible an up-to-date situation of free water and wetlands in the Kamuli district, data from the Terra Aster sensor has been used. This was because Landsat data for the African region were not received after 2001. The Aster scene 4225 km^2 (65 $\times$ 65 km) displays surroundings of Kamuli town, up to the east part of Lake Kyoga and was acquired on 27 December 2006.

20.3.1 Free Water Detection

Both Landsat scenes have been processed by similar methods; the details are listed below. As a first step, free water was detected using a BOMOSA method-

Normalized Difference Vegetation Index approach (Hesslerová, Šíma, & Pokorný, 2009). The accuracy was decreased by processing of whole scenes, which is, due to high heterogeneity of land cover, an unusual procedure. However the purpose was to demonstrate one of many advantages of remote sensing data, processed over large areas.

The method is based on the radiometric enhancement of satellite data using a process called image rationing. In order to distinguish water bodies relevant to the East-African landscape, we employed the Normalized Difference Snow Index – NDSI (Dozier & Marks, 1989), which is calculated using data from spectral channels Landsat TM/ETM+ and adjusted for MultiSpec software in the following way:

$$\mathrm{NDSI} = (\mathrm{TM2} - \mathrm{TM5})/(\mathrm{TM2} + \mathrm{TM5}) \times C.$$

The final map of free water is derived by thresholding of an NDSI image, simultaneously using RGB synthesis of the original Landsat channels 457. Finally, the output NDSI image is recoded into the water bodies' values and background and this is how the final binary mask of water bodies is defined.

The multiplication coefficient used for the 1986 scene was $C = 265$ and recoded mask of water bodies' digital number (DN) $=< 2; 255 >$. The values for the 2001 scene are the following: $C = 247$ and DN values have been thresholded DN $=< 47; 255 >$ as free water and DN $=< 3; 46 >$ as water with emerged macrophytes.

In the case of Terra Aster data a different approach has been used. Aster band 3 N records reflected near-infrared radiation (0.76–0.86 μm) and is essential mainly in vegetation analysis. Dense and green vegetation reflects, in this spectral interval, a significant amount of incident radiation (usually displayed by light hues of grey scale). Water objects appear dark. By very cautious tresholding of this band, it is possible to extract water bodies. The principle of tresholding is the same one, as in the case of an NDSI approach.

A tresholded mask of water bodies, using Terra Aster data is the following: DN $= < 1; 49 >$ free water, DN $=< 50; 69 >$ water with emerged macrophytes.

20.3.2 Wetlands Detection

Information about wetlands was derived by supervised multispectral classification methods applied to both types of satellite data. The objective of image classification is to automatically categorize pixels in an image into thematic categories. Multispectral classification utilizes pixel spectral information as the basis. The analyst attempts to locate specific sites in the remote sensed data that represent homogeneous examples of known objects (in these case wetlands). These areas are commonly referred to as training sites. The spectral characteristics of these sites are used to train the classification algorithm. Multivariate statistical parameters (means, standard deviations, covariance and correlation matrices) are calculated for each training site. Every pixel is evaluated according to these parameters and assigned to the class of which it has the highest likelihood of being a member (Lillesand & Kiefer, 2000).

The maximum likelihood algorithm has been used in most cases of classification. The classifier considers both the variances and covariances of the class signatures. A Gaussian distribution of a training class is assumed, and classes are characterized by the mean vector and the covariance matrix. Each pixel in the image is valued by statistical probability to determine the membership of the pixels to the class. Each pixel is assigned to the class to which it has the highest probability of being a member. The maximum likelihood classifier is considered to give relatively accurate results. In one case Fisher Linear Likelihood was used. The principle is similar to the Maximum Likelihood case; the main difference is in having a common covariance matrix for all classes (Lillesand & Kiefer, 2000).

This advanced processing category was used to distinguish and delineate wetlands. During the classification process there emerged the problem of discrimination of dense aquatic macrophyte vegetation that has a similar spectral response to the other non-forest terrestrial vegetation. However the main purpose was to extract wetlands, with apparent predominance of water. The methods and/or parameters that differed image by image were as follows:

The class of wetlands, detected in 1986, was created by step-by-step classification. Dependent on different degrees of water logging and overgrowing, the wetlands class was classified in four sub-categories. A maximum likelihood algorithm was used, with threshold value 10%, channels used TM 6-4-5-2; in the case of the last sub-category, channel TM7 was added. The same procedure was applied on the scene from 2001. To get a final result, only two sub-classes of wetlands were sufficient. The first sub-class was classified by the Fisher Linear Likelihood algorithm, thresholded by 20%, channels used ETM+ 7-4-5-2. In the second case, maximum likelihood was employed; threshold 15%, channels used ETM+ 4-5-2. To indicate shadows caused by cumulus clouds, which are incorrectly classified as free water, a separate class was determined. Parameters used: maximum likelihood, threshold 10%, channels ETM+ 4-5-2. In most cases the shape of the cloud helps to indicate incorrect water bodies. Aster image classification from 2006 was done only by one wetland class and with the following parameters: maximum likelihood algorithm, threshold 5%, channels 3-4-5-6.

In all cases final thematic images were composed by adding of individual wetland sub-classes and thematically recoded into two basic categories of wetlands: high wetness (water prevails) and wet vegetation (predominance of vegetation). Complete maps of places where potentially new BOMOSA plots could be established were constructed by adding images of wetlands, gained as results of multispectral classification and free water, derived from NDSI images.

Changes-over time of images (from 1986 to 2001) can be done easily by visual comparison. The other method of changes detection is visualization of individual images in an RG (red-green) colour system. This superimposition method was used to capture changes in free water. The interpretation is based on a combination of red and green colours that indicate the presence of water in selected terms. Because of different characteristics of Terra Aster data and scene size, it was not possible to concurrently analyse results from 2006.

20.4 Results

There are several ways to use remote sensing data for evaluation of wetlands in the Lake Kyoga region. Some possibilities from the easiest, available to anybody to more sophisticated are listed below.

20.4.1 Visual Interpretation of RGB Synthesis

The first analyses of wetlands can be done through simple RGB image visualization and interpretation. Recommended channels combination is 4-5-7 that enhance vegetation and water; nearly black colour indicates water, dark blue wetlands, where water prevails and hues of red and orange display vegetation cover, cyan bare grounds. We assume areas displayed by dark hues of blue might be suitable for cage fish farming as open water has been seen there recently. Visualization in appropriate geographical software offers a possibility to find the geographical coordinates of selected localities (if the image has been georeferenced), followed by usage of GPS technology enabling one to search the proper places in the terrain.

Side-by-side visualization of RGB 4-5-7 synthesis of both scenes 1986 (Fig. 20.1a) and 2001 (Fig. 20.1b) displays an evident decrease of wetlands between both years. According to satellite images, it seems that large areas have dried out and the landscape has lost water. This is evident in nearly all parts of the area assessed, despite the fact that the region was affected by heavy rains and floods in 1997–1998. The reason for drainage could be population pressure and increasing demand on arable land. The population growth rate is estimated at 5.1% per year in the Kamuli region. A discussion including local inquiries and GPS measurement of the current state would be helpful for the verification of results. The same situation is apparent from the Terra Aster scene in 2006 (Fig. 20.2). In this case RGB synthesis of channels 3-4-5 was used. There is an evident structure of seasonal water courses on all images that may be flooded during the rainy season; however neither of the images recorded this situation.

Multispectral images also very clearly display the extent of dense macrophyte vegetation. Shorelines are fringed with papyrus and other swamps, sometimes forming a belt of several kilometres width between land and the open water. As it was mentioned above, because of similar spectral responses, it is very difficult to separate dense shoreline beds of papyrus and other macrophytes from the rest of non-forest vegetation. However this situation can be monitored through the shape of the stands. In RGB synthesis, 4-5-7 are displayed as reddish-orange-brownish fringes of watercourses and lakes, or greyish green areas. The shape is important, because some have the same colour as other vegetation.

The increase of water level due to heavy rains at the end of the 1990s is still apparent mainly in the fingers of certain lakes, from which large mats of macrophytes were dislodged and cumulated by the Lake Kyoga outlet. This situation is detectable from the 2001 image; unfortunately the Aster image from 2006 does not capture this area. Enlargements of free water areas in main inflows into Lake Kyoga (especially in its eastern parts) and the Victoria Nile are also evident on both Landsat images.

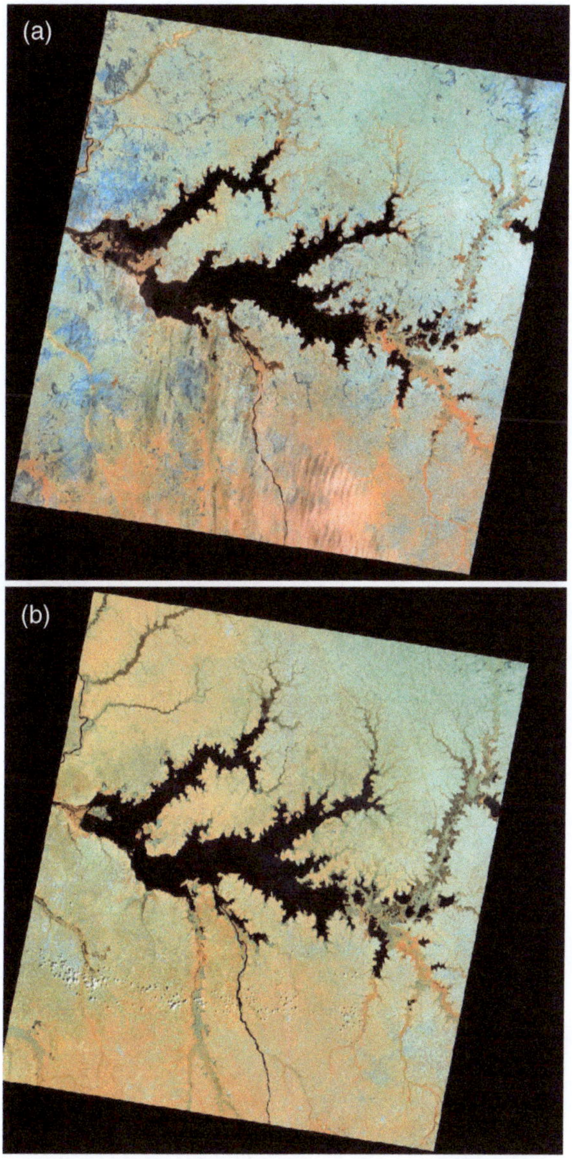

Fig. 20.1 Land cover of Lake Kyoga region on (**a**) 10 January 1986 and (**b**) 27 January 2001. RGB synthesis of Landsat channels 4-5-7. Scene size 189 × 189 km. Nearly black colour – water, dark blue – wetlands, where water prevails, red, orange – vegetation, cyan – bare grounds

20.4.2 Multispectral Classification and Free Water Detection via Normalized Difference Snow Index

Thematic information about wetlands with predominance of free water was gained from the original multispectral image by supervised classification. While the RGB synthesis of original multispectral channels displays the whole land cover, the

Fig. 20.2 Land cover of Kamuli district, Ndolwa and Kasolwe Dams surroundings. Terra Aster image acquired on 27 December 2006. RGB synthesis of channels 3-4-5. Scene size 65 × 65 km

classification results enhance only selected features. Figure 20.3a represents the situation in water and wetlands distribution in 1986, Fig. 20.3b in 2001 and Fig. 20.4 in 2006. It is very easy to compare the situation in the past and today. Changes are significant, many water reservoirs are overgrown now, or during 15 years nearly all wetlands dried out. Objections can be addressed based on different weather conditions, however due to nearly the same date of acquisition, one cannot expect dissimilarities. It would be conceivable to consider dissemination of BOMOSA plots into such places, where wetlands still occur, however the degree of overgrowing has increased. In order to restore these water bodies for BOMOSA cage fish farming, wetland vegetation should be harvested. In addition it will be essential to verify up-to-date conditions of selected localities by ground observation and GPS measurements. The selection of suitable localities is a task for local authorities. As it was mentioned above, finding potential geographical positions is possible from all images in geographical information systems software.

The issue of Lake Kyoga's hydrological regime is very complicated and ambiguous. The morphology of the flat area between Lakes Victoria and Kyoga plays a substantial role in flooding and wetlands dynamics. Lake Kyoga is very prone to fluctuations of water level and has a quick response to precipitation amounts. In addition, the rise of water level can be caused by accumulation of floating mats of aquatic vegetation. Because of nearly zero level change tolerance, even a slight increase of the water level of Lake Kyoga can produce flooding of vast areas. Due to these facts, the extension of wetlands in the area is very changeable and unpredictable.

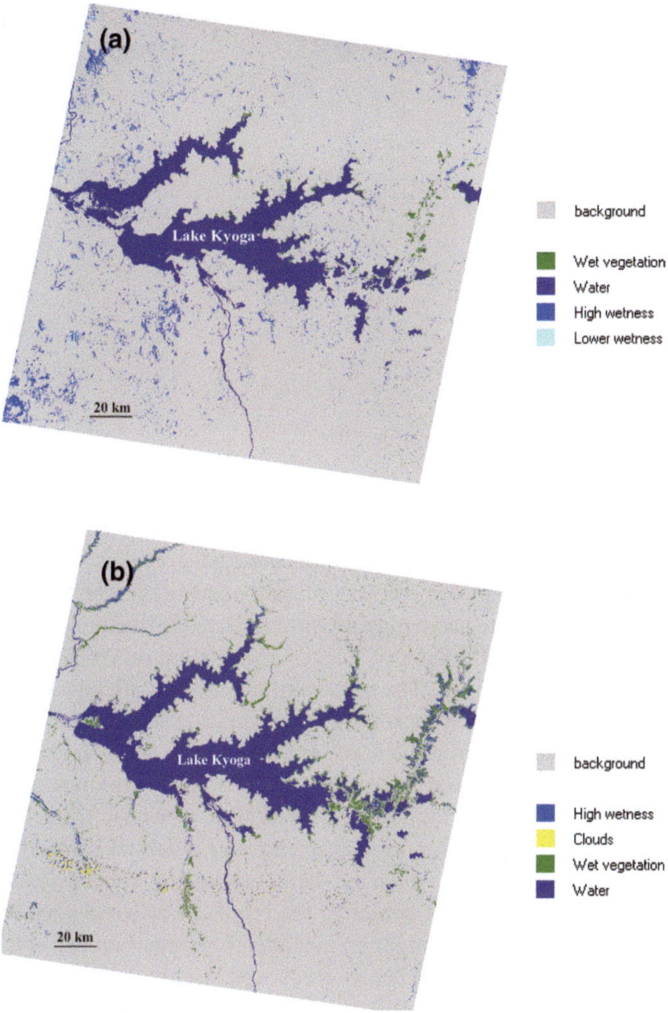

Fig. 20.3 Thematic image derived from Landsat multispectral data by supervised classification methods. Water and wetlands distribution in Lake Kyoga region in (**a**) 1986 and (**b**) 2001

20.5 Conclusions

The comparison of three satellite scenes from the years 1986, 2001 and 2006 enabled us to evaluate changes in occurrence of water reservoirs and wetlands and overgrowing by macrophytes. Unfortunately, because of a dense cloud cover, good quality multispectral data for tropical areas are usually available only for the dry season. The method used as a mapping tool of water bodies was developed for the purposes of BOMOSA fish farming. In addition, methods of supervised classification provided information about wetlands distribution in the Lake Kyoga

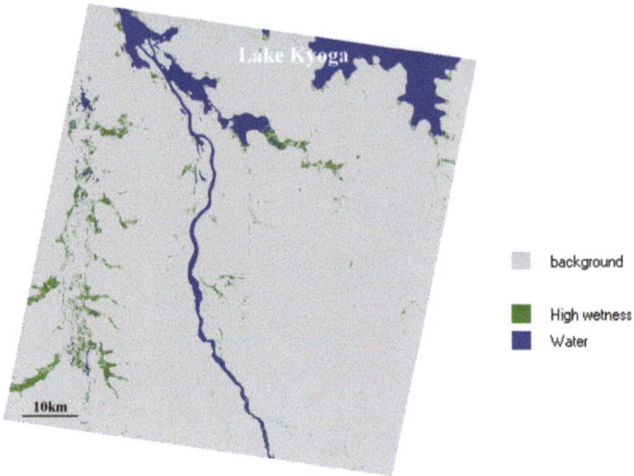

Fig. 20.4 Thematic image derived from Terra Aster data by supervised classification methods. Water and wetlands distribution in Kamuli district in 2006

region. The combination of methods contributes to evaluation of "water potential" of a selected region. The methods use all advantages that generally provide remote sensing data; they make it possible to detect water bodies in huge areas (several thousands of sq km) in one moment of data acquisition, and by one objective method can observe remote and barely accessible areas. In addition, due to the large data archives and repeated regime of images acquisition, one is able to look back and monitor changes-over-time. The outputs of the method are full-valued GIS layers (Geographical Information System), thus integration with GPS is also possible. What are the constraints of the method? The size of the water bodies that can be potentially detected by this method depends on the spatial resolution of the satellite data. In case of Landsat data, the smallest discernible areas had the size of 0.08 ha. As it has been mentioned above, clouds are another potential problem, especially the shadows cast by the clouds which are then mistaken for water bodies. But shadows are also a problem when it comes to the relief. Even some very steep slopes and bottoms of canyon valleys might be mistaken for water bodies, which makes it difficult to detect small water bodies in areas with highly variable relief. Another factor potentially distorting the results is thick vegetation. The multispectral satellite systems scan the land cover only and cannot penetrate below the thick cover of plants. This makes it very difficult to detect water bodies in some areas. Urban units present a potential problem as well as they normally contain a huge amount of mixels (mixed elements; pixels with mixed spectral information from different objects), and in order to eliminate them, other more sophisticated correction procedures need to be employed.

The aim of this research was not to provide exact and targeted information about localities, i.e., where to establish new BOMOSA sites. The main contribution was to introduce the tools and basic methods that enable one to look back in landscape development and that can help to make decisions. However, field verification of the results is essential. Satellite images (water and wetlands mask and multispectral image) can also be used for discussion with local and wetland protection authorities on future dissemination of BOMOSA fish farming in Eastern Africa.

Acknowledgement Many thanks to our African partners involved in the BOMOSA project for their support during ground truthing, for comments and reviews – Kenya Marine & Fisheries Research Institute, Sangoro Aquaculture Station, Kenya Ethiopian Institute of Agricultural Research, NFLARR, Sebetha, Ethiopia, Egerton University, Egerton, Kenya,Ministry of Agriculture, Animal Industry and Fisheries, Dept of Aquaculture, Entebbe, Uganda, Moi University, Dept of Zoology, Eldoret, Kenya, Kenyan Ministry of Livestock and Fisheries Development, Sagana Fish Farm, Kenya, as well as our European partners from University of Natural Resources & Applied Life Sciences, Vienna, Austria, University of Bologna, Italy, Austrian Academy of Sciences, Institute for Limnology, IPGL-Office, Mondsee Austria. Financial support was provided by BOMOSA (INCO-CT-2006-032103) and by the Ministry of Education, Youth and Sports of the Czech Republic NPV 2B06023 projects. Special thanks to Prof. Patrick Denny for valuable comments on the issue and for correction of the paper. Source for free Landsat data set was the Global Land Cover Facility (GLCF), www.landcover.org.

References

CIA. (2007). The library of congress country studies and the CIA world factbook. Available June 20, 2009 from, http://www.photius.com/countries/uganda/geography/uganda_geography_lakes_and_rivers.html

Dozier, J., & Marks, D. (1989). Snow mapping and classification from Landsat Thematic Mapper data. *Annals of Glaciology*, *9*, 97-103.

Hesslerová, P., Šíma, M., & Pokorný, J. (2009). *BOMOSA remote sensing method for small water bodies detection.* ENKI, o.p.s. and Ministry of Agriculture: Uganda.

ILEC. (2001). International Lake Environment Committee. World lakes database. Available November 17, 2009 from, http://www.ilec.or.jp/database/afr/afr-15.html

ILM. (2004). Support to the management of sudd blockage on Lake Kyoga. Produced for the integrated lake management project by environmental impact assessment centre of Finland, EIA Ltd. 81 p. Available June 18, 2009 from http://www.eia.fi/kyoga/report/Kyoga_Model_report-EIA_Ltd.pdf

Lillesand, T. M., & Kiefer, R. W. (2000). *Remote Sensing and Image Interpretation.* New York, NY: Wiley.

Chapter 21
Factors Affecting Metal Mobilisation During Oxidation of Sulphidic, Sandy Wetland Substrates

Gijs Du Laing, Tom Hanssen, Geert Bogaert, and Filip M.G. Tack

Abstract Most metals accumulate as sulphides under anoxic conditions in wetland substrates, reducing their bioavailability due to the solubility of metal sulphides. However, upon oxidation of these sulphides when the substrate is occasionally oxidised, metals can be released from the solid phase to the pore water or overlaying surface water. This release can be affected by the presence of carbonates, organic matter and clay. We compared changes of Cd, Cu and Zn mobility ($CaCl_2$ extraction) during oxidation of a carbonate-rich and a carbonate-poor sulphidic, sandy wetland substrate. In addition, we studied how clay with low and high cation sorption capacity (bentonite and kaolinite, respectively) and organic matter (peat) can counteract Cd, Cu and Zn release during oxidation of both carbonate-rich and carbonate-poor sulphidic sediments. $CaCl_2$-extractability of Cu, a measure for its availability, is low in both carbonate-poor and carbonate-rich substrates, whereas its variability is high. The availability of Cd and Zn is much higher and increases when peat is supplied to carbonate-poor substrates. A strong reduction of Cd and Zn extractability is observed when clay is added to carbonate-poor substrates. This reduction depends on the clay type. Most observations could be explained taking into account pH differences between treatments, with kaolinite resulting in a lower pH in comparison to bentonite. These pH differences affect the presence and characteristics of dissolved organic carbon and the metal speciation, which in turns affects the interaction of metals with the solid soil phase. In carbonate-rich substrates, Cd and Zn availability is lower and the effects of peat and clay amendment are less clear. The latter can also be attributed to the high pH and lack of pH differences between treatments.

Keywords Cadmium · Copper · Mobilisation · Sulphides · Zinc

G. Du Laing (✉)
Laboratory of Analytical Chemistry and Applied Ecochemistry, Ghent University, Ghent, Belgium
e-mail: Gijs.DuLaing@UGent.be

J. Vymazal (ed.), *Water and Nutrient Management in Natural and Constructed Wetlands*, DOI 10.1007/978-90-481-9585-5_21,

21.1 Introduction

Wetland soils and sediments consist of a mixture of organic and inorganic compounds. In the pore water, metals can be present as free ion, or as organic or inorganic complex. In the solid phase, adsorption sites for metals can be clay minerals, Fe-, Mn- or Al-(hydr)oxides, organic matter and carbonates. Metals can also be included in the mineral structure or precipitate with formation of metal carbonates and metal sulphides. Sulphide formation is mediated by sulphate reducing bacteria in anoxic conditions. Accordingly, they are normally found in sulphate-rich wetland substrates. According to e.g. Kornicker and Morse (1991), Morse and Arakaki (1993), Huerta-Diaz, Tessier, and Carignan (1998) and Billon, Ouddane, Laureyns, and Boughriet (2001), iron monosulphide (FeS) is often considered the most reactive of the sulphide phases in sediments. Trace metals entering anoxic sediment environments can be expected to react to form sulphide phases according to: $M^{2+} + FeS(s) \leftrightarrow MS(s) + Fe^{2+}$, where M = e.g., Cd, Cu, Ni, Pb, Zn, Co, Hg (Simpson, Rosner, & Ellis, 2000). Because the solubility of these metal sulphides is very low, sediments with an excess of reactive sulphide to trace metals will exhibit very low dissolved metal concentrations in their pore water (Du Laing et al., 2008). The reactive sulphide fraction may be estimated by measuring the acid-volatile sulphide (AVS) content of the sediment, whereas the simultaneously extracted metal ($SEM = \Sigma\ Cd, Cu, Ni, Pb, Zn$) content approximates the reactive trace metal fraction (Di Toro et al., 1992; Allen, Fu, & Deng, 1993).

When sulphide-containing wetland substrates are oxidised upon contact with water containing dissolved oxygen or with air, the sulphides will tend to be gradually oxidised to sulphate, and metals will be released into the pore water. This can occur occasionally in constructed wetlands used for wastewater treatment or more frequently in tidal marshes, and result in an increased mobility and bioavailability of metals retained within the system (Du Laing, Rinklebe, Vandecasteele, Meers, & Tack, 2009b). Once the reactive sulphide phase, which is determined as AVS (predominantly FeS), has been exhausted in the absence of other binding phases, the metals are expected to appear in the pore waters in order of decreasing metal sulphide solubility, that is Ni, Zn, Pb, Cd and then Cu (Di Toro et al., 1992). Berry et al. (1996) provided evidence for this relationship for each of the metal ions Cd, Cu, Ni, Pb and Zn (Simpson et al., 2000). Next to the direct release of metals precipitated as sulphides, the pH can decrease during sulphide oxidation due to the following reaction (presented for FeS):

$$4\ FeS_{(s)} + 9\ O_2 + 10\ H_2O \rightarrow 4\ Fe(OH)_{3(s)} + 4\ SO_4^{2-} + 8\ H^+ \qquad (1)$$

For each molecule of FeS oxidized, two protons are produced. In un-buffered sediments, these reactions will cause a drop in pH which will limit transfer of trace metals from the water phase to the wetland substrate and/or causing a desorption from the soils or sediments (e.g., Salomons, de Rooij, Kerdijk, & Bril, 1987; Gambrell, Wiesepape, Patrick, & Duff, 1991; Calmano, Hong, & Förstner, 1993).

The presence of carbonates in calcareous wetland substrates constitutes an effective buffer against such pH decrease and thus hampers metal release (Satawathananont, Patrick, & Moore, 1991; Tack, Callewaert, & Verloo, 1996). Moreover, carbonates may also directly precipitate metals (e.g., Gambrell, 1994; Guo, DeLaune, & Patrick, 1997; Charlatchka & Cambier, 2000). They can be geochemically or biogenically formed and be deposited as part of the sediments in tidal flats, floodplains or marshes. However, decalcification occurs as carbonates are consumed upon oxidation of sulphides and concurrent release of protons: $CaCO_3 + H^+ \leftrightarrow Ca^{2+} + HCO_3^-$. Evidently, the intensity of decalcification due to sulphide oxidation varies with the amount of sulphide previously formed in the wetland substrate, and thus on the duration of waterlogging, and the availability of sulphate and organic matter. Also the increased Ca concentrations in the pore water during partial decalcification can cause an enhanced release of metals in the calcareous wetlands' substrate layer (Du Laing, Vanthuyne, Vandecasteele, Tack, & Verloo, 2007).

The SEM/AVS ratio is used to assess potential sediment toxicity. Some authors (Di Toro et al., 1992; van den Hoop, den Hollander, & Kerdijk, 1997) state that no lethal toxic effects on organisms should be expected if the SEM/AVS ratio is smaller than one. Di Toro et al. (1990) initially showed that lethal toxic effects of cadmium spiked marine sediment on amphipods can successfully be predicted by considering the Cd/AVS ratio, whereas estimation of the toxicity based on the total metal content of the sediment failed. Lethal toxic effects were found to be absent for Cd/AVS ratios smaller than one, but appeared to be present at Cd/AVS ratios exceeding one. Comparable results were afterwards also observed for freshwater sediments (e.g. Ankley et al., 1991), sediments contaminated with Cu, Pb, Ni and Zn (e.g., Di Toro et al., 1992; Ankley, Mattson, Leonard, West, & Bennett, 1993; Casas & Crecelius, 1994), and also for other organisms (Carlson, Phipps, Mattson, Kosian, & Cotter, 1991; van den Hoop et al., 1997). However, simultaneously extracted metals may also include metals which are associated with alternative binding phases when these are present, e.g. metals associated with carbonates, organic matter or clay. When metals are associated with these compounds, the risk for metal toxicity and bioavailability may be much lower than expected from the fact that SEM exceeds AVS.

To assess the importance of the presence of carbonates in determining metal mobility and availability when SEM starts to exceed AVS during oxidation of sulphidic sediments, we monitored Cd, Cu and Zn mobility ($CaCl_2$ extraction) during oxidation of both a carbonate-rich and a carbonate-poor sulphidic, sandy wetland substrate. In addition, the effects of clay with low and high cation sorption capacity (bentonite and kaolinite, respectively) and organic matter (peat) on counteracting Cd, Cu and Zn release were studied.

21.2 Materials and Methods

A carbonate-poor, metal-contaminated sulphidic sand was sampled from the bottom of the river Dommel (Belgium). A carbonate-rich sand consisted of commercially available estuarine sand which was enriched in sulphides by waiting for 7 weeks after addition of a growth medium for sulphate-reducing micro-organisms. This induced reduction of sulphates to sulphides. Meanwhile, metals were regularly spiked to the substrate, inducing metal sulphide precipitation. Parts of the carbonate-rich and the carbonate-poor sulphidic sand were also mixed with peat (2% w/w), bentonite (10% w/w) and kaolinite (10% w/w). This resulted in 8 different substrates to be tested.

Glass recipients (ø = 15 cm, height = 6 cm) were filled with 525 g of each substrate type. The initially wet substrates were left to dry and oxidise at room temperature (20°C). Every 2 days, the substrates were mixed well with a spoon to eliminate the gradient caused by drying and oxidation. The experiment was conducted in triplicate. Dry weight percentage, pH, AVS contents, HCl- and $CaCl_2$-extractable metal contents of the substrates were measured regularly. In the beginning of the experiment, measurements were conducted daily, and less frequently afterwards.

To measure dry weight percentage, 10 g of substrate was dried in an oven at 200°C for 16 h. The pH was measured with a pH electrode in a suspension of substrate in distilled water (1:5 ratio) after equilibration for 18 h. Acid volatile sulphide (AVS) was determined on fresh substrate samples by conversion of sulphide to H_2S and absorption in a Zn-acetate solution, followed by back titration (Tack, Lapauw, & Verloo, 1997). Therefore, three pyrex gas washing bottles were connected in series. The first bottle was filled with 100 mL of deionized water, whereas the next two bottles were filled with 95 mL of deionized water. All flasks were deoxygenated by bubbling N_2 through for 30 min. Five mL of 2 M ZnOAc was added to each of the second and third bottles. Subsequently, 10 g of substrate and 20 mL of 6 M HCl were added to the first bottle and N_2 was re-bubbled through it for 30 min. The acid induced the conversion of sulphides to H_2S in the first bottle, which was transported by the N_2 carrier gas to the second and third bottle and absorbed in the Zn-acetate solution. The Zn-acetate solution was subsequently acidified, and KI and KIO_3 were added, which induced the formation of I_2. As a result, the collected sulphides (S^{2-}) were oxidized to elemental sulphur (S^0) by an equivalent amount of I_2, which was reduced to I^-. The excess I_2 was titrated with $Na_2S_2O_3$. HCl-extractable metals were defined as the metals released from the substrate upon addition of the HCl during AVS analysis (also defined as the simultaneously extracted metals or SEM). To measure $CaCl_2$-extractable metal contents, 50 mL of 0.01 mol L^{-1} $CaCl_2$ was added to 10 g of substrate and shaken for 2 h. The suspension was filtered over a 0.45 μm filter (Chromafil RC 45/25, Macherey – Nagel) and the filtrate was analysed using ICP-MS (Elan DRC-e, Perkin Elmer SCIEX). Trace metal speciation changes in the extracts as function of pH were modelled using the software Visual MINTEQ ver. 2.40 (http://www.lwr.kth.se/English/OurSoftware/vminteq/) using as input data: pH measured in water extract, 0.01 mol L^{-1} Ca^{2+} and 0.02 mol L^{-1} Cl^-

(i.e., background cation and anion concentrations in 0.01 mol L^{-1} $CaCl_2$ extractans) and Cd, Cu and Zn concentrations measured in the $CaCl_2$ extract (assumed to be present as Cd^{2+}, Cu^{2+}, Zn^{2+}). Absence of dissolved organic carbon was assumed.

21.3 Results and Discussion

21.3.1 Carbonate-Rich Versus Carbonate-Poor Substrate

The dry weight percentage of the substrates increased from between around 96.5% in the beginning of the experiment to around 100% from day 7 onwards. The carbonate-rich substrate had a higher initial dry weight and reached its maximum dry weight percentage slightly faster compared to the carbonate-poor substrate. The initial pH was very high (around 10) in the carbonate-rich substrate. This should probably be attributed to the presence in both substrates of carbonates and/or sulphides which remove protons released into solution, in combination with a lack of other compounds and processes which may release these protons into solutions (e.g. organic matter). The presence of carbonates constitutes an effective buffer against a decrease in pH. Whereas the pH of the carbonate-poor substrate decreased from 6.3 to below 5 during the first 5 days of drying and oxidation (Fig. 21.1, blank), the pH of the carbonate-rich substrate fluctuated within the range 9.5–10.5 (data not shown).

Due to oxidation of sulphides to sulphates, AVS contents decreased from above 7 mmol kg^{-1} DM to below 1 mmol kg^{-1} DM during the first 3 days (Fig. 21.2, blank). Minimum values were reached faster in the carbonate-rich substrate, which also reached its maximum dry weight percentage a bit faster. From the moment AVS contents decreased below 1 mmol kg^{-1} DM, $CaCl_2$-extractable Cd and Zn contents, indicating a very mobile metal fraction, started to increase (Fig. 21.3, blank). They reached their maximum from around day 7 onwards, which coincided with reaching

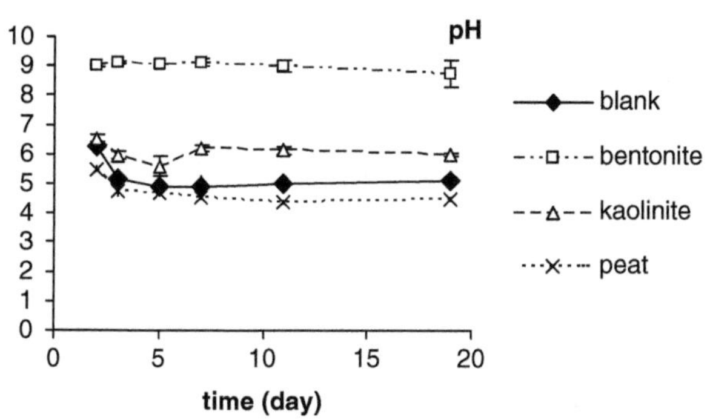

Fig. 21.1 pH during drying and oxidation of a sulphidic, carbonate-poor sandy substrate to which bentonite, kaolinite and peat are added (blank = no amendment)

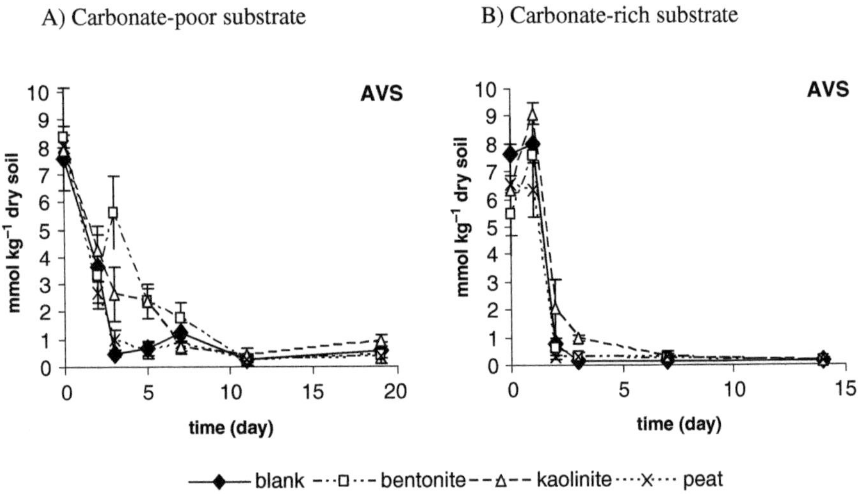

Fig. 21.2 Acid-volatile sulphide (AVS) contents (mmol kg^{-1} DM) during drying and oxidation of a sulphidic, carbonate-poor (**a**) and carbonate-rich (**b**) sandy substrate to which bentonite, kaolinite and peat are added (blank = no amendment)

the minimum moisture content. The $CaCl_2$-extractable levels mostly persisted afterwards, indicating minimum readsorption of the metals to the solid sediment phase. When $CaCl_2$-extractable contents were expressed relative to HCl-extractable contents (Table 21.1), it is clear that the metals are more mobile in the carbonate-poor substrate. This can be attributed to the lower pH, which prevents transfer of most trace metals from the water to the solid phase or causes desorption from the wetland soil and sediments (Calmano et al., 1993). At low pH, the negative surface charge of organic matter, clay particles and Fe and Al oxides is reduced, and various compounds such as carbonates and sulphides become more soluble. Carbonates may also directly precipitate metals (Charlatchka & Cambier, 2000). Presence of carbonates can result in $CuCO_3$, $CdCO_3$ and $ZnCO_3$ occurring in solution. $CuCO_3$ is most easily formed, but especially $ZnCO_3$ is expected to precipitate. Compared to Cd and Zn, Cu is less mobile and also less affected by the presence of carbonates. Copper mobility, represented by the $CaCl_2$-extractable contents in relation to the HCl-extractable contents, is initially low. In contrast to Cd and Zn, $CaCl_2$-extractability of Cu also does not increase much during oxidation of the sulphides (e.g., data for day 7 presented in Table 21.1). This points towards a strong association of Cu to the solid phase before as well as after oxidation of the sulphides.

21.3.2 Role of Organic Matter and Clay

Both at low and high carbonate contents, the variability of $CaCl_2$-extractable Cu contents was high. At low (Fig. 21.3) and high (Fig. 21.4) carbonate contents, Cu was somewhat more mobile when peat was applied due to its affinity to form

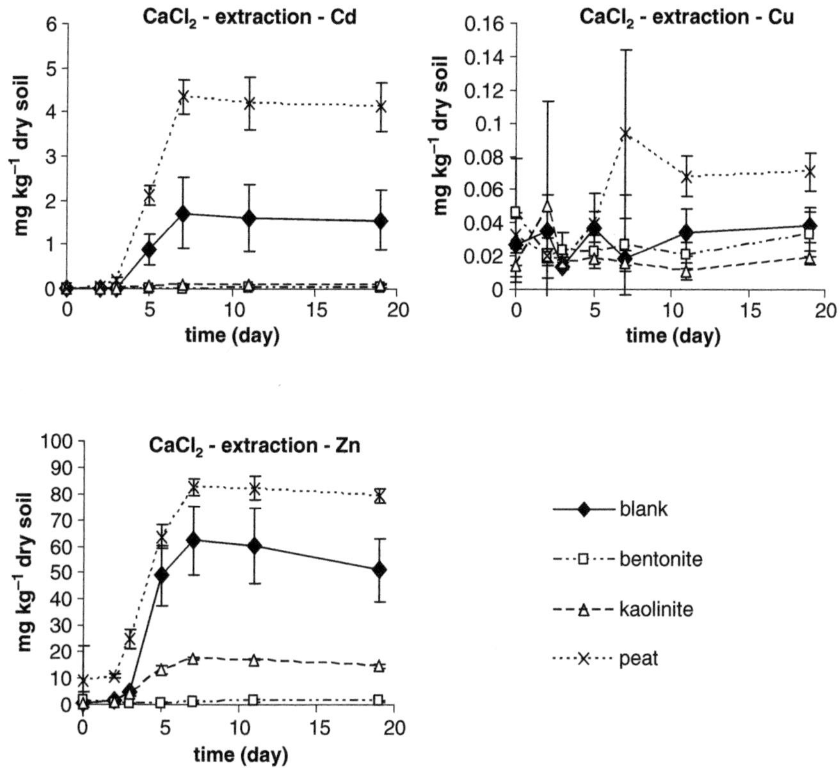

Fig. 21.3 $CaCl_2$ – extractable Cd, Cu and Zn contents (mg kg^{-1} DM) during drying and oxidation of a sulphidic, carbonate-poor sandy substrate to which bentonite, kaolinite and peat are added (blank = no amendment)

Table 21.1 $CaCl_2$ – extractable Cd, Zn and Cu contents (mg kg^{-1} DM) relative to HCl-extractable contents (%) in the carbonate-poor and carbonate-rich substrates as affected by amendment at day 7 of the experiment ($n = 3$)

	Cd	Zn	Cu
Carbonate-poor			
Blank	12.3 ± 5.6	37.6 ± 6.1	0.38 ± 0.08
Peat	31.4 ± 3.4	42.3 ± 7.5	1.77 ± 1.05
Kaolinite	0.7 ± 0.1	10.3 ± 1.0	0.29 ± 0.09
Bentonite	0.1 ± 0.0	0.7 ± 0.2	0.58 ± 0.66
Carbonate-rich			
Blank	2.6 ± 2.3	1.6 ± 0.7	0.58 ± 0.39
Peat	1.7 ± 1.4	2.9 ± 1.5	0.70 ± 0.54
Kaolinite	1.5 ± 1.9	4.1 ± 3.5	0.32 ± 0.17
Bentonite	1.8 ± 1.0	2.0 ± 0.8	0.70 ± 0.42

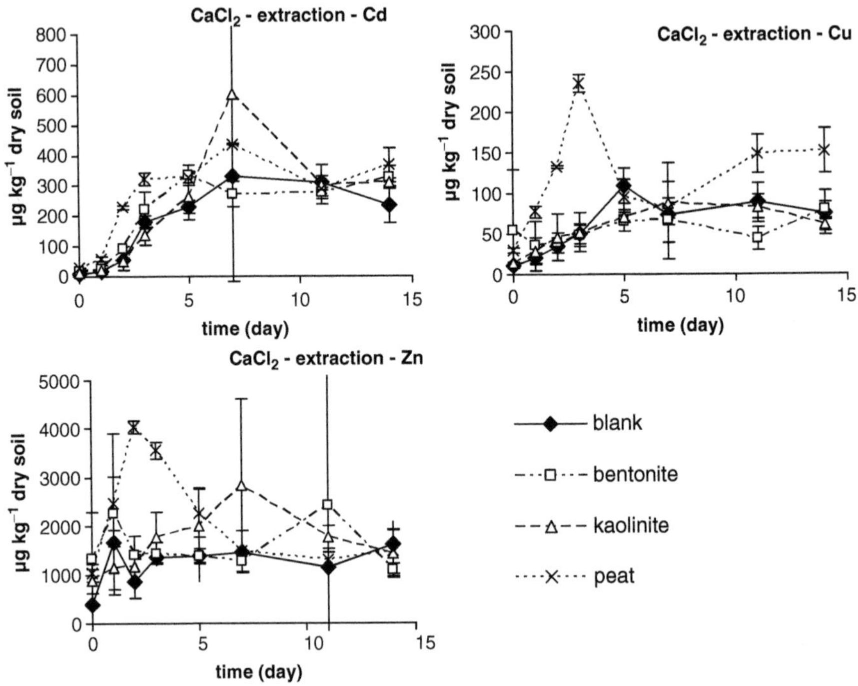

Fig. 21.4 $CaCl_2$ – extractable Cd, Cu and Zn contents (μg kg^{-1} DM) during drying and oxidation of a sulphidic, carbonate-rich sandy substrate to which bentonite, kaolinite and peat are added (blank = no amendment)

complexes with soluble organic molecules (Ashworth & Alloway, 2008; Du Laing et al., 2009a), released from the peat. After oxidation of the sulphides, the mobility was slightly lower in the presence of kaolinite or bentonite as compared to the blank (Fig. 21.3, Table 21.1). Bentonite reveals a lower capacity to reduce Cu mobility as compared to kaolinite, although bentonite has a higher cation exchange capacity (CEC). The CEC is expected to be especially high as the pH of the bentonite-amended substrate did not decrease below 9 during the oxidation (Fig. 21.1). Moreover, speciation modelling indicated that tenorite (CuO) also starts to precipitate above pH 7.7, so additional removal of Cu from solution could be expected at these higher pH levels. However, the differences in pH between substrates amended with kaolinite and those amended with bentonite may have affected the speciation of Cu in solution which might explain the apparent anomaly that sorption is weaker on bentonite than on kaolinite. According to thermodynamical equilibrium calculations, Cu^{2+} is expected to be the main species in solution at pH 6, whereas $Cu(OH)^+$ and $Cu(OH)_2$ are most abundant in solution at a pH of about 9. The speciation shift from Cu^{2+} in solution at pH 6 to $Cu(OH)^+$ and $Cu(OH)_2$ in solution at pH levels close to 9 may have reduced the affinity of Cu for sorption to the clay. Moreover, the pH-dependent presence and characteristics of organic compounds in solution probably have affected Cu concentrations in solution, especially due to the strong association of Cu with organic matter as previously reported in literature (Du Laing et al., 2009a). When pH increases from 6 to above 8, the negative charge on organic molecules is increased. The negatively charged molecules are repelled into the soil solution, thereby increasing dissolved organic matter concentrations and Cu concentrations in solution (Ashworth & Alloway, 2008). Moreover, the negative charge promotes additional association between these dissolved organic molecules and metals, such as Cu. In the carbonate-rich substrates, clear differences in Cu extractability between bentonite and kaolinite cannot be observed, which may be attributed to the quite stable pH, not depending on the type of amendment and varying between 9.4 and 10.5 (data not shown).

Extractability of Cd and Zn by $CaCl_2$ is affected to an important extent by the supply of bentonite, kaolinite and peat in the carbonate-poor substrates (Fig. 21.3, Table 21.1). Peat increases Cd and Zn extractability, which should be attributed to the presence of dissolved organic molecules in the extracts. The effect is slightly higher for Cd. A strong reduction of Cd and Zn extractability is observed when kaolinite and bentonite are added. Bentonite results in the lowest extractability, probably due to its higher cation exchange capacity. Differences between bentonite and kaolinite are smaller for Cd (factor 5) as compared to Zn (factor 15). This may again be due to pH differences between these treatments (Fig. 21.1) and the effect of pH on the speciation of Cd and Zn. At a pH below 7.5 (kaolinite amended substrate), almost all Zn occurs as Zn^{2+} in solution whereas Cd occurs as Cd^{2+} or $CdCl^+$ (both about 48%) in solution. At pH 9 (bentonite amended substrate), 80% of the Zn is present as neutral $Zn(OH)_2$ in solution, whereas less than 20% is still positively charged (Zn^{2+} or $Zn(OH)^+$ in solution). This speciation shift may have changed the tendency of Zn to adsorb onto solid phases. In comparison to Cu, organic matter is expected to play a lesser role as the affinity of Zn for dissolved organic matter at higher pH is lower (Ashworth & Alloway, 2008). Moreover, at the observed Zn concentration, Zn was found to start precipitating as zincite (ZnO) when pH 8.7 is exceeded. By contrast, Cd still mainly occurs as Cd^{2+} or $CdCl^+$ (both about 47%) and only 0.1% occurs as $Cd(OH)_2$ at pH 9. No precipitation occurs and the formation of $Cd(OH)_2$ only starts to become significant (>5%) at pH 10. Accordingly, differences in Cd and Zn mobility between amendments are not observed in the amended carbonate-rich substrates (Fig. 21.4), as they do not differ significantly in pH. The extractability of Cd and Zn by $CaCl_2$ is highly variable in these substrates, probably due to I) the impact of Ca released from the solid soil phase upon dissolution of the carbonates on the Ca and $CO_3{}^{2-}$ concentration in the liquid phase and II) the heterogeneous distribution of small shells, and thus carbonates, in the solid phase of the estuarine sand used as substrate.

21.4 Conclusion

$CaCl_2$-extractability of Cu, a measure for its availability, is low in both carbonate-poor and carbonate-rich substrates, whereas its variability is high. The availability of Cd and Zn is much higher and increases when peat is supplied to carbonate-poor substrates. A strong reduction of Cd and Zn extractability is observed when clay is added to carbonate-poor substrates. This reduction depends on the clay type. Most observations could be explained taking into account pH differences between treatments, with kaolinite resulting in a lower pH in comparison to bentonite. These pH differences affect the presence and characteristics of dissolved organic carbon and the metal speciation, which in turns affects the interaction of metals with the solid soil phase. In carbonate-rich substrates, Cd and Zn availability is lower and the effects of peat and clay amendment are less clear. The latter can also be attributed to the high pH and lack of pH differences between treatments.

References

Allen, H. E., Fu, G., & Deng, B. (1993). Analysis of acid-volatile sulphide and simultaneous extracted metals for the estimation of potential toxicity in aquatic sediments. *Environmental Toxicology and Chemistry, 12*, 1441–1453.

Ankley, G. T., Mattson, V. R., Leonard, E. N., West, C. W., & Bennett, J. L. (1993). Predicting the acute toxicity of copper in freshwater sediments: evaluation of the role of acid volatile sulfide. *Environmental Toxicology and Chemistry, 12*, 315–320.

Ankley, G. T., Phipps, G. L., Leonard, E. N., Kosian, P. A., Cotter, A. M., Dierkes, J. R., et al. (1991). Acid volatile sulfide as a factor mediating cadmium and nickel bioavailability in contaminated sediments. *Environmental Toxicology and Chemistry, 10*, 1299–1307.

Ashworth, D. J., & Alloway, B. J. (2008). Influence of dissolved organic matter on the solubility of heavy metals in sewage-sludge-amended soils. *Communications in Soil Science and Plant Analysis, 39*, 538–550.

Berry, W. J., Hansen, D. J., Mahony, J. D., Robson, D. L., Di Toro, D. M., Shipley, B. P., et al. (1996). Predicting the toxicity of metal-spiked laboratory sediments using acid-volatile sulphide and interstitial water normalizations. *Environmental Toxicology and Chemistry, 15*, 2067–2079.

Billon, G., Ouddane, B., Laureyns, J., & Boughriet, A. (2001). Chemistry of metal sulfides in anoxic sediments. *Physical Chemistry Chemical Physics, 3*, 3586–3592.

Calmano, W., Hong, J., & Förstner, U. (1993). Binding and mobilisation of heavy metals in contaminated sediments affected by pH and redox potential. *Water Science and Technology, 28*, 223–235.

Carlson, A. R., Phipps, G. L., Mattson, V. R., Kosian, P. A., & Cotter, A. M. (1991). The role of acid volatile sulphide in determining cadmium bioavailability and toxicity in freshwater sediments. *Environmental Toxicology and Chemistry, 10*, 1309–1319.

Casas, A. M., & Crecelius, E. A. (1994). Relationship between acid volatile sulfide and the toxicity of zinc, lead and copper in marine sediments, Environmental *Toxicology and Chemistry, 13*, 529–536.

Charlatchka, R., & Cambier, P. (2000). Influence of reducing conditions on solubility of trace metals in contaminated soils. *Water, Air, and Soil Pollution, 118*, 143–167.

Di Toro, D. M., Mahony, J. D., Hansen, D. J., Scott, K. J., Carlson, A. R., & Ankley, G. T. (1992). Acid Volatile Sulfide predicts the acute toxicity of cadmium and nickel in sediments. *Environmental Science and Technology, 26*, 96–101.

Di Toro, D. M., Mahony, J. D., Hansen, D. J., Scott, K. J., Hicks, M. B., Mayr, S. M., et al. (1990). Toxicity of cadmium in sediments: the role of acid volatile sulphide. *Environmental Toxicology and Chemistry, 9*, 1487–1502.

Du Laing, G., De Meyer, B., Meers, E., Lesage, E., Van de Moortel, A., Tack, F. M.G., et al. (2008). Metal accumulation in intertidal marshes along the river Scheldt: role of sulphide precipitation. *Wetlands, 28*, 735–746.

Du Laing, G., Meers, E., Dewispelaere, M., Vandecasteele, B., Rinklebe, J., Tack, F. M. G., et al. (2009a). Heavy metal mobility in intertidal sediments of the Scheldt estuary: Field monitoring. *Science of the Total Environment, 407*, 2919–2930.

Du Laing, G., Rinklebe, J., Vandecasteele, B., Meers, E., & Tack, F. M. G. (2009b). Heavy metal mobility and availability in estuarine and riverine floodplain soils and sediments: a review. *Science of the Total Environment, 407*, 3972–3985.

Du Laing, G., Vanthuyne, D. R. J., Vandecasteele, B., Tack, F. M. G., & Verloo, M. G. (2007). Influence of hydrological regime on pore water metal concentrations in a contaminated sediment-derived soil. *Environmental Pollution, 147*, 615–625.

Gambrell, R. P., Wiesepape, J. B., Patrick, W. H.Jr., & Duff, M. C. (1991). The effects of pH, redox, and salinity on metal release from a contaminated sediment. *Water, Air, and Soil Pollution, 57–58*, 359–367.

Gambrell, R. P. (1994). Trace and toxic metals in wetlands – a review. *Journal of Environmental Quality, 23*, 883–891.

Guo, T., DeLaune, R. D., & Patrick, W. H. Jr. (1997). The influence of sediment redox chemistry on chemically active forms of arsenic, cadmium, chromium, and zinc in estuarine sediment. *Environment International, 23*, 305–316.

Huerta-Diaz, M. A., Tessier, A., & Carignan, R. (1998). Geochemistry of trace metals associated with reduced sulphur in freshwater sediments. *Applied Geochemistry, 13*, 213–233.

Kornicker, W. A., & Morse, J. W. (1991). Interactions of divalent cations with the surface of pyrite. *Geochimica Cosmochimica Acta, 55*, 2159–2171.

Morse, J. W., & Arakaki, T. (1993). Adsorption and coprecipitation of divalent metals with mackinawite (FeS). *Geochimica Cosmochimica Acta, 57*, 3635–3640.

Salomons, W., de Rooij, N. M., Kerdijk, H., & Bril, J. (1987). Sediments as a source for contaminants? *Hydrobiologia, 149*, 13–30.

Satawathananont, S., Patrick, W. H. J., & Moore, P. A. J. (1991). Effect of controlled redox conditions on metal solubility in acid sulfate soils. *Plant and Soil, 133*, 281–290.

Simpson, S. L., Rosner, J., & Ellis, J. (2000). Competitive displacement reactions of cadmium, copper, and zinc added to a polluted, sulfidic estuarine sediment. *Environmental Toxicology and Chemistry, 19*, 1992–1999.

Tack, F. M., Callewaert, O. W. J. J., & Verloo, M. G. (1996). Metal solubility as a function of pH in a contaminated, dredged sediment affected by oxidation. *Environmental Pollution, 91*, 199–208.

Tack, F. M., Lapauw, F., & Verloo, M. G. (1997). Determination and fractionation of sulphur in a contaminated dredged sediment. *Talanta, 44*, 2185–2192.

van den Hoop, M. A. G. T., den Hollander, H. A., & Kerdijk, H. M. (1997). Spatial and seasonal variations of acid volatile sulphide (AVS) and simultaneously extracted metals (SEM) in Dutch marine and freshwater sediments. *Chemosphere, 35*, 2307–2316.

Chapter 22
The Presence of Mycorrhiza in Different Habitats of an Intermittent Aquatic Ecosystem

Nataša Dolinar, Nina Šraj, Paula Pongrac, Marjana Regvar, and Alenka Gaberščik

Abstract Environmental conditions in wetlands were long thought to suppress mycorrhizal fungi and it has been assumed that mycorrhiza in wetlands is limited and is of little significance. This contribution summarises the presence of arbuscular mycorrhizal (AM) fungi and dark-septate endophytes (DSE) in different habitats of the intermittent Lake Cerknica. Mycorrhizal colonisation of wetland plants from the following wetland habitats was analysed: (1) frequently inundated small depressions in the lakebed colonised by amphibious plants; (2) *P. australis* stands covering large parts of the lake area; and (3) wet meadows surrounding the lake area colonised by different *Schoenus* species. All of the examined amphibious species were found to be mycorrhizal, however terrestrial shoots were more mycorrhizal than aquatic shoots. In the former mycorrhizal frequency F% was up to 100% and mycorrhizal intensity M% up to 67%, while F% was up to 80% and M% up to 16% in the latter. Mycorrhizal colonisation of *P. australis* with AM and DSE fungi was confirmed in plants growing in different soils and water regimes (17% < F% < 50%; 0.6% < M% < 3%). In the two *Schoenus* species, F% was up to 99% and M% was up to 14%. According to the literature this is the first detailed report on the presence of AM and DSE mycorrhiza in *S. ferrugineus* and *S. nigricans*.

Keywords Amphibious plants · Intermittent lake · Mycorrhiza · *Phragmites australis*

A. Gaberščik (✉)
Department of Biology, Biotechnical faculty, University of Ljubljana, Ljubljana, Slovenia
e-mail: alenka.gaberscik@bf.uni-lj.si

J. Vymazal (ed.), *Water and Nutrient Management in Natural and Constructed Wetlands*, DOI 10.1007/978-90-481-9585-5_22,

22.1 Introduction

22.1.1 Mycorrhiza and Its Benefits for Plants

Mycorrhizas are mutualistic associations between plant roots and fungi found in more than 90% of land plants and often play an important role in plant nutrition and survival of abiotic stress (Smith & Read, 1997). One of the best noted effects of mycorrhizal symbiosis is the improved phosphorus uptake by plant hosts: by enlarging the absorption area of plant roots and/or by active metabolic processes with specific enzymes that release the P otherwise unavailable to the plant (Gurevitch, 2002). The increased P uptake can lead to larger plant growth unless some other nutrient is then in limitation (Smith & Read, 1997). Mycorrhizal plants often have improved water status compared to nonmycorrhizal plants. This effect was often ascribed to the larger size of the plant; however improved water status of plants in drying soils was evident even when improved nutrition was taken into account (Augé, 2004).

Recently a lot of attention has been given to dark-septate endophytes (DSE) that are often found when inspecting plant roots for mycorrhiza. DSE are a diverse group of ascomycetous anamorphic fungi that colonise root tissues inter- and intracelullarly without causing apparent negative effects (Jumpponen & Trappe, 1998). During intracellular colonisation, DSE might form clusters of inflated, rounded, thick-walled cells within cortical cells called microsclerotia (Jumpponen & Trappe, 1998). According to some authors, DSE could, similar to arbuscular mycorrhiza, play a role in plant nutrient acquisition, in water uptake and in improved tolerance to environmental pressures (Mandyam & Jumpponen, 2005). The symbiosis between plants and DSE can be considered as true mycorrhiza, at least under some conditions (Jumpponen, 2001).

22.1.2 Mycorrhiza in Wetlands

The factors most often influencing the presence and abundance of mycorrhiza in plants are soil characteristics (such as nutrient status and water content) and plant characteristics (Smith & Read, 1997). It has been assumed that arbuscular mycorrhiza (AM) in wetlands is limited and is of little significance because of specific environmental conditions in wetlands, such as anoxic soil due to flooding, water level fluctuations, high or low concentrations of soil minerals and available nutrients (Bohrer, Friese, & Amon, 2004). However, more recently, many wetland species have been found colonised with AM and DSE fungi (Cornwell, Bedford, & Chapin, 2001; Weishampel & Bedford, 2006) including submerged plants (Šraj-Kržič et al., 2006; Šraj-Kržič, Pongrac, Regvar, & Gaberščik, 2009).

The occurrence of mycorrhiza in wetlands can vary. Seasonal dynamics of mycorrhizal colonisation, often described in other habitats, could be under a strong influence of water level changes in wetlands. Flooding of wetland soils has usually been reported to lead to a decrease in mycorrhizal colonisation in wetland plants (Stevens & Peterson, 1996; Turner & Friese, 1998; Miller, 2000; Bajwa,

Yaqoob, & Javaid, 2001), although this is not always the case (Bauer, Kellogg, Brigham, & Lamberti, 2003). On the other hand, mycorrhizal colonisation can also reflect plant phenological phases; with more mycorrhizal fungi present when the plant's need for nutrients are higher (Bohrer et al., 2004).

22.1.3 Mycorrhiza in Different Habitats of an Intermittent Wetland

The aim of this contribution is to discuss the colonisation with AM fungi and DSE in plants from different habitats of the intermittent Lake Cerknica. Lake Cerknica forms on the bottom of a karstic polje when large seasonal water runoff exceeds the capacity of the underground drainage system in carbonate rocks (Gaberščik, Urbanc-Berčič, Kržič, Kosi, & Brancelj, 2003). As a result, a variety of wetland habitats with spatial and/or temporal variability can be found on the seasonally flooded lakebed, ranging from occasionally flooded wet meadows to a range of habitats with frequent water level fluctuations. The importance of such ecosystems is in the numerous ecosystem services they provide and because they are under threat due to global changes (Dolinar, Rudolf, Šraj, & Gaberščik, 2010).

Root colonisation of different plant species with AM fungi and DSE from three different wetland habitats of Lake Cerknica was surveyed: (1) frequently inundated small depressions in the lakebed colonised by amphibious plants; (2) reed stands covering large parts of the lake area; and (3) wet meadows with frequent water level changes surrounding the lake area.

22.2 Material and Methods

22.2.1 Site Description

Terrestrial shoots of amphibious plants were collected on the slopes of small depressions in the lakebed, while aquatic shoots were collected at the bottom of the depressions. The rhizosphere soil was poor in plant available P content which ranged from 0.7 to 1.6 mg L^{-1}, while organic matter content was about 1.6% (Šraj-Kržič et al., 2006, 2009).

Reed stands from different parts of the lake differ considerably in the environmental characteristics of their habitat. Mycorrhizal colonization of three reed stands was investigated. The locations of these reed stands, which differ in the extent of water level fluctuations and organic matter content, are described in detail by Urbanc-Berčič and Gaberščik (2003).

Accordingly, only a short description of characteristics of the survey locations (with geographical names in parenthesis) is provided: Location 1 (Dujice) is characterised by a thick organic layer of soil and consequently high organic matter content (45%); P content in soil is 10 mg 100 g^{-1}. This location is never flooded, however the soil is often water saturated. Reed stands from Location 2 (Gorenje Jezero) grow alongside a stream and are frequently flooded with up to 0.5 m deep water. The soil at this location is also rich in organic matter (23%) and contains 24 mg 100 g^{-1} of P. Location 3 (Zadnji kraj) lies on the edge of the lake and is frequently in up to 1 m deep water. The soil at this location is poor in organic matter (8%) and P (7 mg 100 g^{-1}).

Wet meadows in the wider area of Lake Cerknica are colonised by two *Schoenus* species: *S. ferrugineus* grows on wet meadows characterised by high organic matter content (45%) and P content of 10 mg 100 g^{-1} soil; and *S. nigricans* grows on wet meadows with lower organic matter content (8%) and P content (1.6 mg 100 g^{-1}).

22.2.2 Mycorrhizal Quantification

Plant root samples were collected approximately in the middle of the growing season. Fine roots were carefully separated from the soil, cleared with 10% KOH and stained in 0.05% Trypan blue, according to the procedure described by Philips and Hayman (1970). Mycorrhizal colonisation of each plant sample was determined on thirty 1 cm long fine root segments. The colonisation of root segments by arbuscular mycorrhiza was assessed according to Trouvelot, Kough, and Gianinazzi-Pearson (1986) and mycorrhizal frequency (F%) – the proportion of root segments colonised; intensity (M%) – the abundance of mycorrhizal colonisation in a root segment; and the densities of arbuscules (A%), vesicles (V%), hyphal coils (C%) and microsclerotia (MS%) were calculated.

22.3 Results and Discussion

The survey of mycorrhiza in the area of Lake Cerknica has revealed that mycorrhiza is abundant in plants growing under different hydrological conditions, such as frequent flooding and draining, which were long thought to have an adverse effect on mycorrhizal fungi. Flooding of plant roots often results in anoxic conditions that could suppress the growth of mycorrhizal fungi (Khan, 2004). However, possible mechanisms that enable the survival of mycorrhizal fungi under anoxic conditions could be that they require less oxygen than previously thought, that they concentrate nearer the plant roots where soil is oxygenated due to root radial oxygen loss and/or some fungal species tolerate flooding better than others (Miller & Bever, 1999).

22.3.1 Mycorrhiza in Amphibious Plants

Frequent flooding and draining can present stress for many wetland plants and therefore habitats with such characteristics are often colonised with amphibious plants. These are characterised by adaptations which enable survival in both flooded and un-flooded conditions (aerenchyma, heterophylly, different growth forms) (Braendle & Crawford, 1999). In our survey, mycorrhizal colonisation of *Alisma plantago-aquatica*, *Glyceria fluitans*, *Gratiola officinalis*, *Mentha aquatica*, *Miosotis scorpioides*, *Oenanthe fistulosa*, *Ranunculus lingua*, *Sium latifolium*, *Sparganium emersum*, *Teucrium scordium*, and *Veronica anagallis-aquatica* is presented. Both aquatic and terrestrial shoots of these species were investigated and almost all were mycorrhizal. F% and M% were usually lower in aquatic than terrestrial shoots of the same species or in some cases they were nonmycorrhizal (*G. fluitans* and *S. emersum*) (Fig. 22.1). The most abundant structures observed in the root samples were arbuscules and vesicles and relative abundances of these structures are presented in Fig. 22.2. Furthermore, terrestrial shoots with high M% (*O. fistulosa*, *R. lingua*, *S. latifolium*, *T. scordium* and *V. anagalis-aquatica*) also harboured higher abundances of arbuscules in relation to vesicles. Opposite to that, in terrestrial shoots with low M% (*A. plantago-aquatica*, *G. fluitans*, *G. officinalis*, *M. aquatica*, *M. scorpioides* and *S. emersum*) vesicles prevailed over arbuscules. In general, the abundances of arbuscules and vesicles in aquatic shoots were very low, and in most cases vesicles were the most abundant structure, except in *V. anagalis-aquatica*.

The finding that terrestrial shoots of amphibious plants are more mycorrhizal than aquatic shoots is in line with previous studies of mycorrhiza in flooded plants

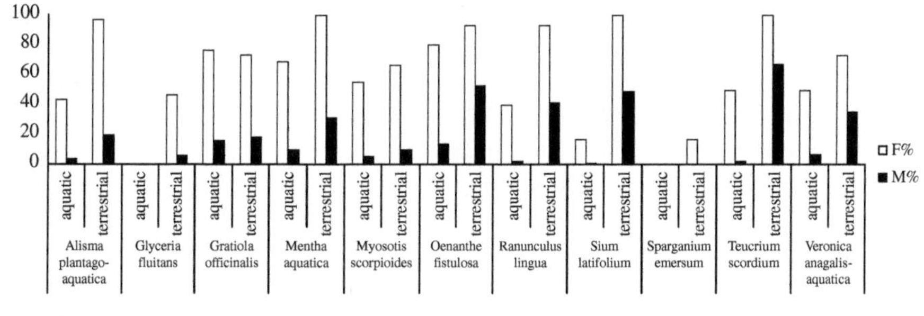

Fig. 22.1 Mycorrhizal frequency (F%) and intensity (M%) in different amphibious plants of Lake Cerknica

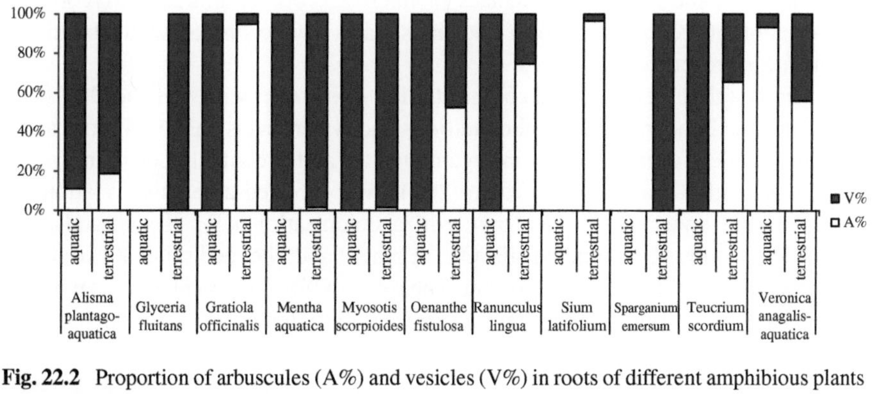

Fig. 22.2 Proportion of arbuscules (A%) and vesicles (V%) in roots of different amphibious plants of Lake Cerknica

(Ray & Inouye, 2006). Šraj-Kržič et al. (2006) have already established a positive correlation between AM and plant available phosphorus for the majority of amphibious plants included in this study. There were apparent differences in the abundance of arbuscules and vesicles between terrestrial and aquatic shoots, with arbuscules being more abundant in terrestrial shoots. Arbuscules in the plant roots are active nutrient transport and exchange sites between the plant and the fungus, while vesicles are considered to be fungal storage organs (Smith & Read, 1997). We presume that flooding of amphibious plants and consequent development of aquatic shoots resulted in reduced abundance of mycorrhizal fungi and triggered the replacement of arbuscules (nutrient exchange sites) with vesicles (storage organs) providing nutrients for future growth under drained conditions.

22.3.2 Mycorrhiza in Phragmites Australis

Phragmites australis is a worldwide occurring emergent macrophyte characteristic of the ecotone between terrestrial and aquatic environments in freshwater to brackish wetlands (Mauchamp & Méthy, 2004). It exhibits wide tolerance to flooding and drainage (Hudon, Gagnon, & Jean, 2005) with the ability of phenotypic plastic responses (Vretare, Weisner, Strand, & Graneli, 2001). There are only a few recent reports on the presence of AM in *P. australis* (e.g. Cooke & Lefor, 1998; Oliveira, Dodd, & Castro, 2001), while some authors found AM to be absent from *P. australis* under flooded conditions (Wirsel, 2004). *P. australis* is often used in wastewater treatment wetlands, where its high primary production presents an effective means of nutrient retention (Vymazal & Kröpfelová, 2005). Further knowledge on ecology of mycorrhizal symbiosis in *P. australis* could contribute to better understanding of the functioning of such systems.

In roots of *P. australis* AM structures (arbuscules, vesicles and hyphal coils) as well as DSE structures (microsclerotia) were observed (Fig. 22.3). Mycorrhizal colonisation was generally very low, but differed considerably between the locations investigated. Reed samples from Location 1 had the lowest abundances of AM structures and the highest abundances of microsclerotia (Fig. 22.4). Mycorrhizal colonisation of reed samples from Locations 2 and 3 were similar, with a bit higher abundances present at Location 3.

Even though the mycorrhizal colonisation of *P. australis* with AM and DSE fungi was confirmed under various soil conditions and water levels, AM was more abundant in samples from soils poor in organic matter and P content (Location 3). The same observation was made by Khan (2004), who noticed a negative relationship between soil P concentrations and mycorrhizal colonisation in P-rich soils and a positive relationship in P-deficient soils. On the other hand, the reed stand at Location 1 had more of the DSE structures present. The role of DSE fungi in plants has not been established but is considered to have a similar role to that of arbuscular mycorrhiza (Jumpponen, 2001).

Locations 2 and 3 were flooded at the time of sampling, but the water level was higher at Location 3, as described in Section 2.1. However, the capacity of

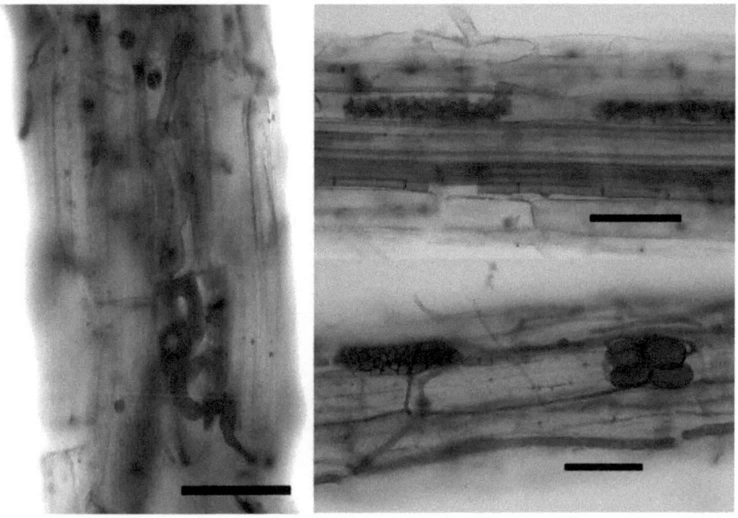

Fig. 22.3 Mycorrhizal structures observed in *Phragmites australis*: hyphal coils (*left*), arbuscules (*upper right picture*), vesicles (*lower right picture – right*) and microsclerotia (*lower right picture – left*). Scale bar is 50 μm

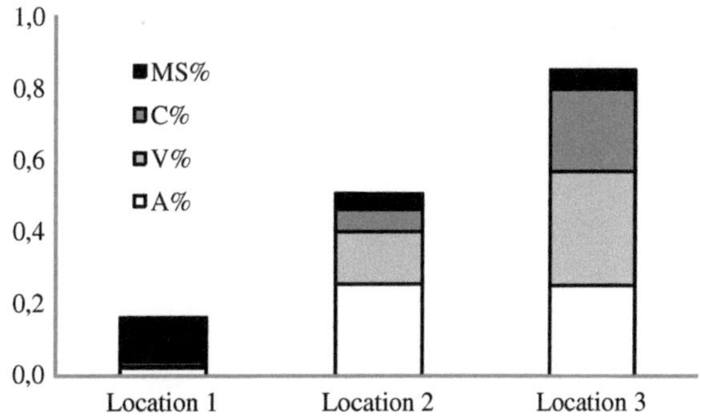

Fig. 22.4 Abundances of arbuscules (A%), vesicles (V%), hyphal coils (C%) and microsclerotia (MS%) in roots of *Phragmites australis* at three locations on Lake Cerknica

P. australis to vent its underground tissues by pressurized through-flow (Brix, Sorrell, & Schierup, 1996) could provide sufficient oxygen levels in the soil surrounding the roots of *P. australis* to sustain mycorrhizal colonisation.

22.3.3 *Mycorrhiza in Schoenus Ferrugineus and Schoenus Nigricans*

Plant species from the family Cyperaceae are not often investigated for mycorrhizal colonisation since they were long assumed to be nonmycorrhizal (Muthukumar, Udaiyan, & Shanmughavel, 2004). In our study, both species of *Schoenus* exhibited relatively high F% (many fragments were colonised with hyphae), but

Fig. 22.5 Abundances of arbuscules (A%), vesicles (V%), hyphal coils (C%) and microsclerotia (MS%) in roots of *Schoenus ferrugineus* and *S. nigricans* from two wet meadows on Lake Cerknica

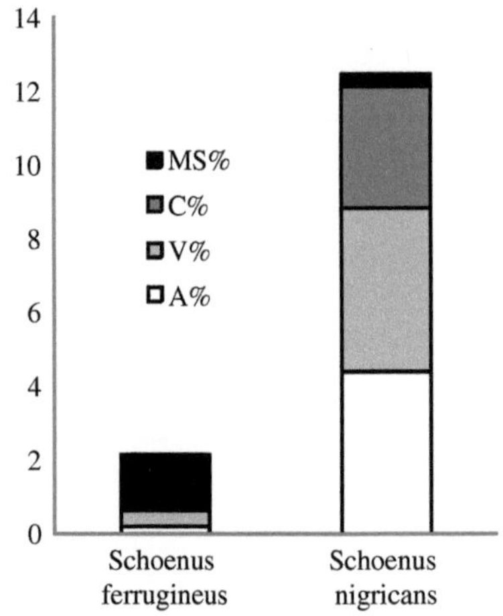

not many hyphal structures were present (Fig. 22.5). However differences between mycorrhizal colonisation of the two species were apparent, *S. ferrugineus* harboured abundant DSE microsclerotia and only a few arbuscules and vesicles were detected, while *S. nigricans* had abundant hyphal coils, vesicles and arbuscules.

According to the literature this is the first detailed report on the presence of AM and DSE mycorrhiza in *S. ferrugineus* and *S. nigricans* since Mejstrik (1972) found vesicular-arbuscular mycorrhiza in this two species, although wetland *Schoenus* species were recently reported as nonmycorrhizal (Fontenla, Puntieri, & Ocampo, 2001). According to our results, *S. ferrugineus* growing in organic soil showed similar mycorrhizal colonisation as *P. australis*, with abundant microsclerotia, while *S. nigricans* had abundant AM structures.

22.4 Conclusions

In intermittent wetlands, the extent of mycorrhizal colonisation in plant roots is directly or indirectly influenced by a variety of factors, such as soil characteristics, soil/water relations and plant characteristics. Which of these factors plays the major role in such ecosystems is hard to discern and probably varies from one wetland to another. Our research revealed that colonisation with AM fungi and DSE is common in wetland plants growing under varying water level and soil conditions. Mycorrhizal colonisation, facilitating the nutrient uptake and mitigating the effects of water stress, could be one of the factors aiding wetland plants in these habitats to withstand environmental changes occurring due to water level fluctuations. Moreover, mycorrhizal colonisation of *P. australis* could play an important role in constructed wetlands contributing to biomass accumulation and overall fitness of plants.

References

Augé, R. M. (2004). Arbuscular mycorrhizae and soil/plant water relations. *Canadian Journal of Soil Science, 84*, 373–381.

Bajwa, R., Yaqoob, A., & Javaid, A. (2001). Seasonal variation in VAM in wetland plants. *Pakistan Journal of Biological Sciences, 4*, 464–470.

Bauer, C. R., Kellogg, C. H., Brigham, S. D., & Lamberti, G. A. (2003). Mycorrhizal colonization across hydrologic gradients in restored and reference freshwater wetlands. *Wetlands, 23*, 961–968.

Bohrer, K. E., Friese, C. F., & Amon, J. P. (2004). Seasonal dynamics of arbuscular mycorrhizal fungi in differing wetland habitats. *Mychorrhiza, 14*, 329–337.

Braendle, R., & Crawford, R. M. M. (1999). Plants as amphibians. *Perspectives in Plant Ecology, Evolution and Systematics, 2*, 56–78

Brix, H., Sorrell, B. K., & Schierup, H.-H. (1996). Gas fluxes achieved by in situ convective flow in *Phragmites australis*. *Aquatic Botany, 54*, 151–163.

Cooke, J. C., & Lefor, M.,W. (1998). The mycorrhizal status of selected plant species from Connecticut wetlands and transition zones. *Restoration Ecology, 6*, 214–222.

Cornwell, W. K., Bedford, B. L., & Chapin, C. T. (2001). Occurrence of arbuscular mycorrhizal fungi in a phosphorus-poor wetland and mycorrhizal response to phosphorus fertilization. *American Journal of Botany, 88*, 1824–1829.

Dolinar, N., Rudolf, M., Šraj, N., & Gaberščik, A. (2010). Environmental changes affect ecosystem services of the intermittent Lake Cerknica. *Ecological Complexity, 7*, 403–409. doi:10.1016/j.ecocom.2009.09.004.

Fontenla, S., Puntieri, J., & Ocampo, J. A. (2001). Mycorrhizal associations in the Patagonian steppe, Argentina. *Plant and Soil, 233*, 13–29.

Gaberščik, A., Urbanc-Berčič, O., Kržič, N., Kosi, G., & Brancelj, A. (2003). The intermittent lake Cerknica: Various faces of the same ecosystem. *Lakes and Reservoirs: Research and Management*, 8, 159–168.

Gurevitch, J. (2002). *The ecology of plants*. Sunderland, MA: Sinauer Associates.

Hudon, C., Gagnon, P., & Jean, M. (2005). Hydrological factors controlling the spread of common reed (*Phragmites australis*) in the St. Lawrence River (Québec, Canada). *Écoscience, 12*, 47–357.

Jumpponen, A. (2001). Dark septate endophytes – are they mycorrhizal? *Mycorrhiza, 11*, 207–211.

Jumpponen, A., & Trappe, J. M. (1998). Dark septate endophytes: A review of facultative biotrophic root-colonizing fungi. *New Phytologist, 140*, 295–310.

Khan, A. G. (2004). Mycotrophy and its significance in wetland ecology and wetland management. In M. H. Wong (Ed.), *Wetlands ecosystems in Asia: Function and management* (pp. 95–114). Amsterdam: Elsevier.

Mandyam, K., & Jumpponen, A. (2005). Seeking the elusive function of the root-colonising dark septate endophytic fungi. *Studies in Mycology, 53*, 173–189.

Mauchamp, A., Méthy, M. (2004). Submergence-induced damage of photosynthetic apparatus in *Phragmites australis*. *Environmental and Experimental Botany, 51*, 227–235.

Mejstrik, V. K. (1972). Vesicular-arbuscular mycorrhizas of the species of a Molinietum coeruleae L.I. association: the ecology. *New Phytologist, 71*, 883–890.

Miller, S. P. (2000). Arbuscular mycorrhizal colonization of semi-aquatic grasses along a wide hydrologic gradient. *New Phytologist, 145*, 145–155.

Miller, S. P., & Bever, J. D. (1999). Distribution of arbuscular mycorrhizal fungi in stands of the wetland grass *Panicum hemitomon* along a wide hydrologic gradient. *Oecologia, 119*, 586–592.

Muthukumar, T., Udaiyan, K., & Shanmughavel, P. (2004). Mycorrhiza in Sedges – an overview. *Mycorrhiza, 14*, 65–77.

Oliveira, R. S., Dodd, J. C., & Castro, P. M. L. (2001). The mycorrhizal status of *Phragmites australis* in several polluted soils and sediments of an industrialised region of Northern Portugal. *Mycorrhiza, 10*, 241–247.

Philips, J. M., & Hayman, D. S. (1970). Improved procedures for clearing roots and staining parasitic and vesicular–arbuscular mycorrhizal fungi for rapid assessment of infection. *Transactions of British Mycological Society, 55*, 158–160.
Ray, A. M., & Inouye, R. S. (2006). Effects of water-level fluctuations on the arbuscular mycorrhizal colonization of *Typha latifolia* L. *Aquatic Botany, 84*, 210–216.
Smith, S. E., & Read, D. J. (1997). Mycorrhizal Symbiosis. London: Academic.
Stevens, K. J., & Peterson, R. L. (1996). The effect of a water gradient on the vesicular-arbuscular mycorrhizal status of *Lythrum salicaria* L. (purple loosestrife). *Mycorrhiza, 6*, 99–104.
Šraj-Kržič, N., Pongrac, P., Klemenc, M., Kladnik, A., Regvar, M., & Gaberščik, A. (2006). Mycorrhizal colonisation in plants from intermittent aquatic habitats. *Aquatic Botany, 85*, 331–336.
Šraj-Kržič, N., Pongrac, P., Regvar, M., & Gaberščik, A. (2009). Photon-harvesting efficiency and arbuscular mycorrhiza in amphibious plants. *Photosynthetica, 47*, 61–67.
Trouvelot, A., Kough, J. L., & Gianinazzi-Pearson, V. (1986). Mesure de taux de mycorhization VA d'un systeme radiculaire. Recherche de méthodes d'estimation ayant une signification fonctionnelle. In V. Gianinazzi-Pearson & S. Gianinazzi (Eds.), *Physiological and genetical aspects of mycorrhizae* (pp. 217–221). Paris: INRA.
Turner, S. D., & Friese, C. F. (1998). Plant-mycorrhizal community dynamics associated with moisture gradient within a rehabilitated prairie fen. *Restoration Ecology, 6*, 44–51.
Urbanc-Berčič, O., & Gaberščik, A. (2003). Microbial activity in the rhizosphere of common reed (*Phragmites australis*) in the intermittent lake Cerkniško jezero. In J. Vymazal (Ed.), *Wetlands – nutrients, metals and mass cycling* (pp. 179–190). Leiden: Backhuys.
Vretare, V., Weisner, S. E. B., Strand, J. A., & Graneli, W. (2001). Phenotypic plasticity in *Phragmites australis* as a functional response to water depth. *Aquatic Botany, 69*, 127–145.
Vymazal, J., & Kröpfelová, L. (2005). Growth of *Phragmites australis* and *Phalaris arundinacea* in constructed wetlands for wastewater treatment in the Czech Republic. *Ecological Engineering, 25*, 606–621.
Weishampel, P. A., & Bedford, B. L. (2006). Wetland dicots and monocots differ in colonization by arbuscular mycorrhizal fungi and dark septate endophytes. *Mycorrhiza, 16*, 495–502.
Wirsel, S. G. R. (2004). Homogenous stands of a wetland grass harbour diverse consortia of arbuscular mycorrhizal fungi. *FEMS Microbiology Ecology, 48*, 129–138.

Chapter 23
Comparison of Reflected Solar Radiation, Air Temperature and Relative Air Humidity in Different Ecosystems: From Fishponds and Wet Meadows to Concrete Surface

Hanna Huryna and Jan Pokorný

Abstract Incoming and reflected shortwave solar radiation, temperature and relative humidity of air at 0.3 and 2 m were measured continuously during the vegetation season (from April 1 to September 30, 2008) at six different sites: fishpond, wet meadows, pasture, barley field, village and concrete surface. All six sites are located in the basin of Třeboň Biosphere Region, Czech Republic. For data evaluation the 183 days of the vegetation season were divided into three classes according to the amount of incoming solar energy they received. The albedo of water was two times lower than the albedo of the field and three times lower than the albedo of the concrete surface, which also had the highest average temperature. The lowest average temperature was measured at the wet meadows. The lowest average daily amplitude in temperature (difference between daily maximum and minimum) was measured at the fishpond. The highest difference (3.58°C) in average temperature at 0.3 m among sites was found for wet meadows and concrete surface on clear days and 2.49°C for all days of the vegetation season. Daily time courses of relative air humidity on sunny days show the ability of vegetation to buffer air humidity extremes. We conclude that changes of land cover results in changes of average temperature.

Keywords Albedo · Local climate · Solar energy income · Wetlands

23.1 Introduction

The role of wetlands and vegetation in Earth's energy budget and in climate change is mostly considered through the greenhouse effect (IPCC, 2007). Of all natural wetland types, peatlands represent a highly important sink for carbon. In addition to live peatlands, littoral wetlands with abundant plant cover, such as reed-dominated marshes, can be important sinks for carbon (Brix, Sorrell, & Lorenzen, 2001).

J. Pokorný (✉)
ENKI, o.p.s., Třeboň, Czech Republic
e-mail: pokorny@enki.cz

J. Vymazal (ed.), *Water and Nutrient Management in Natural and Constructed Wetlands*, DOI 10.1007/978-90-481-9585-5_23,

Floodplains can play an important role for accumulating organic matter and carbon (Čížková et al., 2010). Radiative forcing, caused by greenhouse gases produced since the start of the industrial revolution in 1750, is estimated to range from 1 to 3 W m^{-2}. According to the Intergovernmental Panel on Climate Change (IPCC), radiative forcing will continue to increase over the next 10 years by 0.2 W m^{-2} due to the greenhouse effect.

Vegetation affects solar energy distribution and the energy budget of the Earth's surface directly through its influence on surface reflectivity (albedo) and dissipation of solar energy by evapotranspiration (Ripl, 2003; Pokorný & Rejšková, 2008). The latent heat of water is equal to 2.45 MJ L^{-1}, i.e., 0.68 kWh L^{-1} at 20°C. Wetlands evaporate several liters of water per square meter a day, i.e. several hundreds of W m^{-2} of solar energy are transformed into kinetic energy of water vapour. Under suitable conditions, and if the water supply is sufficient, the prevailing part of the incoming energy can be transformed into latent heat flux. This energy is later released at cooler places when water vapour condensates back to liquid. If water is not present, solar energy is transformed mostly into turbulent sensible heat flux, which increases temperature near the surface.

Although the effects of vegetation on local as well as regional climate through changing energy fluxes and thus temperatures near the surface have been acknowledged – though not always well understood (e.g. Kabat et al., 2004) – continuous data comparing solar energy fluxes in various ecosystems located in the same climatic region are still missing.

For the Project of National Research Program of the Czech Republic entitled "Development of a method for determination of energy fluxes in ecosystems", twelve different sites were chosen and equipped with meteorological stations in order to follow solar energy fluxes and temperature differences of different ecosystems. Incoming and reflected solar radiation, air temperature, air humidity, soil temperature, precipitation, wind direction and wind speed were monitored continuously. In this study we evaluated incoming and reflected solar radiation, air temperature and air humidity at six sites in order to compare albedo, temperature and relative humidity changes in ecosystems with markedly different vegetation cover and water abundance throughout a growing season. We compared a barley field, a graminoid wetland, a concrete area, the open water surface of a fishpond, a pasture and a partly built up area in a village. In addition to the temperature averages, maximum temperatures among the sites were evaluated to study the potential of land cover to mitigate temperature extremes. In the first approach in this article, the sites were compared on the basis of data from the whole vegetation season.

23.2 Methods and Site Descriptions

23.2.1 Meteorological Data

Micro-meteorological stations monitored and recorded air temperature and relative air humidity at two height levels (0.3 and 2.0 m), incident and reflected global solar radiation in the shortwave region (0.31–2.8 μm by CM3 pyranometer Kipp@Zonen, The Netherlands), soil temperature and humidity, precipitation, and the direction and speed of wind at a height of 2 m at 10-min intervals. Additionally, soil temperature was measured at every 10 mm for the first 100 mm of depth and at 200 and 250 mm of depth, atmospheric pressure and incident and emitted IR radiation 4.5–45 μm and the temperature of the detector were monitored at one of the 12 sites. The collected data from all stations were automatically transferred by a GPRS network three times a day to a server, where they were stored in a database accessible for all researchers participating in the project (Jirka, Hofreiter, Pokorný, & Novák, 2009). The accuracy of the temperature sensors was 0.1°C, and the accuracy of the air's relative humidity measurements was 2.5% in the range from 5 to 95%. The pyranometer sensors CM3 were tested at one site before installation at individual sites; their variability was within 1%.

23.2.2 Data Processing

The data were collected for 183 days during the growing season of 2008, from April 1st to September 30th. In this paper, data on incoming and reflected solar radiation, air temperature and relative air humidity (0.3 and 2.0 m) were evaluated. The data were analyzed using SPSS software version 17.0.

The data were divided according to the amount of total incoming shortwave solar irradiance per day into three groups: overcast (0–3000 Wh m^{-2}), cloudy (3000–6000 Wh m^{-2}) and clear (over 6000 Wh m^{-2}). Numbers of overcast, cloudy and clear days in the measuring period for each site are given in Table 23.1. The lowest number of evaluated days was for cloudy days at the field (42), the highest number of evaluated days was for sunny days at the village (82).

Mean hourly values were plotted in figures for each meteorological parameter monitored during the whole vegetation period ($n = 26352$ observation). For Tables we analyzed daily average values ($n = 183$ observation). The meteo-station at the fishpond (Ruda) did not function in periods 10.7–19.7 and 10.8–22.8, 2008. The missing data were replenished by data from a station located in the littoral of the same fishpond.

Linear analyses were used to identify the relations between the average hourly values of incoming solar radiations mean reflected solar radiation, mean air temperature, mean humidity, mean net radiation, mean albedo and time at

Table 23.1 Number of overcast, cloudy and clear days evaluated

Sites	Overcast	Cloudy	Clear
Field	42	78	63
Wet meadows	45	80	58
Concrete surface	47	73	63
Fishpond	44	77	62
Pasture	44	73	66
Domanín village	45	82	56

different sites. A confidence interval of 95% was used and p values less than 0.05 were considered significant.

23.2.3 Site Descriptions

The studied sites were located in the Třeboň Biosphere Reserve (TBR) within a distance of several km (Fig. 23.1). Standard hydro-meteorological data for the TBR were provided by the Czech Hydrometeorological Institute (Třeboň and Borkovice

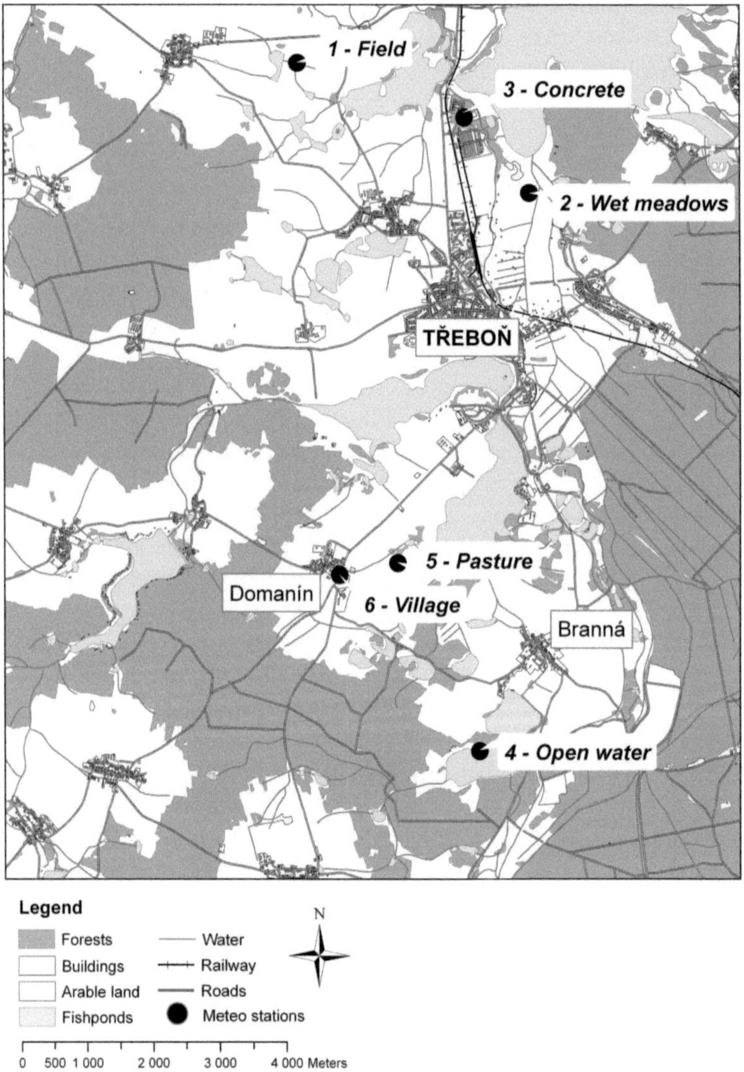

Fig. 23.1 Maps of monitored sites in Třeboň Biosphere Reserve

stations). Long term meteorological data from the TBR used were adapted from Pokorný and Kučerová (2000) and from Kovářová and Pokorný (2005).

The data were collected at the following six sites:

1. Autumn barley field (area of 22 ha)
2. Wet meadows (area of c. 200 ha) in the Rožmberk fishpond (450 ha) floodplain. Dominant macrophyte species: high sedges (*Carex gracilis, Carex vesicaria*), *Calamagrostis canescens, Phalaris arundinacea, Urtica dioica.* The area surrounding the meteorological station is not managed; tall vegetation is regularly cut only in the close vicinity of the meteorological station. The dryer parts of the wet meadows are mowed once a year.
3. Concrete surface (area of 400 m^2) within the area of the Wastewater Treatment Plant of the city of Třeboň.
4. Open water surface of the Ruda fishpond (72 ha, depth c. 2 m)
5. Pasture with a relatively high underground water level with dominant macrophytes *Alopecurus pratensis, Ranunculus repens, Phleum pratense, Dactylis glomerata, Bellis perennis, Poa palustris, Trifolium repens, Taraxacum sect. Ruderalia, Veronica chamaedrys etc.*
6. Built-up area within the Domanín village (the meteo-station is situated on a small lawn between a small family house, small pond and a street)

23.3 Results

The daily courses of the incoming solar radiation were similar at all sites (Fig. 23.2). The average maximum solar radiation fluxes reached about 250, 600 and 890 W m^{-2} on overcast, cloudy and clear days, respectively. The incoming solar radiation on the wet meadows was on clear days slightly lower than at other sites, perhaps due to higher fog formation in the terrain depression of this habitat. Mean fluxes of incoming solar radiation (W m^{-2}) for all sites are given in Table 23.2. The mean daily fluxes ranged from 80.4 to 83.5 W m^{-2}, from 191.3 to 293.9 W m^{-2} and from 286.3 to 293.9 W m^{-2} on overcast, cloudy and clear days, respectively.

The mean daily sums of incoming solar energy calculated from the mean fluxes are given in Table 23.3. The mean daily values of incoming solar energy at six sites on clear days ranged from 7.06 kWh m^{-2} in the field to 6.87 kWh m^{-2} in the wet meadows.

The amount of reflected shortwave radiation differed markedly between the sites (Fig. 23.3). The reflection of the solar radiation was always lowest at the open water surface, not exceeding 50 W m^{-2} even on clear days. On clear days the highest reflection, as high as 220 W m^{-2}, was measured on the concrete surface.

Mean values of reflected solar radiation are given in Table 23.4. On clear days, the highest mean flux of reflected solar radiation (76.93 W m^{-2}) was at the concrete surface. The fishpond showed the lowest value of mean reflected solar radiation (24.66 W m^{-2}). The mean fluxes of reflected solar energy at the remaining sites were close to each other (from 57.37 to 63.63 W m^{-2}).

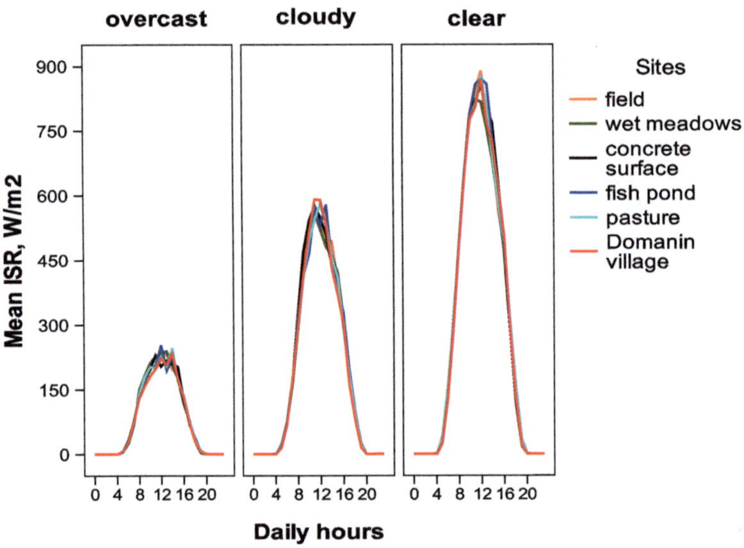

Fig. 23.2 Daily mean time courses of incoming shortwave solar radiation (ISR) for overcast, cloudy and clear days

Table 23.2 Mean incoming solar radiation (W $m^{-2} \pm$ SD) monitored at six sites during vegetation season from April 1 to October 31, 2008

Sites	Overcast	Cloudy	Clear
Field	82.75 ± 31.19	192.97 ± 44.61	294.27 ± 37.02
Wet meadows	80.37 ± 32.84	191.52 ± 41.78	286.25 ± 34.05
Concrete surface	83.05 ± 32.68	193.19 ± 47.73	290.44 ± 40.99
Fishpond	83.52 ± 32.50	195.05 ± 45.49	291.98 ± 42.36
Pasture	82.91 ± 32.02	194.28 ± 39.74	293.45 ± 37.40
Domanín village	82.46 ± 30.19	193.07 ± 42.16	293.85 ± 37.14

Table 23.3 Mean daily income of solar energy kWh $m^{-2} \pm$ SD, calculated from mean incoming solar radiation (see Table 23.2)

Sites	Overcast	Cloudy	Clear
Field	1.99 ± 0.75	4.59 ± 1.11	7.06 ± 0.89
Wet meadows	1.93 ± 0.79	4.68 ± 1.00	6.87 ± 0.81
Concrete surface	1.99 ± 0.78	4.64 ± 1.15	6.94 ± 0.98
Fishpond	2.00 ± 0.78	4.68 ± 1.09	7.00 ± 1.02
Pasture	1.99 ± 0.77	4.66 ± 0.94	7.04 ± 0.90
Domanín village	1.98 ± 0.75	4.63 ± 1.07	7.05 ± 0.86

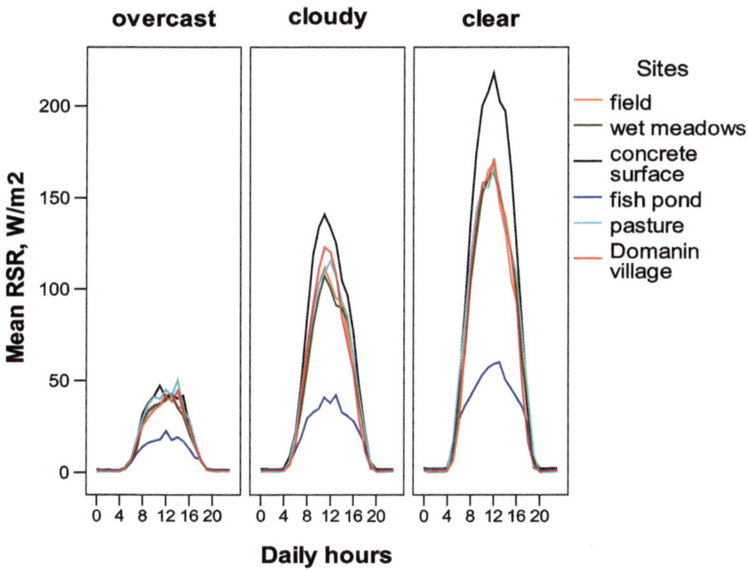

Fig. 23.3 Daily mean time courses of reflected shortwave solar radiation (RSR) for overcast, cloudy and clear days. For details see Fig. 23.2

Table 23.4 Mean reflected solar radiation (W $m^{-2} \pm$ SD) monitored at six sites during vegetation season from April 1 to October 31, 2008

Sites	Overcast	Cloudy	Clear
Field	15.82 ± 6.50	39.43 ± 14.14	58.99 ± 8.65
Wet meadows	14.35 ± 6.96	38.19 ± 10.13	63.63 ± 15.07
Concrete surface	18.09 ± 8.91	49.09 ± 14.35	76.93 ± 14.91
Fishpond	8.56 ± 3.93	16.83 ± 6.24	24.66 ± 8.74
Pasture	17.36 ± 6.79	41.55 ± 8.59	62.09 ± 7.98
Domanín village	15.27 ± 6.46	39.51 ± 9.86	57.37 ± 6.98

The total daily amounts of solar radiation reflected from the surface at the different sites are given in Table 23.5. The highest daily average of reflected solar radiation on clear days was 1.85 kWh m^{-2} at the concrete surface and the lowest average value of the daily reflected solar radiation was 0.59 kWh m^{-2} at the fishpond.

Daily mean time courses of albedo, which is defined as the ratio of reflected to incoming solar radiation, are shown in Fig. 23.4. Average albedo of the water surface was always the lowest from all the evaluated sites (about 10%). Albedo of the concrete surface was about 25% on both cloudy and clear days. On overcast days it was about 20% at this site. Albedo for other sites was about 20% under all irradiance conditions. The high values of albedo in the early morning and late afternoon hours were caused by dividing very small numbers of both variables of

Table 23.5 Mean daily reflected amount of solar energy kWh m^{-2} ± SD, calculated from mean daily flux (Table 23.3)

Sites	Overcast	Cloudy	Clear
Field	0.38 ± 0.16	0.94 ± 0.34	1.42 ± 0.21
Wet meadows	0.34 ± 0.17	0.93 ± 0.24	1.53 ± 0.27
Concrete surface	0.43 ± 0.21	1.18 ± 0.34	1.85 ± 0.36
Fishpond	0.20 ± 0.09	0.40 ± 0.15	0.59 ± 0.21
Pasture	0.42 ± 0.16	1.00 ± 0.21	1.49 ± 0.19
Domanín village	0.37 ± 0.156	0.95 ± 0.24	1.38 ± 0.17

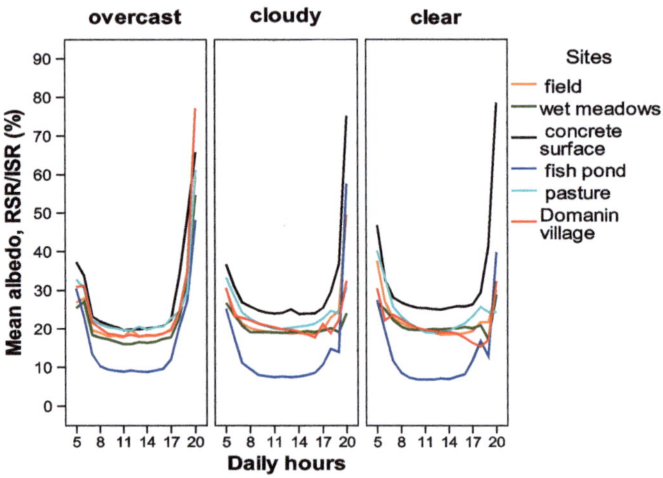

Fig. 23.4 Daily mean time courses of albedo (reflected solar radiation/incoming solar radiation) for overcast, cloudy and clear days. For details see Fig. 23.2

the ratio, i.e. incoming and reflected solar radiation, and thus do not provide much useful information.

Net shortwave radiation (Rsn) (Fig. 23.5) was calculated as a difference of incoming and reflected shortwave radiation. No correction for IR was made. The fishpond had the highest value of net shortwave radiation; the concrete surface and the wet meadows had the lowest values of Rsn. Mean maximum net shortwave solar radiation reached 800 W m^{-2} at the fishpond site on clear days whereas on the concrete surface it reached only 650 W m^{-2} and on other sites Rsn about 700 W m^{-2}.

Mean daily fluxes of Rsn can be calculated as the difference between mean incoming and reflected radiation from Tables 23.2 and 23.3. On sunny days, the mean fluxes of Rsn were 267.32 W m^{-2} at the fishpond site, 222.6 W m^{-2} at the wet meadows site and 213.51 W m^{-2} at the concrete surface site. The wet meadows had relatively lower incoming radiation and, among vegetated sites, the wet meadows showed relatively higher reflection of shortwave radiation.

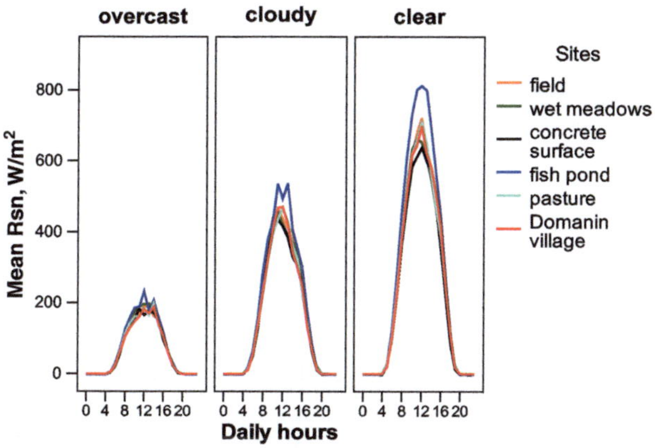

Fig. 23.5 Daily mean time courses of net shortwave radiation (Rsn) for overcast, cloudy and clear days. For details see Fig. 23.2

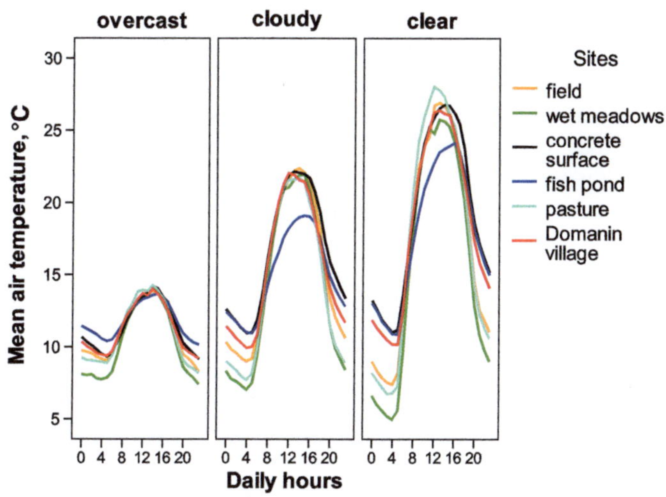

Fig. 23.6 Daily mean time courses of air temperature above ground (30 cm) for overcast, cloudy and clear days. For details see Fig. 23.2

Daily courses of the air temperature above ground (30 cm) for overcast, cloudy and clear days are plotted in Fig. 23.6. On clear days at 30 cm, the lowest average midday temperature was at the fishpond; the other sites had similar midday temperatures. On overcast days the mean temperature was similar at all sites during the whole day. Low early morning temperature in the wet meadows can be explained by the terrain depression of the flood area of Rožmberk fishpond. The early morning temperature was on average highest in the fishpond due to the high heat capacity of its water. Table 23.6 shows average daily temperatures at 30 cm at all sites. On overcast days the difference of mean temperature between the sites ranged from

Table 23.6 Mean temperature (°C ± SD) at 30 cm for all sites from April 1 to October 31, 2008

Sites	Overcast	Cloudy	Clear
Field	10.94 ± 3.99	15.29 ± 4.75	17.09 ± 4.38
Wet meadows	10.21 ± 3.72	14.08 ± 3.94	15.58 ± 3.90
Concrete surface	11.33 ± 4.26	16.52 ± 4.64	19.16 ± 4.44
Fishpond	11.80 ± 4.40	14.96 ± 4.81	17.77 ± 4.25
Pasture	10.95 ± 4.22	14.42 ± 4.76	17.34 ± 4.30
Domanín village	11.25 ± 4.16	15.59 ± 4.58	18.42 ± 4.08

10.21 (wet meadows) to 11.80°C (fishpond). On cloudy and clear days, the lowest mean temperature was measured in the wet meadows (14.08 and 15.58°C, respectively) and the highest mean temperature at 30 cm was measured at the concrete surface (16.52 and 19.16°C, respectively). On clear days, the average difference of temperature between the wet meadows and the concrete surface was 3.58°C. The average temperature on clear days at the field was 17.09°C, at the pasture 17.34°C, the difference with the concrete site was 2.07 and 1.82°C.

The average air temperatures at 2 m during overcast, cloudy and clear days are plotted in Fig. 23.7. The pasture and the concrete surface sites showed the highest values of the mean midday temperature at 2 m. The lowest air temperature at midday was on both cloudy and clear days at the fishpond. The early morning temperatures were lowest in the wet meadows. Table 23.7 shows mean daily temperatures at all sites on the overcast, cloudy and clear days. The average temperature difference between the sites on the overcast days was 1.05°C (10.52°C at the wet meadows and 11.57°C at the fishpond). The average temperature difference between the sites on clear days was 2.1°C (16.7°C at the wet meadows and 18.74°C at the village site).

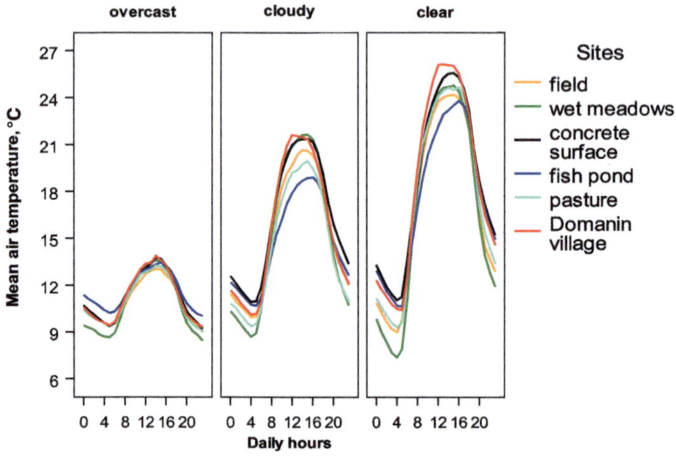

Fig. 23.7 Daily mean time courses of air temperature at 2 m for overcast, cloudy and clear days. For details see Fig. 23.2

Table 23.7 Mean temperature (°C ± SD) at 2 m for all sites from April 1 to October 31, 2008

Sites	Overcast	Cloudy	Clear
Field	10.86 ± 3.96	15.18 ± 4.79	17.12 ± 4.65
Wet meadows	10.52 ± 4.17	15.13 ± 4.49	16.70 ± 4.59
Concrete surface	11.21 ± 4.23	16.12 ± 4.80	18.64 ± 4.52
Fishpond	11.57 ± 4.49	14.64 ± 5.02	17.54 ± 4.31
Pasture	10.96 ± 4.21	14.46 ± 4.84	17.58 ± 4.27
Domanín village	11.21 ± 4.24	15.60 ± 4.84	18.74 ± 4.15

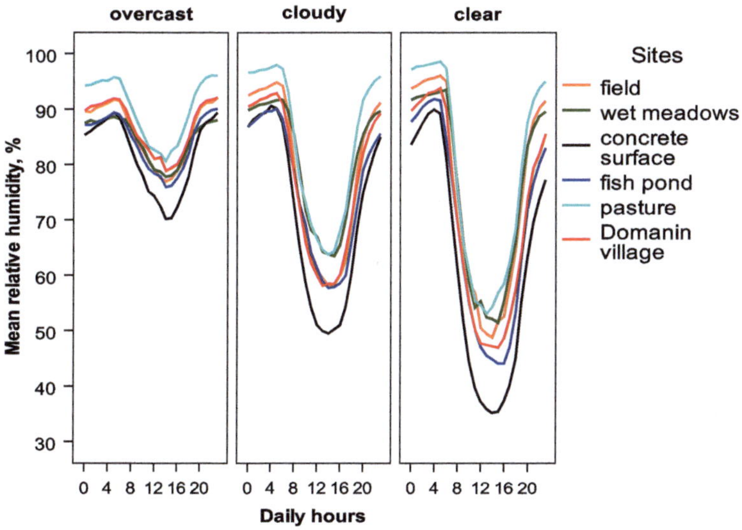

Fig. 23.8 Daily mean time courses of air relative humidity above ground (30 cm) for overcast, cloudy and clear days. For details see Fig. 23.2

The village Domanín and the concrete surface had on clear days the highest average temperature at 2 m (18.74°C and 18.64°C, respectively). The wet meadows had the lowest average daily temperature (16.70°C).

The time courses of the average values of the air's relative humidity at 30 cm above ground are plotted in Fig. 23.8. The relative humidity values of air differed even on cloudy days. The lowest values of the relative air humidity were, under all irradiance conditions, measured at the concrete surface. On the contrary, the highest values of the mean relative humidity were measured at the pasture and at the wet meadows. Rather low values of relative humidity were surprisingly recorded above the open water surface of the fishpond site on overcast and cloudy days. On clear days during midday, the relative humidity of air at the concrete surface was about 20% lower than at wet meadows and in pasture.

Mean daily values of the relative air humidity are given in Table 23.8. The pasture, the wet meadows and the field had on clear days the highest mean air humidity

Table 23.8 Mean relative humidity of air (% ± SD) at 30 cm for all sites from April 1 to October 31, 2008

Sites	Overcast	Cloudy	Clear
Field	86.26 ± 7.56	78.98 ± 7.99	75.45 ± 8.27
Wet meadows	84.80 ± 6.35	80.70 ± 6.81	75.64 ± 8.32
Concrete surface	81.79 ± 7.44	71.39 ±.6.67	61.81 ± 5.86
Fishpond	84.36 ± 6.44	75.74 ± 6.41	68.23 ± 5.39
Pasture	90.27 ± 5.65	84.11 ± 6.24	78.95 ± 6.52
Domanín village	87.06 ± 5.46	77.35 ± 5.26	70.13 ± 5.24

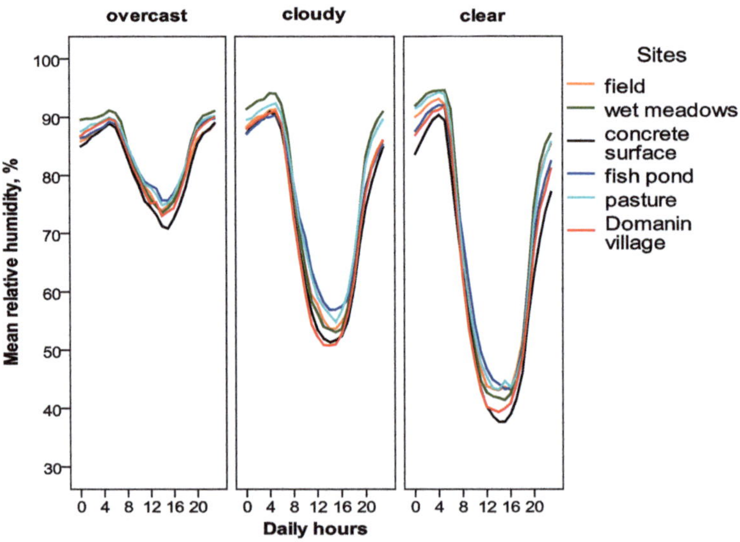

Fig. 23.9 Daily mean time courses of air relative humidity (2 m) for overcast, cloudy and clear days. For details see Fig. 23.2

at 30 cm (78.95, 75.64 and 75.45 % respectively). The concrete surface site showed the lowest mean values of air humidity (61.81% on clear days).

The values of mean relative humidity of the air at 2 m throughout the clear, cloudy and overcast days are plotted in Fig. 23.9. The differences in relative air humidity at 2 m are smaller than the differences in relative air humidity at 30 cm above ground. The highest values of relative air humidity at 2 m were, on all studied days, recorded above the water surface of the fishpond. The village site and the concrete surface sites showed the lowest values both early in the morning and during midday. The relative air humidity at midday was about 10% lower at wet meadows and pasture than at the concrete surface.

The wet meadows, the pasture and the barley field showed on clear days the highest mean air humidity at 2 m (68.92, 68.57 and 68.24%, respectively) (Table 23.9). The concrete surface showed the lowest mean value of the air humidity (62.95%).

Table 23.9 Mean relative humidity of air (% ± SD) at 2 m for all sites from April 1 to October 31, 2008

Sites	Overcast	Cloudy	Clear
Field	83.06 ± 7.15	74.40 ± 6.65	68.24 ± 5.88
Wet meadows	84.62 ± 6.63	76.06 ± 5.26	68.92 ± 5.12
Concrete surface	81.73 ± 7.17	72.19 ± 6.55	62.95 ± 6.01
Fishpond	84.01 ± 6.57	75.44 ± 6.74	67.57 ± 5.61
Pasture	84.26 ± 5.72	76.07 ± 6.09	68.57 ± 4.65
Domanín village	83.54 ± 6.47	72.33 ± 5.89	64.49 ± 5.37

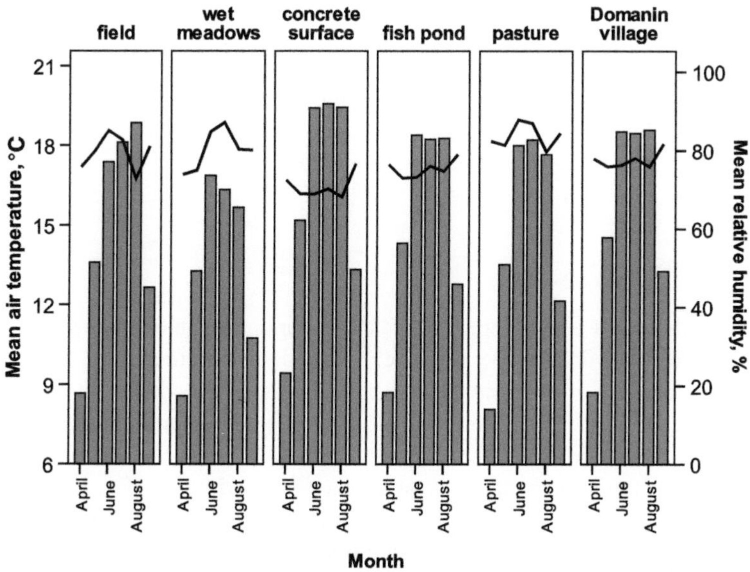

Fig. 23.10 Daily means of relative humidity and air temperature at 30 cm calculated for individual months of vegetation season. The lines indicate mean daily relative humidity measured at the surface of plant stand, the columns indicate mean daily temperature per month measured at 30 cm

Monthly means of air relative humidity and air temperature measured at 30 cm height, calculated per each month of vegetation season (from April to September), are plotted in Fig. 23.10. The lowest average temperatures were during all months measured in the wet meadows; the highest temperatures were measured at the concrete surface. The temperatures in the field were relatively low in May and in June when crop (barley) formed a high active biomass.

Mean temperature at 30 cm for all days during the vegetation season at all six sites is given in Table 23.10. The concrete surface in the area of a wastewater treatment plant has the highest daily average temperature at 30 cm for all days during the vegetation period (16.09°C). The wet meadows has the lowest value of daily average temperature equal to 13.60°C. Mean temperature at 2 m for all days during

Table 23. 10 Mean temperature (°C) at 30 cm for all days from April 1 to October 31, 2008

	Field	Wet meadows	Concrete surface	Fishpond	Pasture	Domanín village
Temperature	14.91	13.60	16.09	15.15	14.64	15.39
SD	5.00	4.37	5.38	5.04	5.07	5.06

Table 23.11 Mean temperature (°C) at 2 m for all days from April 1 to October 31, 2008

	Field	Wet meadows	Concrete surface	Fishpond	Pasture	Domanín village
Temperature	14.85	14.49	15.73	14.89	14.74	15.48
SD	5.11	5.02	5.37	5.16	5.14	5.26

the vegetation season at all six sites are given in Table 23.11. The concrete surface in the area of the wastewater treatment plant has the highest average air temperature at 2 m for all days during the vegetation period (15.73°C). The wet meadows has the lowest value of air temperature equal to 14.49°C.

23.4 Discussion

The aim of the paper is to compare the fate of solar energy in two types of wetlands (wet meadows and fishpond) with a field, sealed surface (concrete), pasture and village. The amount of shortwave incoming solar radiation which contains most of the incoming solar energy is similar at all sites: average mean fluxes on clear days ranged from 286.25 W m^{-2} (wet meadows) to 294.27 W m^{-2} at the village. Lower incoming solar radiation at the wet meadows is caused by fog formed during the night; the lowest morning temperature minimum can occur on overcast, cloudy or clear days. The differences in average incoming solar radiation among the other five sites (except wet meadows) ranged from 290.44 to 294.22 W m^{-2}.

Daytimes in the vegetation season were divided into three classes according to amount of incoming shortwave radiation. As expected, the differences in distribution of solar energy were most pronounced on clear days when daily income of solar energy was over 6 kWh m^{-2}. Marked differences among sites were found in the reflection of solar energy. The open water of the fishpond showed the lowest reflection of all monitored sites. On clear days the mean flux of reflected shortwave solar radiation from the fishpond (24.66 W m^{-2}) was two times lower than radiation reflected from the field (58.99 W m^{-2}) and three times lower than radiation reflected from the concrete surface (76.99 W m^{-2}). In terms of albedo, fishpond water reflects c. 10% of incoming shortwave solar radiation on overcast, cloudy and clear days whereas albedo of concrete is about 25%. Albedo at other sites was around 20%.

Net shortwave radiation (Rsn) inversely relates to albedo. The highest mean Rsn on clear days showed in the fishpond water (6.41 kWh m^{-2} d^{-1}), the lowest Rsn appeared on the concrete surface (5.09 kWh m^{-2} d^{-1}), and the Rsn of the field was 5.63 kWh m^{-2} d^{-1}. A relatively low Rsn of wet meadows (5.34 kWh m^{-2} d^{-1}) in comparison with the pasture (5.55 kWh m^{-2} d^{-1}) and the village (5.67 kWh m^{-2} d^{-1}) results both from lower income of solar radiation (due to morning fog) and slightly higher reflection in spring caused by last year's dead biomass. The data on incoming and reflected radiation of water monitored at fishpond Ruda were compared with data monitored by the same type of meteo-station on the other two fishponds of TBR (Rod, Naděje). The albedo of water at the other two fishponds was also c. 10%, the same as at fishpond Ruda, and differs markedly from the albedo of terrestrial sites (20–25%).

On overcast days the mean temperature was similar at all sites during the whole day. On clear days, the temperature at 30 cm height is influenced by ground surface and type of vegetation. The series of daytime air temperatures on overcast days were similar at all sites both at 30 cm and 2 m. During the night and early morning, wet meadows showed the lowest temperature and the fishpond the highest temperature. We explain the low temperature at wet meadows by location of the site in a relatively low part of Třeboň basin where, during the night, cold air flows from the several meters higher surroundings. Early morning fog at wet meadows indicates that effect. The highest difference (3.58°C) in temperature among the sites at 30 cm was found for wet meadows and the concrete surface on clear days. Both sites are located close together near a large fishpond Rožmberk (see Fig. 23.1). The average temperature on clear days at the field was 17.09°C, at the pasture 17.34°C; the difference with the concrete site was 2.07 and 1.82°C.

The highest midday air temperature at 2 m was reached at the pasture and the concrete surface. It can be explained by absence of water and vegetation on the concrete surface and by an average low amount of vegetation and trampled soil on pasture. The lowest air temperature at midday was on both cloudy and clear days at the fishpond site. The midday temperature values at the fishpond in comparison with values at the concrete and pasture sites differs by about 3°C. On overcast days the midday air temperature was similar at all sites. On clear days, the average temperature difference at 2 m between the sites was 2.1°C (16.7°C at wet meadows, 18.74°C at the village site, and 18.64°C at concrete). The highest mean daily temperatures at village and concrete surfaces are associated both with high midday values and relatively high early morning values. Fishpond water evidently damps temperature extremes between night and day hours.

Particularly remarkable are high differences of early morning temperature among sites. The differences are highest on clear days at 30 cm (c. 10°C). The early morning temperatures at wet meadows, pasture and village are lower on clear days than on cloudy or overcast days. At low solar radiation income, daily maxima of temperature are similar at all sites whereas early morning minima differ. Infrared radiation fluxes play an important role in temperature dynamics, particularly at night. Jirka et al. (2009) described a method for measurement and evaluation of IR fluxes developed for our project. IR fluxes at monitored sites will be evaluated in another paper.

Daily series of relative air humidity at 30 cm differ on overcast, cloudy and clear days. The sites differ in relative humidity within a range of 20% during a clear midday. The lowest values of relative air humidity are shown at the concrete surface, i.e. air moving from a dry surface has the capacity to take and transport most of its water vapour. wet meadows and pasture show the highest air humidity. Daily time courses of relative air humidity on sunny days (Fig. 23.8) show the ability of vegetation to buffer air humidity extremes – the concrete surface shows 20% lower air humidity than wet meadows. The relative air humidity above the water level of the fishpond (68.23%) is lower than expected. Similar values were measured also at two other open water sites. The values of relative air humidity above water level should be verified and explained on the basis of detailed on-site measurements next season.

At 2 m height, the relative humidity of air is highest at the fishpond and the values at all sites were in a range of 10%. On clear days at the concrete surface, relative air humidity fluctuates from 35% at midday up to the night dew point, whereas at wet meadows relative humidity of air does not fall under 50%. Evaluation of the length of the dew period during night hours will be evaluated for individual sites in a special paper.

Average temperature calculated from data measured for the whole vegetation season was highest at the concrete surface, both at 30 cm (16.09°C) and 2 m (15.73°C). The concrete surface is the site of highest albedo (25%). It was not albedo which decreased temperature. Low temperature was measured at wet meadows and other vegetated sites. The fishpond, the site of lowest albedo, shows at 2 m height similar to average temperature of other vegetated sites. The temperature at all sites was controlled by evapotranspiration, ground heat flux (highest in water) and by location in terrain where cold air flows during the night.

Our data shows that absence of water and vegetation is linked with high temperature. In the relatively small region of Třeboň basin where the differences in altitude are within several meters, the differences among wetlands and drained area in average seasonal temperature were several °C.

23.5 Conclusions

On clear days, the amount of shortwave reflected solar energy was 1.85 kWh m^{-2} d^{-1} at the concrete surface, 1.53 kWh m^{-2} d^{-1} at wet meadows, 1.49 kWh m^{-2} d^{-1} at the pasture and 0.59 kWh m^{-2} d^{-1} at the fishpond.

Shortwave radiation albedo of water during vegetation season was two times lower than albedo of the field and three times lower than albedo of the concrete surface. Although of high albedo, the highest average temperature was measured at the concrete surface.

Other monitored sites had lower albedo and lower average temperature than the concrete surface. The lowest average temperature was measured at wet meadows. Daily averages of temperature on clear days are markedly influenced by early morning minima. The lowest average daily amplitude in temperature (difference between daily maximum and minimum) was measured at the fishpond site.

The highest difference (3.58°C) in average temperature at 30 cm among sites was found for wet meadows and the concrete surface on clear days and 2.49°C for all days of the vegetation season.

The lowest values of relative air humidity were shown at the concrete surface, i.e. air moving from a dry surface has the capacity to take and transport most of its water vapour. Daily time courses of relative air humidity on sunny days show the ability of vegetation to buffer air humidity extremes – the concrete surface shows 20% lower air humidity than wet meadows.

At 2 m height, the relative humidity of air is highest at the fishpond and the values at all sites were within a range of 10%. On clear days at a concrete surface, relative air humidity fluctuates from 35% up to the dew point whereas at wet meadows relative humidity of air does not fall under 50%.

On sunny days, water and vegetation buffer midday extremes of low air humidity of more than 10% at 2 m height.

Land cover influences average temperature of several °C at the surface of a stand (30 cm height). It means that changes of land cover result in changes of average temperature.

Acknowledgements The work was supported by Ministry of Education, Youth and Sports of the Czech Republic (NPV 2B06023). The authors thank RNDr. Milena Kovářová for advice in elaboration of meteorological data, Alžběta Rejšková, PhD, for consultation on text and results, Jakub Brom PhD, for advice and data on sites. The authors express their gratitude to colleagues from ENKI, o.p.s. and to the staff of the company Fiedler and Mágr who maintain and manage the network of meteorological stations including the database.

References

Brix, H., Sorrell, B.K., & Lorenzen, B. (2001). Are *Phragmites*-dominated wetlands a net source or net sink of greenhouse gases? *Aquatic Botany*, *69*, 313–324.

Čížková, H., Květ, J., Comín, F., Laiho, R., Pokorný, J., & Pithart, D. (2010). Actual state of European Wetlands and their possible future in the context of global climate change. *Aquatic Sciences – Research Across Boundaries* (in press).

IPCC (2007). Summary for Policymakers. In S. Solomon, D. Qin, M. Manning, Z. Chen, M. Marquis, K. B. Averyt, M. Tignor, & H. L. Miller (Eds.), *Climate Change 2007: The Physical Science Basis*. Contribution of Working Group I to the Fourth Assessment Report of the Intergovernmental Panel on Climate Change (996 pp). Cambridge University Press, Cambridge, United Kingdom and New York, NY, USA.

Jirka, V., Hofreiter, M., Pokorný, J., & Novák M. (2009, September). *Method for estimation of surface fluxes of solar energy in landscape based on remote sensing*. Paper presented at the 6th International Conference on Environmental Hydrology, Cairo.

Kabat, P., Claussen, M., Dirmeyer, P. A., Gash, J. H. C., Bravo de Guenni, L., Meybeck, M. et al. (Eds.). (2004). *Vegetation, water, humans and the climate. A new perspective on an interactive System*. Berlin: Springer.

Kovářová, M., & Pokorný, J. (2005). Hodnocení dlouhodobých data teploty, vlhkosti vzduchu a srážek z Mokrých luk u Třeboně (Evaluation of long term data of temperature, air humidity and precipitatin from Wet Meadows at Třeboň). *Meteorologické zprávy (Prague)*, *58*, 51–60 (in Czech with English summary).

Pokorný, J., & Kučerová, A. (2000). Monitoring klimatu a atmosférických depozic v ChKO Třeboňso (Monitoring of climate and atmospheric depositions in Třeboň Landcape Protected Area In J. Pokorný, J. Šulcová, M. Hátle, & J. Hlásek (Eds.), *Třeboňsko 2000. Ekologie a ekonomika Třeboňska po dvaceti letech.* (*Ecology and Economy of Třeboňsko Region after twenty years*) (pp. 87–99). Třeboň, Czech Republic: UNESCO MaB and ENKI, o.p.s.

Pokorný, J., & Rejšková, A. (2008). Water cycle management. In E. Jørgensen & B. Fath (Eds.), *Encyclopedia of ecology* (Vol. 5, pp. 3729–3737). Oxford: Elsevier.

Ripl, W. (2003). Water: The bloodstream of the biosphere. *Philosophical Transactions of the Royal Society London, B358*, 1921–1934.

Subject Index

J. Vymazal (ed.), *Water and Nutrient Management in Natural and Constructed Wetlands,* DOI 10.1007/978-90-481-9585-5,

MIX
Papier aus verantwortungsvollen Quellen
Paper from responsible sources
FSC® C105338

If you have any concerns about our products,
you can contact us on
ProductSafety@springernature.com

In case Publisher is established outside the EU,
the EU authorized representative is:
Springer Nature Customer Service Center GmbH
Europaplatz 3, 69115 Heidelberg, Germany

Printed by Libri Plureos GmbH
in Hamburg, Germany